The British Steam Railway Locomotive
1925-1965

By the same Author

The Locomotives of Sir Nigel Gresley
The British Locomotives at Work
The Railways of Britain
Kings and Castles of the G.W.R.
Scottish Railways
British Locomotives from the Footplate
The Great Western Railway: An Appreciation
British Trains, Past and Present
The Boy's Book of British Railways
The Premier Line
Four Thousand Miles on the Footplate
Locomotives of the North Eastern Railway
Locomotives of R. E. L. Maunsell
Fifty Years of Western Express Running
The Railway Engineers
British Railways in Action
Steam Locomotive
Branch Lines
The Great Northern Railway
Father of Railways
Historical Steam Locomotives
The Railway Race to the North
Main Lines Across the Border
The London and North Western Railway
British Steam Railways
The South Eastern and Chatham Railway
The Caledonian Railway
Fifty Years of Railway Signalling
The Great Western Railway in the Nineteenth Century
Continental Main Lines
Sir William Stainer—An Engineering Biography
The Great Western Railway in the 20th Century
Pocket Encyclopedia of British Steam Locomotives
The Midland Compounds
Railway Holiday in Austria
The Highland Railway
Britain's New Railway
Steam Railways in Retrospect
Single-Line Railways
The London and South Western Railway
The L.N.W.R. Precursor Family
Historic Railway Disasters

A magnificent example of British locomotive construction: one of the giant 4-8-4 + 4-8-4 *Beyer-Garratt locomotives built in* 1952 *for New South Wales Government Railways.*

The British Steam Railway Locomotive 1925-1965

O. S. NOCK
B.Sc., C.Eng., F.I.C.E., F.I.Mech.E, M.I.Loco. E., M.I.R.S.E.

LONDON
IAN ALLAN LTD

First published 1966
Fourth impression 1988

ISBN 0 7110 0125 1

Published by Ian Allan Ltd, Shepperton, Surrey;
and printed in the United Kingdom by
Richard Clay Ltd, Chicester, Sussex

Contents

O. S. Nock, like the celebrated author of the earlier volume dealing with the first 100 years of the British Steam Railway Locomotive, was trained as a mechanical engineer; but whereas Ahrons in middle age relinquished his professional engineering work and devoted himself entirely to literary work the present author is still actively engaged in engineering work. Nock's early training was at the City and Guilds (Engineering) College, under Professor W. E. Dalby —himself the author of classic works on " Steam Power ", " Valve Gears ", and the " Balancing of Engines ". After graduating he joined the Westinghouse Brake and Signal Company, and his entire professional work has been with that firm. After experience in various departments he was appointed Chief Draughtsman, Brake Department, in 1945 and Chief Draughtsman of the Company in 1949. In January 1960 he became Chief Mechanical Engineer, Signal and Colliery Division, and in November 1965 he took up his present appointment of Planning Manager, Signal and Colliery Division.

His first literary work was a short article in " The Railway Magazine " in January 1932, and his first full length book, " The Locomotives of Sir Nigel Gresley " appeared in 1945. Since that time he has written extensively on many facets of railway engineering and operation, in addition to a number of historical works. The present book is his forty-second. He is also Honorary Editor of the Institution of Railway Signal Engineers, and a Member of Council of this Institution.

The Author on "A4" Pacific, No. 4482 Golden Eagle, *working the down Flying Scotsman,* 1945.

Preface

" Of all those who worship at the shrine of the steam locomotive there has never been one more devout and more learned than the late E. L. Ahrons . . ." So wrote Loughnan Pendred in opening a Foreword to Ahrons' classic work: " The British Steam Railway Locomotive: 1825 to 1925." That book was the republication, in book form, of a series of articles specially commissioned by *The Engineer* in celebration of the Railway Centenary. It was a tremendous task, superbly done. As Pendred said, in 1927: " Take it for all in all, it may now be said without the slightest fear of contradiction that it is the most complete, the most accurate, the most detailed history of the British steam locomotive between 1825 and 1925 which has ever been written."

When Ian Allan asked me to consider taking up the story from the year 1925, and carrying it through to the end of steam in Great Britain, or very nearly to the end, I must admit that the pleasure of receiving such an invitation was mingled with some apprehension. Ahrons remains unique in the field of railway literature; and although the scholarly work that I was to follow did not include any of those delightfully humorous touches that so characterised the equally famous series of articles in " The Railway Magazine " there can never be a second " Ahrons ". In another respect however my task has been easier. Throughout the period covered by this second book I have been a professional engineer closely connected with railway work, and have the clearest personal recollections of the first appearances of most of the locomotives concerned. Ahrons on the other hand traced his story from a time more than 40 years before his birth.

My period covers most of the " grouping " era on the British railways, and the subsequent time of nationalisation. In consequence I have had far fewer different locomotive classes to describe. It has been nevertheless a period of great significance in design, construction and testing, and I have been fortunate in seeing a great deal of the work at first hand. I made my first runs on the footplate in 1934 and there is no express passenger locomotive class built since 1925 of which I have not had first hand experience—except, of course, the unconventional and experimental designs described in Chapter Nine. There is nothing like the footplate for getting the true " feel " of locomotive work, and to my experience in this respect have been added many opportunities of travelling in the dynamometer car on some exacting and exciting test runs.

Over the years, through my membership of professional institutions, and through my literary work I have come to know many of the leading figures in the locomotive world, not only on the home railways but also in the locomotive building industry; and in several of my chapters I have sought to pay tribute to the great firms, steeped in history, who have done so much to foster the export trade of this country in steam locomotives. It is a sign of the times that, one by one, these famous firms have begun to drop out of the picture: Kitsons, Hawthorns, Robert Stephenson, and most poignant of all, North British. Now, at the time of writing, Beyer-Peacock also have ceased the manufacture of locomotives, leaving, among the major works only Vulcan and Hunslet in the field, to carry the old traditions into the new age.

Many locomotive men by their enthusiasm and readiness to " talk shop " at any time have contributed indirectly to this book, and I would like to acknowledge the help I have received on many occasions from R. C. Bond, K. J. Cook, E. S. Cox, Sam Ell, J. F. Harrison, F. W. Hawksworth, H. Holcroft, H. G. Ivatt, R. G. Jarvis, T. Matthewson-Dick, R. A. Riddles, C. T. Roberts and B. Spencer, while among the contractors I must mention with gratitude John Alcock of Hunslet, A. Black of North British, and W. Cyril Williams and Maurice Crane of Beyer-Peacock. Nor must I forget those who are no longer with us, who by their interest and encouragement have helped me at many times: Sir Nigel Gresley, R. A. Smeddle, T. S. Finlayson, Edward Thompson, and, to me the most treasured memory of all, Sir William Stanier.

I began this preface with a quotation from that great Editor Loughnan Pendred of *The Engineer*. Now, among journalists and authors, editors are apt to be regarded as a race of bogey-men, vigorously wielding the blue pencil—that is when they are not cheerfully attaching a rejection slip to one's most cherished MSS. But Pendred wrote most charmingly of Ahrons in that preface, and in following Ahrons perhaps I may add a word about Pendred, who in 1931 reached the pinnacle of his profession with his election as President of the Institution of Mechanical Engineers. It was through the commissions I received from him to write in *The Engineer*, and the help and encouragement he was always so ready to give that I gained much of the first hand experience of steam locomotives that provides the background to this book. I am no less indebted to his son, Benjamin, who succeeded him as Editor-in-Chief of *The Engineer*, and who has continued the association in the same delightful way, even though my more recent contributions have been of the diesel and electric age. Thus the name of Pendred, in many ways the architect of the first volume, has been indirectly, but no less effectively associated with the second..

Loughnan Pendred always regretted that the exigencies of the press hampered Ahrons in painting the majestic picture of 100 years' development, against the background of so vast a canvas. I, on the other hand, am most grateful to my own publisher, Ian Allan, for giving me a virtually free hand, over the shorter period of 40 years, from 1925 to 1965.

I must also express my thanks to the various Regions of British Railways, and to member firms of the Locomotive Manufacturers Association for supplying between them many hundreds of photographs from which the book has been illustrated. The Councils of the Institution of Mechanical Engineers and of the Institution of Locomotive Engineers have kindly permitted me to reproduce certain drawings that have appeared in the proceedings of those Institutions, and I am also much indebted to Mr. B. W. C. Cooke, Editor of *The Railway Gazette* for allowing me to reproduce a number of drawings from that journal and from *The Railway Engineer*. My thanks are no less due to Mr. C. R. H. Simpson who read the proofs, and made a number of suggestions that I was glad to adopt.

Lastly I am, as always, much indebted to Olivia my wife, for coping so well and so quickly with the typing, in this case of a long and unusually intricate manuscript.

O. S. Nock.

Silver Cedars,
High Bannerdown,
Batheaston,
Bath.

March 1966.

Chapter One — A Survey in 1925

At the time of the Railway Centenary the British steam railway locomotive was on the threshold of one of the most interesting periods in its history. Two years earlier the old railway companies of Great Britain had been merged into no more than four "Groups"; and this piece of legislation had not only set in motion the gradual elimination of the marked individualities in practice that had been an established feature of many of the old companies, but the process of amalgamation had caused considerable movement and regrouping of personnel. For many locomotive engineers it was a time of uncertainty, upheaval, and frustration. That period had not ended by the summer of 1925, when the Centenary celebrations took place at Darlington. Nevertheless, through this time of transition some important factors in locomotive design, which hitherto had been the result of isolated and largely unconnected effort by individual engineers began to take shape as marked trends towards a general accepted future practice. These factors were valve gears, boiler design, and balancing.

Another old controversy was revived as a result of the grouping of the railways of Great Britain. If one took a survey of the locomotive stock of the various companies in the year 1922 it would certainly have seemed that compound propulsion found no favour in modern practice. There was still a considerable number of Webb 4 cylinder compounds in service on the London and North Western Railway, mostly on freight duties; but the system was discredited and no compounds had been built at Crewe for nearly 20 years. There were isolated engines, compounded on W. M. Smith's system on the North Eastern and Great Central Railways; these, although doing good work, were regarded as non standard, and it was only on the Midland Railway that any form of compound locomotive was established as a standard design (Fig. 1). Even so, a class of no more than 45 medium-powered 4-4-0s would not ordinarily have been considered likely to affect the future policy of a huge combine like the newly formed London Midland and Scottish Railway, which included in its locomotive stud some 500 Crewe-built simple, superheater engines of the 4-4-0 and 4-6-0 types.

Two factors combined to revive, in the most intense form, the old controversy between simple and compound propulsion. The first was the emergence of the Midland as the predominant partner in the L.M.S. combine, with the result that its locomotive practice came to receive a consideration that was certainly out of all proportion to the numerical strength and nominal tractive capacity of its most powerful locomotives. In the No. 4 class for example, there were only 55 Midland engines in all, against 246 on the London and North Western—not to mention the 130

Fig. 1.—*Midland Railway, 3-cylinder superheated compound 4-4-0 No. 1000, as now preserved.*

locomotives in No. 5 class which the latter company possessed. This is not to suggest that the consideration given to ex-Midland designs was unjustified. Events proved it to have ample justification; but in other circumstances the Midland compounds could have been passed over as easily as the Smith compounds that were in the ownership of the London and North Eastern Railway from the beginning of 1923. In another respect however the practice of compounding claimed a renewal of attention by British engineers. Just as the work of Alfred de Glehn in France had compelled notice some 30 years earlier, so the more recent developments of his practice, particularly as being worked out on the Northern Railway of France, caused many British engineers to reflect upon possible developments.

The grouping into single large concerns of railways which had hitherto pursued different courses in locomotive design led to the running of several series of competitive trials. Both the London Midland and Scottish and the London and North Eastern Railways had dynamometer cars of relatively modern design, and by the year 1925 test results had been obtained, on the one side, from certain London and North Western, Midland, Lancashire and Yorkshire, and Caledonian designs, and on the other from Great Northern, North Eastern and North British locomotives. Some of the data obtained from these trials is discussed in the next chapter because it provides a good general impression of the standards of locomotive performance that prevailed in 1925. But the most immediate outcome of a series of trials conducted in the winter of 1923–24 on the Leeds–Carlisle section of the former Midland Railway was the tremendous fillip it gave to the protagonists of compounding.

Without anticipating the test results which will be discussed later the following salient points may now be mentioned. Three varieties of express passenger locomotive in No. 4 class were tested with the dynamometer car in the running of certain regular service trains which were specially augmented in weight. The engines concerned were the Midland 3-cylinder compound 4-4-0; the Midland "999" class simple 4-4-0, and the L.N.W.R. "Prince of Wales" class 4-6-0. In duty that was very heavy in relation to the size of the locomotives concerned, the Midland compound proved equally suitable as the L.N.W.R. 4-6-0, and more economical in fuel. These results encouraged the management of the L.M.S.R. to embark upon a large programme of compound engine building for general passenger service on all parts of the system, while the locomotive headquarters staff at Derby commenced work on a development of the compound principle towards the production of new designs of locomotives of maximum power for the heaviest main line express work from Euston to the North. The multiplication of the Midland compound 4-4-0 class, and the news of the projected Derby developments provoked a controversy in the technical press of the day that brought memories of the Webb era on the London and North Western Railway; but factors other than those of locomotives engineering were eventually to bring this interested project to a halt.

The exponents of compounding, pointing equally to the results of the L.M.S.R. Leeds–Carlisle trials of 1923–24, and to current work on the Northern Railway of France, claimed that two-stage expansion of the steam provides for greater thermal efficiency. The years following the end of hostilities in Europe in 1918, had been a time of much unrest in the coal industry in Great Britain. There had been a

Fig. 2.—*G.W.R. The basic Churchward 2-cylinder design, represented by engine No.* 2906 Lady of Lynn.

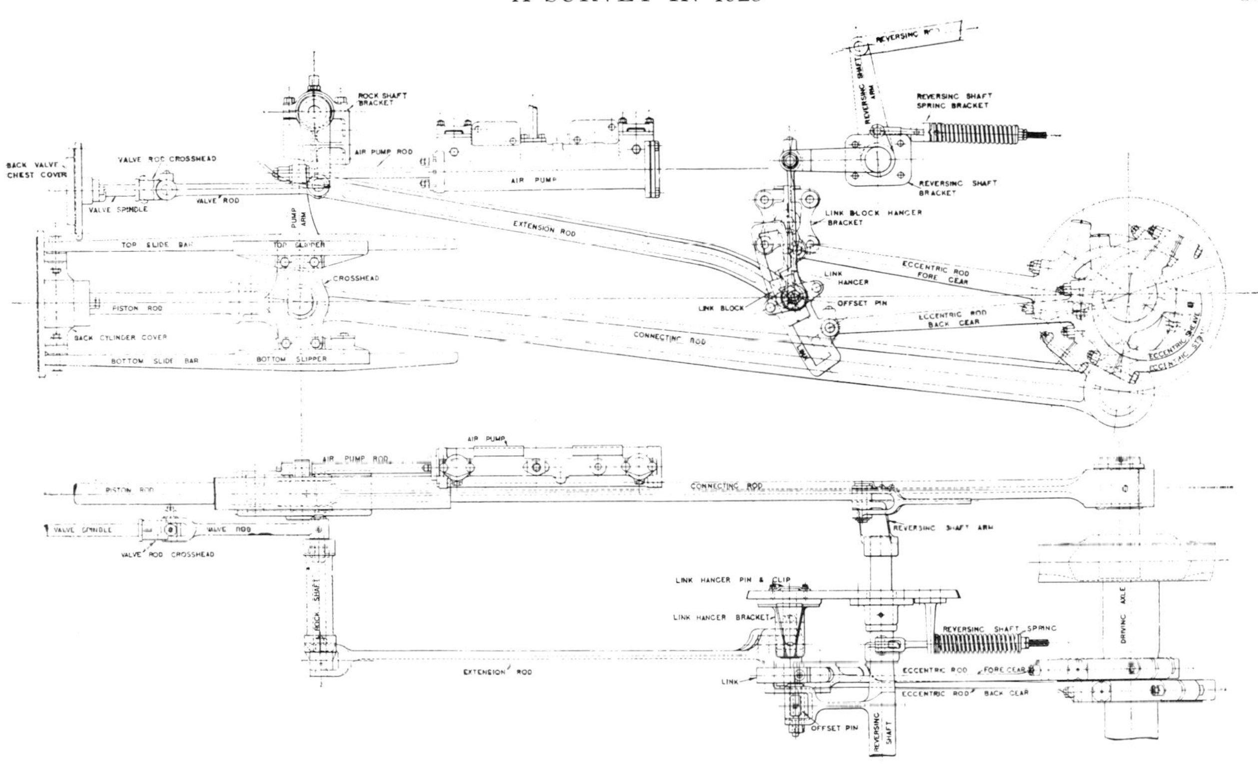

Fig. 3.—*G.W.R. Stephenson's Link Motion, as applied to the Churchward locomotives with two outside cylinders.*

serious strike of mine-workers in the early summer of 1921, while in the year of the present survey the situation had almost reached the stage of a national stoppage in the summer, when Government intervention and a subsidy postponed the crisis until the May of the following year. There was therefore a very live concern among all responsible locomotive officers towards means of reducing coal consumption, though at that stage the problem of dealing with inferior qualities of coal had not seriously arisen. On theoretical grounds the exponents of compounding were unquestionably correct; but as a matter of practical railway working the case that had been worked out on the Great Western Railway by G. J. Churchward, twenty years earlier, had to be gone over again. It is certainly true that in the railway circumstances prevailing in Great Britain up to the time of grouping, and for a few years afterwards, very few engineers had appreciated the full significance of Churchward's work on the Great Western Railway, though its details—if not its thermodynamic results—had been published from time to time, in Churchward's own contributions to the proceedings of the leading engineering institutions.

By his arrangement for the purchase of three de Glehn compound "Atlantics" from France, and the running of these engines in regular express passenger service on the Great Western Railway the merits and disadvantages of the system had become fully understood at Swindon; and to eliminate the complexities of control a sustained attempt had been made to produce a single expansion locomotive that would have a thermal efficiency equal, or nearly equal to that of the compound (Fig. 2). It was argued that in the practical running of a railway a locomotive that is simple in construction, and simple to handle is worth more than a more complicated machine which, under ideal running conditions, will give a slightly lower coal consumption. In working towards this target Churchward had adopted two basic principles: a high working pressure in the boiler, and cylinders in which the stroke was long in relation to the bore. But neither of these factors—either alone or in combination—would have given the desired result had not the utilisation of steam in the cylinders been efficient. The temperature range must be as large as possible, which made essential a high initial pressure, and a low back-pressure.

It was in working towards this target that the valve gear performance with which his name will always be associated was achieved by Churchward. To secure high initial pressure in the cylinder, and low back pressure needed much larger port openings, both to steam and exhaust than were customary at that time, and this required valves with much longer laps, and having a travel in full gear of about 50 per cent. more than conventional arrangements (Fig. 3). His contemporaries were reluctant to adopt such practice, feeling that the advantages in improved thermal efficiency would be offset by increased costs due to wear of valves and liners. On the London and North Western Railway C. J. Bowen-Cooke had, in some measure, gone half-way towards the

Fig. 10.—*L.M.S.R. One of the non-superheater* 4-6-4 *tank engines built for the Furness Railway in* 1920 *by Kitson and Co. Ltd.*

The Midland Railway had been a user of Belpaire fireboxes for some 10 years before grouping took place, but on the London and North Western section of the L.M.S.R. the use of Belpaire fireboxes had been relatively limited. Shortly after grouping however, designs were prepared for new boilers with Belpaire fireboxes that could be used on existing express passenger locomotives of the "George the Fifth" and "Prince of Wales" types, and be interchangeable with the previous standard boilers having round top fireboxes. This was an interesting outward expression of a trend in locomotive practice that was coming to the fore generally on the British railways, to reduce the time that the locomotive was out of traffic through shopping. The existence of interchangeable boilers of both older and newer varieties meant that locomotives did not necessarily have to wait for the repair of the boiler originally fitted if other work of overhaul was completed. On the Southern Railway no single policy as regards boiler design was evolved during Maunsell's time as Chief Mechanical Engineer. Locomotives of the London and South Western origin, and the more recent classes that were derived from them such as the "King Arthurs", retained boilers with round-top fireboxes throughout their existence, and new locomotives like the 3-cylinder 4-4-0s of the "Schools" class, which were built to use existing tools, had boilers with round-top fireboxes. Similarly, engines that were derived from South Eastern and Chatham types had Belpaire fireboxes.

On the Great Western Railway the elaborate shaping of the boiler barrel and the firebox was the result of a very intense study and development that took place at Swindon between 1902 and 1913. In the Great Western boiler the degree of coning of the barrel was much more pronounced than in any other British design of boiler, while the firebox was roughly trapezoidal in plan; also, there were no sharp corners. The top and front were carefully rounded, and every part of the design was carefully studied so as to make provision for rapid and unhindered circulation of the water inside. The largest possible water space was provided at the junction between the barrel and the firebox front, because that is the hottest part of the boiler and circulation of the water inside is most rapid in that region. The boilers were designed so as to give ample steam space above the water line; this minimised any tendency to foam and enabled the steam dome to be dispensed with. Churchward always argued that the provision of a steam dome introduced a point of weakness into the structure of the boiler itself, and in his boilers the steam was collected in a perforated pipe running immediately below the highest part of the boiler barrel. The special shaping and curving which was a speciality of Swindon boiler design and construction did, of course, entail higher first costs than with the simpler and more conventional type with parallel barrel and straight-sided Belpaire firebox; but experience on the Great Western had shown that the higher first costs were more than offset by the greatly reduced maintenance charges, and by freedom from troubles arising out of leaking tubes. It was however, not until W. A. Stanier was appointed Chief Mechanical Engineer of the L.M.S.R. in 1932 that the precepts of Swindon in connection with boiler design were adopted by other administrations.

Chapter Two 1925–27 : The Standards of Locomotive Performance

The period from the time of the grouping until the end of 1926 probably witnessed more testing with dynamometer cars than any other, until the nationalisation of the British railways in 1948. Although very little in the way of official information was published at the time the results of most of these trials have subsequently been made available to the author through the courtesy of the Chief Mechanical Engineers of the Western, London Midland, and North Eastern Regions of British Railways, and it is now possible to study these results collectively, and form a broad overall impression of the standards of performance that were current in Great Britain at the time of the Railway Centenary. Generally speaking, too, the various test runs that were made to compare the performance of dissimilar designs, then merged in one or another of the grouped railways, did represent a fair assessment of the demands made upon locomotives on different routes by the timetables, and the traffic being conveyed at the time. An exception was the Leeds–Carlisle main line of the former Midland Railway, where the test loads were exceptionally heavy compared with the current demands of ordinary business.

The test results now available for study cover locomotives of the Great Western Railway, and the following pre-grouping designs of the constituents of the London Midland and Scottish, and of the London and North Eastern Railways:

L.M.S.R.

Former Railway	Type	Class
L.N.W.R.	4-6-0	" Prince of Wales "
L.N.W.R.	4-6-0	" Claughton "
Midland	4-4-0	3-cylinder compound
Midland	4-4-0	Class 4 simple " 999 "
L.Y.R.	4-6-0	Hughes Class " 8 "
Caledonian	4-4-0	Pickersgill type
Caledonian	4-6-0	Pickersgill " 60 " class

L.N.E.R.

Former Railway	Type	Class
G.N.R.	4-4-2	Ivatt—large boilered Atlantic—" 251 " class
G.N.R.	4-6-2	Gresley Pacific Class " A1 "
N.E.R.	4-4-2	2-cylinder " V " class
N.E.R.	4-4-2	3-cylinder " Z " class
N.E.R.	4-6-2	Raven Pacific
N.B.R.	4-4-2	Reid superheater Atlantic

No dynamometer car test runs were made with engines of the former Great Eastern and Great Central Railways, so far as the former English companies were concerned, nor with Glasgow and South Western, or Highland Railway locomotives in Scotland. But the test results nevertheless are very comprehensive, and the figures included in the respective reports make most interesting comparison with those obtained with the Great Western Railway 4-cylinder 4-6-0s of the " Castle " class (Fig. 11), which made experimental runs on a variety of routes away from their own routes in 1925 and 1926. The details of these " foreign " workings are notably supplemented by the results of some special dynamometer car test runs carried out on the Great Western Railway between Swindon and Plymouth, communicated to the World Power Conference in 1924 by Mr. C. B. Collett, Chief Mechanical Engineer of the G.W.R. There is no doubt that the publication of the coal consumption figures for the 4-6-0 locomotive No. 4074 *Caldicot Castle*, recorded on some particularly heavy workings, created something of a sensation among engineers of the northern companies. This was understandable seeing that the best figures obtained with any locomotive of the L.M.S.R., or of the L.N.E.R. showed coal consumptions of little below 4 lb. per d.h.p., and the figure quoted for the " Castle ", in Collett's paper to the World Power Conference was 2·83 lb. per d.h.p. hour.

It has been demonstrated, in the many trials conducted under strictly controlled conditions by British Railways since 1948, that the actual consumption of coal per drawbar horse-power hour is not constant for any locomotive, but varies according to the rate of evaporation of water in the boiler. It is important to emphasise this before discussing the test results obtained in the various competitive trials of 1923–26. To take an actual example: one could not expect the same overall results from the " Castle " class 4-6-0 when tested on the L.M.S.R. between Euston and Carlisle, in 1926, as were obtained in the tests between Swindon and Plymouth reported in Mr. Collett's paper in 1924. The variations in point-to-point timings, in gradients, and in the fuel used would be enough to introduce considerable divergencies; and it was these divergencies that later testing methods sought to eliminate. So far as the " Castle " engines were concerned, certain duties in working the train service between London and the West of England made demands upon locomotive capacity that were probably more severe than anything regularly worked elsewhere in Great Britain at that time; and the haulage of the Cornish Riviera Express non-stop from Paddington to Plymouth under maximum loading conditions is quoted opposite as exemplifying a maximum demand duty.

The details quoted for engine working are typical of a run with good point-to-point timekeeping throughout. Hard steaming was required on all the adverse lengths of line, with useful respites from Savernake to Westbury, and from Castle Cary to Taunton. But the working of this train required a very high standard of enginemanship and intimate road knowledge, because of the very numerous curves, and slight speed restrictions. Only an experienced driver could negotiate such a route at such high booked speed without loss of time due to over emphasis of restrictions, or excessive coal consumption. On this duty, with the maximum loads quoted the fuel consumption of the "Castle" class locomotives was 40 to 45 lb. per train mile.

On the trials conducted with engine No. 4074 *Caldicot Castle*, in March 1924, the test loads were made up to the maximum tare tonnages permitted over each stretch of line, as follows:

DOWN LINE	
Swindon to Taunton	485 tons
Taunton to Newton Abbot	390 tons
Newton Abbot to Plymouth	288 tons
UP LINE	
Plymouth to Newton Abbot	288 tons
Newton Abbot to Swindon	485 tons

The above figures are the tare loads. Fully loaded with passengers and luggage, trains of the above formations would scale approximately 530, 420 and 310 tons. These latter figures were regularly attained over certain sections of the westbound Cornish Riviera Express. Three return trips were made with the test trains from Swindon to Plymouth and back and the following figures for one round trip were quoted in Mr. Collett's paper to the World Power Conference.

Calorific Value of coal, B.Th.U. per lb.	14,780
Water evaporation per lb. of coal (actual lb.)	9.95
Coal per i.h.p. hour (lb.)	2.1
Water per i.h.p. hour (lb.)	20.9
Coal per d.h.p. (lb.)	2.83
Water per d.h.p. hour (lb.)	28.1
Coal per sq. ft. of grate area per hour running time (lb.)	66.1
Average steam chest pressure with full regulator in per cent. of boiler pressure (inside cyl.)	92.6
Average steam chest pressure with full regulator in per cent. of boiler pressure (outside cyl.)	93.6
Thermal efficiency of boiler (per cent.)	79.8
Thermal efficiency overall (i.h.p. basis)	8.22
Overall efficiency work at drawbar heat in coal	6.1
Drawbar pull sustained at 71 m.p.h. (tons)	2.35

The coal consumption per train mile was 43 lb., despite the heavy work done. At that time there were 10 locomotives of the class in service, and their average consumption on all duties during the first four months of 1924 was as follows:

Month	Coal consumption lb. per engine mile
January	44.7
February	41.2
March	42.8
April	43.0

These were very good figures, seeing that they included all light running and coal used for lighting up. All ten locomotives of the class were then stationed at Old Oak Common. Most of them were engaged on runs between Paddington and Plymouth, but their workings included a hard " single-home " trip from Paddington to Wolverhampton and back. This latter involved the running of one of the most severe of the highly competitive Birmingham–London expresses. The coal consumption figures of the " Castle " class locomotives must be studied against the background of the loads hauled and of the high booked speed, and it is in this

Fig. 11.—*G.W.R. " Castle " class* 4-6-0, *as originally built with small tender, and used in the trials of* 1924-5.

GREAT WESTERN RAILWAY
Working of Cornish Riviera Express

Section	Distance miles	Booked time min.	Average speed m.p.h.	Load tons (gross trailing)	Nature of road	Engine working	
						Reg.	Cut-off %
Reading	36.0	37	58.3	530	Level	3/4	26
Savernake	34.1	36 1/2	56.1	530	Rising	3/4	26–30
Westbury	25.5	24	63.7	530	Downhill	—	—
Castle Cary	19.7	22 1/2	52.6	450	Difficult	3/4	22–25
Taunton	27.6	28	59.2	450	Easy	—	—
Whiteball Box	10.9	13	50.3	385	Steeply Adverse	3/4	25–35
Exeter	19.9	18	66.3	385	Downhill	—	—
Newton Abbot	20.2	24	50.5	310	Level, many restrictions	—	—
Plymouth	31.8	44	43.4	310	Very severe	Full	42 max.

Overall: 225.7 miles; 247 min., 54.7 m.p.h.

context that the figures caused surprise and some incredulity in other railway circles.

Of the competitive trials run within the L.M.S.R. and L.N.E.R. organisations a group run by the latter railway with "Atlantic" engines may be taken first. These, in turn, were in two separate groups. There was a series of tests made in October 1923 to establish the relative performance of locomotives of the Great Northern, North Eastern, and North British design, and then a second series almost a year later to compare the performance of 2 cylinder and 3 cylinder "Atlantics" of North Eastern design, both having identical boilers (Figs. 12 to 15). The first set of trials was carried out mainly on service trains, while the latter was entirely with specials. All the trains concerned were booked to run non-stop between Newcastle and Edinburgh, at average speeds of less than 50 m.p.h. The results obtained from the Great Northern and North British "Atlantics" in 1923 are important as they constitute the only records in existence of these two well-known designs, both of which continued to render excellent service in traffic for many years after grouping. At the same time, in quoting the figures for the G.N.R. engine certain reservations must be made.

L.N.E.R. ATLANTIC TRIALS

Results of the Tests

Railway	G.N.R.	N.E.R.	N.B.R.
Engine No.	1447	733	878
Average load, tons tare			
Service trains	307	311	345
Special trains	406	406	406
Average speed m.p.h.*			
Service trains	49.3	49.0	48.8
Special trains	46.0	47.2	47.5
Average d.h.p.			
Service trains	393	391	529
Special trains	527	529	525
Average steam chest pressure lb./sq. in.			
Special trains	97	132	138
Average cut-off			
Special trains %	40	42	38
Coal per d.h.p. hour lb.			
Service trains	5.08	4.45	4.12
Special trains	4.42	3.63	4.15
Average superheat temp. in deg. Fahr.	624	577	520
Evaporation lb. of water per lb. of coal	7.2	8.6	8.2

* Inclusive of various checks on most runs.

Fig. 12.—*L.N.E.R. Standard G.N.R. superheated "Atlantic" with reduced-height boiler mountings, for running between Newcastle and Edinburgh.*

Fig. 13.—*N.E.R. 3-cylinder 4-4-2 Class "Z" fitted with indicator shelters.*

Performance on Cockburnspath Bank

Railway Engine No.	G.N.R. 1447	N.E.R. 733	N.B.R. 878
Load tons	406.5	406.5	366
Average boiler pressure lb./sq. in.	162	175	175
Steam chest pressure lb./sq. in.	140	147	138
Average cut-off %	40	56	52
Speed on bank m.p.h.			
at foot	47.5	50.5	44.0
minimum	20	22.5	21.5
Drawbar pull, tons			
at foot	2.45	2.20	2.35
maximum actual	4.75	4.80	4.70
maximum corrected for gradient	5.92	6.08	5.97

The North Eastern and North British locomotives were standard examples of their respective classes worked by local enginemen who were thoroughly familiar with the road. On the other hand the Great Northern engine required to have the boiler mountings reduced in height in order to clear the loading gauge of the former North British Railway. The engine concerned, No. 1447, was worked by a particularly good G.N.R. driver from King's Cross shed, A. Pibworth by name, and a man who was not likely to be overawed by a strange and important assignment. On the other hand the alteration to the height of the chimney of the engine may have had an adverse effect upon the steaming. This point is mentioned, because in view of the high reputation of the class in general No. 1447 gave disappointing results in the trials between Newcastle and Edinburgh. While the results of the trials were to some extent inconclusive, as between the North Eastern and North British engines, the Great Northern can be considered only as " a bad third ". A summary of the test results is given in the accompanying table, and a supplementary table gives details of performance in climbing the Cockburnspath Bank on the southbound run—an ascent of four miles at 1 in 96. So far as the actual amounts of fuel used were concerned the following figures of coal per train mile are interesting in comparison with current figures on the Great Western Railway:

Atlantic Locomotive Working—L.N.E.R.

Engine	Load tons	Average Speed m.p.h.	Coal per train mile lb.
G.N.R. No. 1447	307	49.3	40.5
,,	406	46.0	50.7
N.E.R. No. 733	311	49.0	36.5
,,	406	47.2	40.7
N.B.R. No. 878	345	48.8	44.5
,,	406	47.5	46.2

Fig. 14.—*L.N.E.R. ex-N.E.R. 2-cylinder 4-4-2 Class "V" with dynamometer car.*

The above results reflect the variations that can occur in road testing at variable speeds and rates of evaporation. The North British locomotive, for example, showed a relatively high figure for coal per train mile with the lighter trains, but in relation to the work done, as registered in the dynamometer car her performance was the best of the three engines, with a basic consumption of 4·12 lb. per d.h.p. hour, against 5.08 and 4.45 by the Great Northern and North Eastern engines respectively.

In the trials conducted in September and October 1924, between 2- and 3-cylinder "Atlantics" of North Eastern design, loads of approximately 365 tons (empty stock) were conveyed on all occasions. Five return trips from Heaton to Edinburgh were made with each engine, with the following results:

L.N.E.R. "Atlantic" engine trials Sept. – Oct. 1924

Engine No. / Design	701 2-cylinder Class V	729 3-cylinder Class Z
Load of test train, tons	363	365
Average speed, m.p.h.	46.3	45.7
Average d.h.p.	422	454
Average superheat, deg. Fahr.	530	550
Average steam pressure, p.s.i.		
Boiler	170	169
Steam chest	127	125
Coal consumed		
per train mile, lb.	56	45.5
per d.h.p. hr.	6.15	4.6

The figures for the 3-cylinder "Z" class engine correspond fairly well with those obtained with the service trains in the comparative trials a year earlier, but one can only remark that the cylinder performance of the 2-cylinder Class "V" engine must have been very poor to need a coal consumption of 56 lb. per train mile to work a load of 363 tons at an average speed of 46.3 m.p.h., on a non-stop run of 122 miles. The basic consumption of 6.15 lb. per d.h.p. hour registered with this engine was certainly excessive.

Fig. 15.—*L.N.E.R. ex-N.B.R. "Atlantic" as tested against G.N.R. and N.E.R.* 4-4-2*s.*

Turning now to the L.M.S.R. test results, the data available concerns several sets of trials over the former Midland Railway main line between Leeds and Carlisle run at various times between December 1923 and January 1925, and two sets of trials over the former London and North Western main line between Preston and Carlisle in May 1925 and May 1926. All these trials centred upon the way in which the different engines concerned climbed the severe gradients en route. In the Midland case it was the negotiation of Aisgill summit, 1,166 ft. above ordnance datum, and in the North Western case the ascent to Shap, 915 ft. above ordnance datum. The approach gradients are rather longer on the Midland line, and the test runs extended over 113 miles, against 90 miles on the Preston–Carlisle route. The trials on the Midland line are summarised in separate tables for the southbound and northbound runs.

Four different varieties of the well-known Midland compound were concerned in these trials: the standard Midland Railway superheated type with 7 ft. 0 in. coupled wheels and cylinders 19 in. diameter high pressure and 21 in. low pressure (Fig. 16); the first

Fig. 16.—*L.M.S.R. ex-M.R. standard superheated compound* 4-4-0 *as tested against L.N.W.R. locomotives in* 1923-5.

Fig. 17.—*M.R. A superheater non-compound class "4" 4-4-0 "999" series, as used in trials between Leeds and Carlisle in* 1923-4.

L.M.S.R. development with 6 ft. 9 in. coupled wheels and larger cylinders 19 ¾ and 21 ¾ in. diameter; a variant of this latter, with a reduced size blastpipe; and the final L.M.S.R. standard type, with 6 ft. 9 in. coupled wheels, and the original Midland cylinder dimensions (Fig. 18). From the viewpoint of comparative data the L.N.W.R. "Claughton" used in these trials was a poor representative of the class, and gave a performance well below the usual standard. The impressions created by this first trial of the design under L.M.S.R. auspices were largely corrected in later trials. It must also be recorded that the Caledonian 4-4-0 (Fig. 23) was outclassed in these trials. In L.M.S.R. reckoning it was included in the No. 3 power class, whereas the other locomotives engaged were either No. 4 class, or No. 5 in the case of the L.N.W.R. "Claughton" 4-6-0 (Fig. 19). Apart from the Caledonian engine, and the compound No. 1023, which was in poor shape, all the locomotives engaged kept schedule time, other than incidental minutes lost through signal and other out-of-course checks. The performance of the 7 ft. compound engine No. 1008 was exceptionally good. It was not equalled by any other engine of the class on any subsequent tests either on the Midland or on the L.N.W.R. line.

The performance is summarised in three separate tables. During all the tests the locomotives were running non-stop from Carlisle to Leeds on the up-journeys; but on the earlier series the down journeys were also made non-stop while the later series included a stop at Hellifield. With the exception of engine No. 1008 the general level of coal consumption by the Midland compounds was from 4.15 to 4.77 lb. per d.h.p. hour. The North Western "Prince of

L.M.S.R. Leeds–Carlisle Test Runs
Non-Stop Trains

Engine			Load trailing tons	Av. Speed m.p.h.	No. of Tests	Coal lb./ mile	Coal lb./ d.h.p. hr.
Former Rly.	No.	Design					
M.R.	1008	Compound	294	51.1	1	38.1	3.64
M.R.	1008	,,	223	51.5	2	36.7	4.35
M.R.	1008	,,	249	51.1	1	33.0	3.94
M.R.	1008	,,	343	49.5	2	42.3	3.82
M.R.	998	"999" class	296	49.2	6	44.5	4.44
M.R.	998	"999" class	349	48.6	2	52.5	4.78
M.R.	998	"999" class	224	48.6	1	35.7	4.66
L.N.W.	388	"Prince of Wales"	295	50.7	3	45.4	4.83
L.N.W.	388	"Prince of Wales"	343	48.0	2	55.2	4.80

Fig. 18.—*L.M.S.R. The standard superheater compound with left hand drive, as built for general service after* 1925.

Fig. 19.—*L.N.W.R. 4-cylinder 4-6-0 "Claughton" class, engaged in comparative trials on L.M.S.R.* 1923-5.

L.M.S.R. Carlisle—Leeds Test Runs

Engine			Load trailing tons	Av. Speed m.p.h.	No. of Tests	Coal lb./mile	Coal lb./d.h.p. hr
Former Rly	No.	Design					
M.R.	1008	Compound	307	48.3	2	41.5	3.97
M.R.	1008	,,	352	48.0	4	44.0	3.86
M.R.	998	"999" class	296	47.9	7	48.0	4.61
M.R.	998	,,	355	45.6	2	56.5	4.65
M.R.	1023	Compound	348	46.7	2	43.3	4.44
M.R.	1065*	,,	306	48.2	2	39.3	4.42
M.R.	1065	,,	355	47.0	2	45.5	4.39
M.R.	1066*	,,	300	48.5	1	40.2	4.68
M.R.	1066	,,	351	47.1	1	45.1	4.43
M.R.	1060†	,,	300	46.7	2	43.5	4.68
M.R.	1060	,,	348	47.0	1	49.9	4.77
M.R.	1065‡	,,	304	47.2	2	38.8	4.19
M.R.	1065	,,	358	47.5	1	43.0	4.15
L.N.W.	388	"Prince of Wales"	304	47.4	3	46.5	4.42
L.N.W.	388	,,	353	47.2	2	50.0	4.35
L.N.W.	2221	Claughton	302	47.3	2	51.8§	5.56§
L.N.W.	2221	,,	354	47.7	2	56.4§	5.03§
C.R.	124	Pickersgill	301	46.6	1	54.7§	6.53§

* L.M.S. built: 6 ft. 9 in. cyls. 19 3/4 in. dia. H.P.; 21 3/4 in. dia. L.P.
† L.M.S. built: 6 ft. 9 in. reduced size blastpipe.
‡ L.M.S. built: 6 ft. 9 in. cyls. 19 in. dia. H.P.; 21 in. dia. L.P.
§ Coal for round trip Carlisle to Leeds and back.

Wales" class engine (Fig. 20) and the Midland "999" class 4-4-0 (Fig. 17) were little above this, and these results compare closely with the figures obtained on the L.N.E.R. with Atlantic engines between Newcastle and Edinburgh. The L.M.S.R. trials carried out between Preston and Carlisle in 1925 and 1926 confirmed the general impression that a basic consumption of about 4 1/2 lb. per d.h.p. hour was a good average figure for locomotives engaged in express passenger work in 1925–26, other than on the Great Western Railway, and with a very few locomotives on the Southern. The results of the Preston–Carlisle trials are summarised in a separate table. As in the later trials between Carlisle and Leeds the figures for coal consumption relate to the round trip from Carlisle to Preston and back. It is therefore not possible to assess the effect of the much faster timings worked to on the down journeys from Preston to Carlisle. The engines concerned were all thoroughly representative examples, and all did good work. The

Fig. 20.—*L.M.S.R. ex-L.N.W. 2-cylinder* 4-6-0 *"Prince of Wales" class.*

Fig. 25.—*L.N.E.R. The ex-N.E.R. " Pacific " No.* 2400 *at King's Cross during comparative dynamometer car trials in* 1923.

" Pacific " gave results closely similar in coal consumption per d.h.p. hour to those obtained with the 3-cylinder " Atlantics " of the same company, in tests between Newcastle and Edinburgh described in this chapter. By comparison with the locomotives of other railways recorded in the first years after grouping the performance of the Gresley Pacific was very satisfactory, though the design had not then attained its final form.

To summarise, with the exception of the Great Western Railway, the test results of fourteen different designs of express passenger locomotives gave basic coal consumption rates varying between 3.64 and 6.53 lb. per d.h.p. hr. The figures for the Caledonian " Pickersgill " 4-4-0, and for the North Eastern 2-cylinder " Atlantic " of Class " V " were poor by comparison, and it has already been emphasised that some of the figures obtained with the Midland compound No. 1008 were exceptionally low. Neglecting these extremes the remainder of the results lie roughly between 4 and 5 lb. per d.h.p. hour, and this range can be considered to cover the great majority of British express passenger locomotives in the period 1923–25, with the exception of those of the Great Western Railway. It is important to bear these figures in mind in assessing the importance of the work towards the improving of valve gears in the period following the year 1925.

Chapter Three Improvements in Design 1926–30

An event that had the most far-reaching results was the interchange of express passenger locomotives between the Great Western and the London and North Eastern railways in April and May 1925. The locomotives concerned were the G.W.R. "Castle" class 4-6-0, and the Gresley "Pacific". At that time it would have been difficult to find two top-line locomotive designs that differed so markedly in all their basic features. Quite apart from the fundamental difference that one was a 4-6-0 and the other a 4-6-2, the boiler designs differed greatly; the G.W.R. engine had a boiler pressure of 225 lb. per sq. in. and the L.N.E.R. only 180, and the valve gears differed not only in their mechanical layout but in the important items of the length of the lap, and the maximum travel. The G.W.R. engine used a relatively small degree of superheat, though in this respect the actual temperature of the steam entering the cylinders was approximately the same.

Superheater Proportions

Engine	G.W.R. Castle	L.N.E.R. Gresley 4-6-2
Evaporation		
Heating surface, sq. ft.	1857.7	2635
Superheating		
Heating surface, sq. ft.	262.6	525
Boiler pressure, p.s.i.	225	180
Temperature of steam, deg. Fahr.	520	547

By far the hardest duty performed by the rival locomotives was the haulage of the down Cornish Riviera Express. The scheduled running times of this train were analysed in the previous chapter. The Gresley "Pacific" engine No. 4474 kept good time on all these trips, but at the expense of a heavier coal consumption, as one would have expected from a comparison of the relative valve gear designs of the two locomotives. The loads hauled were the same throughout the week, namely 530 tons gross trailing load from Paddington to Westbury; 445 tons forward to Taunton; 390 tons to Exeter, and 310 tons to Plymouth. The coal consumption figures in pounds per train mile were:

Date	Engine		Coal lb./mile
April 27	L.N.E.R.	No. 4474	50.0
,, 28	G.W.R.	No. 4074	44.1
,, 29	L.N.E.R.	No. 4474	48.8
,, 30	G.W.R.	No. 4074	45.6
May 1	L.N.E.R.	No. 4474	52.4
,, 2	G.W.R.	No. 4074	46.8

All runs except that of May 2 were made with close adherence to the schedule point-to-point running times, though on May 1 the L.N.E.R. engine was hampered by a strong contrary wind. But on May 2, the Great Western locomotive was considerably extended, by way of a demonstration, and arrived at Plymouth 15 min. ahead of time. The important point to note is that although the average speed to passing Exeter was increased from the scheduled 58 m.p.h. to 61.5 m.p.h., and the overall average from 54.7 to 58.3 m.p.h. the coal consumption was still lower than the best of the L.N.E.R. figures, working the train to the normal schedule.

The results were widely publicised and discussed at great length in railway and engineering circles generally, and at the time much of the controversy centred around the greatly differing proportions of the boilers and fireboxes. There were also a number of extraneous factors, none very large in themselves, that contributed to the difficulty the L.N.E.R. enginemen found in working the Cornish Riviera Express; but two major points of the design emerged from the tests:

(a) the susceptibility of the Swindon boiler and firebox to inferior fuel.

(b) the superiority of the Great Western valve gear design, in contributing to a free-running engine and low coal consumption.

The first of these two factors was not apparent at the time of the trials, because all the engines concerned were supplied with coal of first-class quality. A year later however there occurred the prolonged coal strike, which necessitated the importing of much foreign fuel. During that emergency the Great Western express passenger 4-6-0s engaged on the heaviest duties were frequently in trouble for steam, whereas the larger boilers and wide fireboxes of the Gresley "Pacifics" coped with emergency conditions more readily. The coal difficulty proved no more than a passing phase, and no change was made in the general design of Great Western boilers and fireboxes, though the difficulties were to recur, with increasing severity during the period of World War II, and then some changes in design were inevitable.

On the London and North Eastern Railway some careful studies were made towards the improvement of valve gear design, to give a freer exhaust. In 1926 engine No. 4477 was fitted with a modified arrangement. In so doing as many as possible of the existing parts were used, but the experiment proved successful in reducing the coal consumption, and providing an

COMPARATIVE TESTS OF LOCOMOTIVES WITH 180 AND 220 LB. PRESSURE

Table I — Dynamometer Car Tests, L.N.E.R. "Pacific" Engine No. 4473, Pressure 180 lb.

Road / Date / Trip	Doncaster to King's Cross and return 13-2-28 1	Doncaster to King's Cross and return 14-2-28 2	Doncaster to King's Cross and return 15-2-28 3	Doncaster to King's Cross and return 16-2-28 4	Doncaster to King's Cross and return 17-2-28 5	Doncaster to King's Cross and return 18-2-28 6
Weight behind tender, tons	(a) 427.0 (b) 495.25 (c) 334.25	(a) 430.5 (b) 495.75 (c) 327.75	(a) 425.5 (b) 494.25 (c) 333.25	(a) 439.75 (b) 488.75 (c) 331.75	(a) 432.25 (b) 488.25 (c) 332.25	(a) 456.25 (b) 489.75 (c) 333.75
Total weight of train including engine, tons	(a) 567.0 (b) 635.25 (c) 474.25	(a) 570.5 (b) 635.75 (c) 467.75	(a) 565.5 (b) 634.25 (c) 473.25	(a) 579.75 (b) 628.25 (c) 471.25	(a) 572.25 (b) 628.25 (c) 472.25	(a) 596.25 (b) 629.75 (c) 473.75
Coal per mile, lb.	40.1	34.95	37.93	40.65	40.61	38.76
Coal per d.h.p. hour, lb.	3.44	3.09	2.89	2.99	3.06	3.00
Coal per ton-mile, lb.	0.072	0.062	0.068	0.072	0.072	0.068
Coal per sq. ft. of grate area per hour, lb.	50.43	44.13	47.73	51.63	52.82	51.0
Water, lb. per lb. of coal	7.97	8.11	8.3	8.46	7.95	8.3
Water per d.h.p. hour, lb.	27.9 26.9	24.7 25.4	23.8 24.2	24.3 26.4	23.8 24.9	24.4 25.5
Gallons per mile	31.65 32.2	28.75 27.9	32.5 30.5	33.7 35.0	31.6 32.8	31.8 32.5
Average speed, m.p.h.	54.9 49.2	54.8 49.7	53.3 50.65	54.83 50.22	56.08 51.6	55.7 53.0
Work done in h.p. hours	1766.8 1867.0	1816.0 1712.6	2125.3 1958.5	2164.6 2070.3	2078.1 2059.6	2034.6 1987.5

NOTE — Coal per ton-mile includes weight of engine, but excludes coal used for shed duties. Rossington coal used throughout. Boiler pressure 180 lb. per square inch.
(a) Doncaster to King's Cross; (b) King's Cross to Peterborough; (c) Peterborough to Doncaster.

Table II — Dynamometer Car Tests, L.N.E.R. "Pacific" Engine No. 2544, Pressure 220 lb.

Road / Date / Trip	Doncaster to King's Cross and return 20-2-28 1	Doncaster to King's Cross and return 21-2-28 2	Doncaster to King's Cross and return 22-2-28 3	Doncaster to King's Cross and return 23-2-28 4	Doncaster to King's Cross and return 24.2-28 5
Weight behind tender, tons	(a) 433.75 (b) 506.75 (c) 348.75	(a) 428.75 (b) 508.25 (c) 347.25	(a) 424.75 (b) 505.75 (c) 346.75	(a) 427.75 (b) 502.25 (c) 346.25	(a) 428.75 (b) 512.25 (c) 355.25
Total weight of train including engine, tons	(a) 577.75 (b) 650.75 (c) 492.75	(a) 572.75 (b) 652.25 (c) 491.25	(a) 568.75 (b) 649.75 (c) 490.75	(a) 571.75 (b) 646.25 (c) 490.75	(a) 572.75 (b) 656.25 (c) 499.25
Coal per mile, lb.	37.18	35.89	35.4	34.39	34.01
Coal per d.h.p. hour, lb.	3.27	3.17	3.24	2.92	2.99
Coal per ton-mile, lb.	0.065	0.063	0.062	0.060	0.059
Coal per sq. ft. of grate area per hour, lb.	49.6	47.92	46.8	45.4	44.32
Water, lb. per lb. of coal	7.67	7.82	8.32	8.75	8.31
Water per d.h.p. hour, lb.	24.35 25.81	24.18 25.51	25.53 28.52	25.91 25.15	23.9 25.76
Gallons per mile	28.7 28.3	28.9 27.18	28.58 30.38	28.76 31.42	27.22 29.28
Average speed, m.p.h.	58.3 52.16	57.18 53.3	57.5 51.8	55.55 53.5	55.7 51.93
Work done in h.p. hours	1837.0 1709.8	1862.8 1658.5	1745.5 1660.7	1729.3 1947.6	1775.1 1773.8

NOTE — Coal per ton-mile includes weight of engine, but excludes coal used for shed duties. Rossington coal used throughout. Boiler pressure 220 lb. per square inch,
(a) Doncaster to King's Cross; (b) King's Cross to Peterborough; (c) Peterborough to Doncaster,

Fig. 26.—*Development of the Gresley " Pacific ": the standard " A3 " class, with* 220 *lb. per sq. in. boiler pressure, built Doncaster,* 1928. *This engine attained a maximum speed of* 108 *m.p.h. on test in* 1935.

engine that was noticeably freer in running. This prompted a more complete re-design of the gear, in which the lap of the valves was increased from the original 1 $^1/_4$ in. to 1 $^5/_8$ in. and the valve travel in full gear from 4 $^9/_{16}$ in. to 5 $^3/_4$ in. This was applied early in 1927 to engine No. 2555, and comparative trials were conducted between Doncaster and King's Cross between this engine and No. 2559, which had the original gear. In the working of the heavy and fast Leeds expresses the coal consumption of No. 2555 was no more than 38–39 lb. per mile, against an average of 50 lb. per mile for the engine with the original gear. In the light of such excellent results the entire stud of Gresley " Pacific " engines was altered to the valve gear layout fitted to engine No. 2555. The accompanying table shows the comparative settings of the original and the modified gear.

From this time onwards the Gresley " Pacific " locomotives, which had always been very satisfactory motive power units, now stepped into the front rank in respect of thermal efficiency, and in a series of trials conducted between Doncaster and King's Cross in February 1928 with engine No. 4473 the average figure for coal consumption per d.h.p. hour, taken over six return trips of 312 miles each, was only 3.08 lb. Taking into account the difference in calorific value between the South Yorkshire " hards " used by express locomotives working from Doncaster shed, and the first-grade Welsh coal then used on the Great Western Railway there was practically no difference between these L.N.E.R. results, and those which has startled the locomotive world in Mr. C. B.

TABLE OF VALVE SETTING (FORWARD GEAR)

For Gresley " A1 " Pacifics with three 20 in. × 26 in. cylinders and 8 in. piston valves.
Original setting: Lead $^3/_{16}$ in.; steam lap 1 $^1/_2$ in.; exhaust lap $^1/_4$ in. (negative).

Cylinder	Nominal cut-off %	Valve opening in.		Cut-off %		Exhaust opens %		Exhaust port opening above full port in.		Exhaust closes %	
		F.	B.	F.	B.	F.	B.	F.	B.	F.	B.
Outside	25	$^{19}/_{64}$	$^{19}/_{64}$	25.9	24	63.4	60.8	$^1/_{16}$	$^1/_{16}$	75.5	77
Centre	25	$^5/_{16}$	$^5/_{16}$	25.9	24.3	64.9	59.9	$^1/_{16}$	$^1/_{16}$	75.7	79.4
Outside	65	$^{31}/_{32}$	1	67	62.9	85.7	83.3	$^{25}/_{32}$	$^{23}/_{32}$	90.8	92.3
Centre	65	$^{29}/_{32}$	1	64.2	62.2	87	85	$^3/_4$	$^{21}/_{32}$	92.3	92.7

Later standard setting: Lead $^1/_4$ in.; steam lap 1 $^5/_8$ in.; exhaust lap = line and line.

Cylinder	Nominal cut-off %	Valve opening F.	Valve opening B.	Cut-off F.	Cut-off B.	Exhaust opens F.	Exhaust opens B.	Exhaust port opening F.	Exhaust port opening B.	Exhaust closes F.	Exhaust closes B.
Outside	15	$^{12}/_{64}$	$^{13}/_{64}$	14.4	15.6	65.6	65.1	$^5/_{64}$	$^4/_{64}$	65.1	65.6
Inside	15	$^{12}/_{64}$	$^{14}/_{64}$	15.3	16.5	65.9	65.3	$^6/_{64}$	$^4/_{64}$	65.3	65.9
Outside	25	$^{19}/_{64}$	$^{20}/_{64}$	24.3	25.9	72.8	71.6	$^{12}/_{64}$	$^{11}/_{64}$	71.6	72.8
Inside	25	$^{18}/_{64}$	$^{23}/_{64}$	24.7	25.7	73	71.1	$^{15}/_{64}$	$^{10}/_{64}$	71.1	73
Outside	65	$^{13}/_{16}$	$^{11}/_{4}$	67.3	63.9	89.9	88.2	$^{11}/_{8}$	$^{11}/_{16}$	88.2	89.9
Inside	65	$^{11}/_{8}$	$^{15}/_{16}$	64.9	63.7	90.6	89.9	$^{13}/_{16}$	1	89.9	90.6

Fig. 27.—*The final development of the* " A3 " *class under Gresley: engine No.* 2500 *with* " *banjo* " *steam collector instead of dome, built Doncaster* 1935.

Collett's paper to the World Power Conference in 1924. The figure of 2.81 lb. per d.p.h. hour then quoted for the " Castle " class engine No. 4074 related to a single round trip from Swindon to Plymouth and back. The best of the runs with the L.N.E.R. engine No. 4473, made on February 15, 1928, gave a figure of 2.89 lb. per d.h.p. hour.

These trials of 1928 on the L.N.E.R. represented a further stage in the evolution of Doncaster locomotive design. The working of the Great Western engine in the interchange trials of 1925 had created such an impression on the L.N.E.R. that consideration was given to the use of higher boiler pressures, with a view to greater economies in working. Furthermore, despite the successful weight-haulage performance of the Gresley " Pacifics ", and the improvement in their runniug by the use of long-lap, long-travel valves, there is no doubt that Mr. Gresley was seeking means of augmenting power, without any appreciable increase in locomotive weight, and in 1927 two standard " Pacific " locomotives were fitted with new boilers carrying a pressure of 220 lb. per sq. in. instead of the previous 180 lb. The increased first costs of boiler construction, and the probability of increased maintenance charges had to be weighed against the enhancement of tractive effort obtained from the higher working pressure. The introduction of long-lap, long-travel valves had made customary the use of an absolutely full regulator opening for all hard running, and in conjunction with cut-offs of 15 to 20 per cent. the range of expansion of steam in the cylinders was long. In lighter steaming conditions it had been found possible to use cut-offs of considerably less than 15 per cent., so that with higher working pressures it was felt that an even greater range of expansion would be a practical reality.

Of the two engines fitted with 220 lb. boilers one, No. 4480, retained the original cylinder dimensions, and the nominal tractive effort was thereby increased from 29,835 lb. to 36,465 lb. As such this engine proved somewhat *too* powerful for its allocated duties, and worked most of the time with partly-closed regulator—thus nullifying some of the advantages of higher steam pressure. The second engine, No. 2544 had the cylinders lined up to 18 1/4 in. diameter to make the nominal tractive effort the same as that of the standard 180 lb. engines, and render possible a direct comparison between engines of equal tractive effort,

Fig. 28.—The Flying Scotsman, *as rebuilt to Class* " A3 " *and running in British Railways livery.*

but with differing boiler pressures. Actually the comparison was not strictly " like for like ", because in one respect the high pressure engine had the considerable advantage. Though the cylinders were lined up from 20 in. to 18 1/4 in. diameter the piston valves remained the same, and the high pressure engine had therefore port openings that were larger in relation to the cylinder volume than on the standard engine, and was consequently still more free in running. This did not however seem to have any appreciable effect on her economy. In February 1928 a series of dynamometer car tests were carried out between engines 4473 (180 lb.) and 2544 (220 lb.), and taking an average of the complete week's running with each engine the coal consumption slightly favoured No. 4473, with a basic figure of 3.08 lb. per d.h.p. hour, against 3.12 lb. for engine No. 2544. But the difference was entirely marginal and the accompanying table shows how the actual coal per train mile, and also the basic consumption varied from day to day.

Although the results of these trials did not show any particular advantage to the high pressure engine in respect of basic fuel consumption there is no doubt that in general performance there were benefits to be gained, and the immediate outcome was the adoption of 220 lb. per sq. in. as standard for future " Pacific " engines on the L.N.E.R. and in a batch of new locomotives turned out from Doncaster in 1928 the cylinder diameter was 19 in., with 8 in. diameter piston valves. The nominal tractive effort of these engines, classified " A3 " was 32,909 lb., an appreciable advance over the original 180 lb. engines. These locomotives (Figs. 26, 27 and 28) can be seen as the ultimate outcome, so far as the L.N.E.R. was concerned, of the historic interchange trial of 1925 with the Great Western Railway—an outcome that was to the great advantage of the L.N.E.R. express passenger motive power stud.

Fig. 29.—*Southern Railway: R. E. L. Maunsell's 4-cylinder 4-6-0 No.* 850 Lord Nelson, *built Eastleigh* 1926.

The years 1925–27 were ones of exceptional activity in the design departments of all four of the " grouped " railways of Great Britain, though these activities stemmed from a diversity of origins. On the London Midland and Scottish Railway, with Sir Henry Fowler succeeding George Hughes as Chief Mechanical Engineer in 1925, it was no more than natural that the Midland influence in the locomotive department should increase, particularly as the 3-cylinder compound 4-4-0 engines of Derby design had done so well in the various sets of dynamometer car test runs that had been conducted since grouping took place. On the L.M.S.R. however, between the departments concerned with the provision and utilisation of motive power there were considerable divergencies of view, and the need for enhanced engine power was not so clearly recognised as on the Southern, and on the G.W.R., where in the years between 1924 and 1926 the locomotive departments had definite directives from the top management as to future haulage requirements. On the L.M.S.R. too, a further point of controversy was whether or not the compound system of propulsion should be adopted as a future standard.

It was a time of intense interest in the history of the steam locomotive, not only in Great Britain but in all the major railway countries of the world. While in Europe as a whole the pre-war addiction towards compounding seemed to be decreasing the principles of Alfred de Glehn continued to be developed with remarkable success in France, and particularly on the Northern Railway where the " Super-Pacifics " of M. Bréville's design were doing outstanding work on the English boat expresses between Paris and Calais. On the L.M.S.R. Fowler and his staff somewhat naturally looked to the compound for their top-line express passenger power, and prior to initiating detail design at Derby the best continental practice had been carefully studied. The design for a large 4-cylinder compound " Pacific " was eventually worked out in full detail, and construction had commenced to the extent that the frames were cut for the first engine. Then other interests supervened.

At that point in British locomotive history the Webb compound era on the London and North Western Railway was still a vivid memory in the minds of many of the older men; and when that era eventually came to an end in 1903 there was among the men responsible for locomotive running a general feeling towards compounds that could be summed up in the two words: "never again". Furthermore there had been the trial of de Glehn compound "Atlantics" on the Great Western Railway, ending in the production of the famous 4-cylinder single expansion 4-4-2 *North Star,* by Churchward, which was generally considered to be superior on all counts to the French compounds. On the London Midland and Scottish Railway the compound *versus* simple controversy was eventually settled more by practical politics than by considerations of engineering design. A Great Western 4-cylinder 4-6-0 of the "Castle" was borrowed, and put through a series of tests, some with the Horwich dynamometer car, between Euston, Crewe and Carlisle; while by way of exchange a Midland 3-cylinder compound 4-4-0 was lent to the Great Western Railway. Although the G.W.R. engine No. 5000 did not reproduce, on the L.M.S.R. metals, the remarkable figures of coal consumption that had startled the locomotive world when Mr. Collett's paper had been presented to the World Power Conference, in 1924, the engine did display a comfortable mastery over most of the duties assigned. The only trouble experienced was in bad weather in the North country, when the G.W.R. arrangement of dry sanding proved ineffective against the combination of persistent side winds and drizzling rain, and the engine was consequently handicapped by much incipient slipping.

But the running of engine No. 5000 *Launceston Castle* convinced the management of the L.M.S.R. that a well-designed 4-6-0 of 32,000 or 33,000 lb. tractive effort would suffice for all the heaviest duties then envisaged, and all work on the 4-cylinder compound "Pacific" was stopped. From this decision there eventuated the well-known "Royal Scot" class of 3-cylinder 4-6-0 locomotive; but before referring to the details of this design, in the interest of true historical sequence, two other large express passenger locomotives of the 1926–27 period must be mentioned. Both were the outcome of demands by the respective traffic departments for greatly enhanced power, and both incorporated some notable advances in design. The Southern 4-cylinder 4-6-0 No. 850 *Lord Nelson* (Fig. 29), was the result of a requirement for the haulage of 500-ton express trains at average speeds of 55 m.p.h. The basic dimensions are given in the table on page 37, but an interesting feature was the setting of the cranks at 135 deg. so as to give eight exhausts per revolution, instead of the usual four. This was a development of the point claimed for a 3-cylinder locomotive with cranks at 120 deg., that the softer exhaust, and consequently less fluctuation in smokebox vacuum, had a beneficial effect upon coal consumption.

At the time of its completion at Eastleigh Works, in 1926, the *Lord Nelson* had the highest nominal tractive effort of any British express passenger locomotive, namely 33,500 lb. at 85 per cent. boiler pressure, though this supremacy was not to remain for long. The *Lord Nelson* incorporated the refinements in front-end design that were an established feature of Maunsell's practice; but while the success of the Great Western "Castle" class had undoubtedly been a factor influencing the use of four cylinders the actual layout differed considerably from standard Swindon practice, in which the cylinders were so disposed as to permit of equal lengths of connecting rod for all four cylinders. This led to rather long steam passages, whereas in the *Lord Nelson* the four cylinders were placed almost in line, and permitted of short and direct passages both for steam and exhaust. With the drive divided between two axles, as in the G.W.R. engine, this layout of cylinders resulted in connecting rods of 11 ft. centres for the outside cylinders and 6 ft. 11 in. for the inside ones. Four sets of valve gear were used. At one time

Fig. 30.—*G.W.R. The largest development of the Churchward 4-cylinder* 4-6-0*: C. B. Collett's "King" class, of* 1927.

Fig. 31.—*L.M.S.R.* 3-*cylinder* 4-6-0 *No.* 6100 Royal Scot, *as originally built* in 1927 *by the North British Locomotive Co. Ltd.*

consideration was given to a conjugated motion to avoid the use of inside valve gear; but the disadvantages of the 3-cylinder motion experienced on 2-6-0 tender, and 2-6-4 tank locomotives of South Eastern and Chatham design were considered to outweigh the undoubted merits of avoiding inside motion altogether.

The second large new express passenger locomotive which preceeded the completion of the " Royal Scot " by a few weeks was the G.W.R. *King George V* (Fig. 30)—a very striking development in size and power of the basic Churchward 4-cylinder 4-6-0 locomotive of 1907, evolved as a direct result of the trials of the de Glehn compound " Atlantics " purchased from France in 1903 and 1905. In the production of a locomotive of more than 40,000 lb. tractive effort on a 4-6-0 chassis the Swindon drawing office had achieved a masterpiece of design, and the performance of these locomotives very soon showed that the tractive force was no theoretical figure only to be realised in the most favourable circumstances. In the haulage of heavy express trains at high speed the work was outstanding, and will be referred to in detail in a later chapter. At this stage it is enough to state that the advance in tractive effort over the " Castle " was achieved in four ways:

1. Boiler pressure increased from 225 to 250 p.s.i.
2. Driving wheel diameter reduced from 6 ft. $8\frac{1}{2}$ in. to 6 ft. 6 in.
3. Cylinder diameter increased from 16 to $16\frac{1}{4}$ in.
4. Piston stroke increased from 26 to 28 in.

The boiler was a very carefully designed enlargement of the classic Churchward " Swindon No. 1 " standard, having in the new design a grate area of 34.3 sq. ft. A notable feature of the " King " class was an increase in maximum axle load from 19·75 tons in the "Castle" to 22.5 tons. The significance of this will be discussed later.

On the L.M.S.R. the trials of the G.W.R. " Castle " class locomotive took place in the late Autumn of 1926, and new locomotives were required for the summer traffic of the following year. When the decision was taken to adopt the 4-6-0 type the railway workshops were not in a position to design and build the locomotives in time, and a contract was placed with the North British Locomotive Company for no fewer than 50, straight off the drawing board, and the contract was to include all the detail designing. The broad specification laid down a 3-cylinder locomotive of approximately 33,000 lb. tractive effort; in view of the extreme urgency of the contract the assistance of Mr. R. E. L. Maunsell of the Southern Railway was obtained, and a complete set of drawings of the *Lord Nelson* were made available to the contractors. In Glasgow constructional work was divided between the Hyde Park and Queens Parks Works, with an allocation of 25 locomotives of the class apiece. The *Royal Scot* (Fig. 31) was not ready for the summer traffic of 1927, but sufficient engines of the new class were available for the introduction of the remarkable innovation, in the autumn of that year, of regular daily non-stop running between Euston and Carlisle, 299.1 miles. The dimensional details of the " Royal Scots " are set out in the table on page 37, in comparison with those of the premier locomotives of the other three " group " railways of Great Britain. It remains to be added that the production of the " Royal Scots " in so short a time was a remarkable tribute to the potentialities of the British locomotive building industry; and that the engines so admirably met the very exacting demands of the new services on the L.M.S.R. was very much to the credit of the coordinating efforts of the design staff at Derby, to whom, incidentally, the cancellation of the compound " Pacific " project must have been a great disappointment.

The improvement in performance made possible by refinements in valve gear design have already been discussed, but the physical implications of the running of locomotives of greater total weight, and because of the great increase in boiler diameter with a higher centre of gravity, needed equally important consideration. The behaviour of the locomotives as "vehicles", was dependent to a large extent upon the balancing of the revolving and reciprocating parts, and in this respect British locomotive engineers had received invaluable assistance from the work of the Bridge Stress Committee set up in March 1923 under the auspices of the Department of Scientific and Industrial Research. The effect of locomotives on bridges led naturally to an examination of the various practices then in vogue for balancing, and some wide divergencies as between the various design centres in Great Britain were revealed. It was also shown that the effects of these divergencies, so far as the track and under-line structures were concerned, were not clearly appreciated by the civil engineers of the day. Indeed one case existed of a locomotive that had been condemned and transferred elsewhere, but which was far less severe in its effect on the track than a subsequent design that had been accepted in substitution.

Certain design practices, although they do not concern locomotives built new in the period now under review, are worth mentioning in view of subsequent changes made. On the former London and North Western and Midland Railways for example, it was the practice, with 2-cylinder engines, to concentrate the whole of the balance of the reciprocating parts into the driving axle, with the result that the hammer-blow from this axle was greater than the combined effect of the entire engine. The varying methods of balancing can be exemplified by hammer-blow effects for three well-known 4-4-0 locomotives of pre-1914 design.

Hammer-blow at 6 revs. per second

Railway .. .. Class	L.N.W.R. "George V"	M.R. 2P	G.W.R. "County"
Speed at 6 r.p.s. m.p.h.	87	90	86
Max. axle load, tons	19.15	17.5	19.4
Hammer-blow at 6 r.p.s.			
Whole engine, tons	9.7	13.1	16.6
Max. axle, tons	14.1	11.8	8.5
Max. combined load at 6 r.p.s., tons	33.2	29.3	27.9

Weights and Hammer-blow Effect in Typical British Locomotives at 6 revs. per sec.

Railway	Engine Class	Type	Max. Axle load tons	Speed at 6 r.p.s. m.p.h.	Hammer-blow at 6 r.p.s.		Max. Combined load at 6 r.p.s.
					Whole engine	Axle	
G.W.R.	"Saint"	4-6-0	18.4	86	17.9	6.9	25.3
,,	"Star"	4-6-0	18.6	86	3.7	2.9	21.5
,,	"Castle"	4-6-0	19.7	86	3.5	3.4	23.1
,,	"King"	4-6-0	22.5	84	2.2	4.0	26.5
Southern	"King Arthur"	4-6-0	19.5	85	20.2	9.1	28.6
,,	"Lord Nelson"	4-6-0	20.65	85	5.2	2.6	23.2
S.E.C.R.	"D1"	4-4-0	17	86	3.4	5.8	22.8
L.N.W.R.	"George V"	4-4-0	19.15	87..	9.7	14.1	33.2
,,	"Claughton"	4-6-0	19.75	87	nil	nil	19.75
Midland	Compound	4-4-0	19.75	87	1.76	3.65	23.4
,,	"2P"	4-4-0	17.5	90	13.1	11.8	29.3
L. & Y.R.	Class "8"	4-6-0	20	80	2.5	1.6	21.6
Caledonian	"60"	4-6-0	19.3	78	23.5	11.7	31.0
Highland	"Castle"	4-6-0	15.15	74	16.3	6.2	21.2
,,	"River"	4-6-0	17.75	77	1.7	4.2	21.4
,,	"Clan"	4-6-0	15.33	77	15.3	7.1	22.4
L.M.S.R.	"Royal Scot"	4-6-0	20.9	87	0.9	0.3	21.2
G.N.R.	"251"	4-4-2	20	86	9.2	10.4	30.4
N.E.R.	"Z"	4-4-2	19.95	88	3.5	2.8	23.1
N.B.R.	"868"	4-4-2	20	87	16.1	11.6	31.6
G.C.R.	"Lord Farringdon"	4-6-0	20	87	5.6	3.3	22.2
L.N.E.R.	"A3"	4-6-2	22	86	2.3	2.6	24.7

In passing it is curious to recall that at Crewe, where there was acceptance of the above arrangements, not only on the "George V" class 4-4-0s but also on the very numerous inside cylinder 4-6-0s of the "Prince of Wales" class, the 4-cylinder 4-6-0 of the "Claughton" class should have been produced, having, by reason of its cylinder and crank arrangements, no hammer-blow at all. The balancing arrangements on the Great Western 4-cylinder 4-6-0 locomotives were not to be commended, and the work of the Bridge Stress Committee drew attention to their ill-effects. An appreciable proportion of reciprocating parts was balanced separately for the inside and outside cylinders, in the leading and middle pairs of coupled wheels respectively. Although the balance applied to the leading axle was opposed to that in the middle one, and the total engine hammer-blow was relatively small, the hammer-blow from each of the individual axles actually exceeded that of the 2-cylinder "Saint" class engines, which was nearly 7 tons maximum on any axle. During the course of the Bridge Stress Committee's work the method of balancing the 4-cylinder engines was revised at Swindon, with the advantageous results to be noted in the table on this page.

In this table particulars have been included for the former Highland Railway "River" class 4-6-0s, which were the locomotives previously referred to as being condemned and transferred elsewhere. These engines were put into service in 1915, and the unfortunate circumstances connected with them were undoubtedly too fresh in memory for any comment to be made in Ahrons' classic work of 1925; in fact they are not even mentioned. While one can appreciate the reason this omission does scant justice to what was undoubtedly the most advanced design of passenger

Fig. 32.—*L.M.S.R. The* Royal Scot *as equipped with electric headlamp for the North American tour of* 1933-4.

Fig. 33.—*L.M.S.R. Derby-built " Royal Scot "* 4-6-0 *with large tender, and smoke-deflecting plates.*

Large Express Passenger Locomotives 1926–28

Year Introduced	1926	1927	1927	1928
Railway	Southern	G.W.R.	L.M.S.R.	L.N.E.R.
Type	4-6-0	4-6-0	4-6-0	4-6-2
Class	" Nelson "	" King "	" Royal Scot "	" A3 "
Cylinders number	4	4	3	3
dia., in.	$16\,^1/_2$	$16\,^1/_4$	18	19
stroke, in.	26	28	26	26
Coupled wheel dia., ft. in.	6–7	6–6	6–9	6–8
Boiler pressure, p.s.i.	220	250	250	220
Heating surfaces, sq. ft.				
Evaporative	1795	2007.5	1892	2736.6
Superheater	376	289	416	706
Firebox	194	193.5	189	215
Combined total	2365	2490	2497	3657.6
Grate area, sq. ft.	33	34.3	31.2	41.25
Valve gear details				
Max. travel, in.	$6\,^1/_2$	$7\,^1/_4$	$6\,^3/_{16}$	$5\,^3/_4$
Steam lap, in.	$1\,^1/_2$	$1\,^3/_8$	$1\,^7/_{16}$	$1\,^5/_8$
Exhaust clearance, in.	Nil	Nil	Nil	Nil
Lead, in.	$^1/_4$	$^3/_{16}$	$^1/_4$	$^1/_4$
Cut-off full gear, %	75	76.5	75	65
Total engine wt., tons	$83\,^1/_2$	89	84.9	96.2
Adhesion weight, tons	62	$67\,^1/_2$	$62\,^1/_2$	66.1
Tractive effort lb. at 85% boiler pressure	33,500	40,300	33,150	32,909

locomotive to be built for service on any of the former independent Scottish railways. The particulars so far as balancing is concerned are given in the table on page 36, and although the maximum axle load was considerably higher than anything previously used on the Highland Railway the effect on the track was no more severe than that of the much less powerful locomotives already in service on that line. Furthermore, the engines of the " Clan " class, which were subsequently accepted by the civil engineer, and took up the work intended for the " Rivers " had a very much more severe effect on the track so far as hammer blow was concerned.

So far as future practice in Great Britain affected the balancing of locomotives one has only to study the hammer-blow figures for the four powerful locomotives of the " King ", " Lord Nelson ", " Royal Scot " and Gresley " Pacific " classes to appreciate the very great progress made in this respect. As will be shown in later chapters the principles established as a result of the cooperation of locomotive and civil engineers during the work of the Bridge Stress Committee were of subsequent value towards the introduction of larger and more powerful locomotives on certain secondary routes where the limitation in maximum axle load had hitherto been severe.

Fig. 38.—*Great Indian Peninsular Railway: one of the largest non-articulated freight engines introduced prior to the setting up of the Locomotive Standards Committee; a* 2-10-0 *of* 1924.

Fig. 39.—*South Indian Railway: combined rack and adhesion* 0-8-2 *for the Siliguri line.*

away from the war zone, to French designs, and 300 of the well-known Great Central "04" 2-8-0, which was adopted by the Ministry of Munitions as a national standard.

The table on page 39 includes particulars of some typical locomotives supplied to the railways of India, South Africa, South America, and elsewhere in the period 1910–1925, and as indicated, a number of these locomotives are illustrated herewith. With the exception of the Mallet and other articulated types, which are the subject of a separate chapter, it will be seen from the illustrations that all have a characteristically "British look". The table, studied in conjunction with the illustrations, gives some impression of the varied conditions of service for which these locomotives were built.

Turning now to the commencement of the true period of this book one of the most significant trends of the time was the establishment, in 1924, of the first Indian Locomotive Standards Committee, which had the duty of drawing up designs and specifications for a range of standard locomotives that should be suitable for service on any of the Indian railways. By that time the majority of the railways in India were Government owned, though retaining their former names and continuing to be operated as separate concerns. The need for economy and the use of lower grades of coal prompted the decision to use the "Pacific" type generally, because this type permitted the adoption of wide fireboxes and large grate areas

Fig. 40.—*Bengal Nagpur Railway: narrow gauge* 2-8-2 *for* 2*ft.* 6*in. gauge lines.*

Fig. 41.—*New Zealand Government Railway: a 3ft. 6in. gauge* 4-6-2 *with Vanderbill tender.*

Fig. 42.—*South African Railways:* 4-6-2 *passenger locomotive.*

Fig. 43.—*Benguela Railway: Wood burning* 4-8-0.

Fig. 44.—*Buenos Aires Pacific Railway: an early* 4-6-2 *express passenger locomotive.*

Fig. 45.—*Sao Paulo Railway:* 4-6-2 *passenger locomotive.*

Fig. 46.—*Central Argentine Railway:* 4-8-0 *heavy freight locomotive.*

Fig. 47.—*Pekin-Kalgan Railway: the first British built Mallet articulated compound locomotive* 0-6-6-0 *type* (1907).

required to burn the lower grades of coal economically. Nevertheless while the specifications for the new designs were being drawn up, large orders were placed in England for 4-6-0 locomotives of the H.P.S. design: a simple, straightforward machine of typically British appearance, with two cylinders 20 1/2 in. diameter by 26 in. stroke; 6 ft. 2 in. coupled wheels, and a boiler pressure of 180 lb. per sq. in. Despite the marked trend towards larger locomotives the "H.P.S." has withstood the test of time, and further large orders were executed by the Vulcan Foundry in 1947 for locomotives but little modified from the original design of 1924 (Fig. 48).

The original standard "Pacific" designs for the broad gauge Indian railways were of three classes:

"XA", for branch lines, maximum axle load . . . . 13.1 tons

"XB", light main lines, maximum axle load . . . . 17 tons

"XC", heavy main lines, maximum axle load . . . . 19.8 tons

Common to all three varieties was an improved front-end, with large smokebox volume, free exhaust passages and good draughting; the steam and exhaust ports were straight and direct, and a moderately long valve travel was adopted. It was a guiding principle to use a large firebox heating surface and grate area. Certain basic ratios were adopted in calculating the leading dimensions of all three classes as follows: diameter of cylinders to give a tractive effort at 85 per cent. of the boiler pressure, of the adhesion weight divided by 4.25; the grate area was to be the tractive effort divided by 600; the evaporative and superheater heating surfaces were obtained by multiplying the grate area by 50 and 12 respectively. All three classes were to use a boiler pressure of 180 lb. per sq. in. The basic dimensions were thus evolved for the three classes as follows:

Class	"XA"	"XD"	"XC"
Illustration ref.	Fig. 49	Fig. 50	Fig. 51
Cylinders (2), dia., in.	18	21 1/2	23
stroke, in.	26	28	28
Coupled wheel dia., ft. in.	5 1 1/2	6 2	6–2
Grate area, sq. ft.	32	45	51
Heating surfaces, sq. ft.			
tubes	1277	1642	2222
firebox	122	198	207
Total evaporative	1399	1840	2429
superheater	348	463	636
Nom. tractive effort at 85% B.P., lb.	20,960	26,760	30,625
Adhesion weight, tons	39	51	59
Total engine weight, tons	66.5	90.25	97
Total engine and tender, tons	108 75	155.25	175
Coal capacity, tons	7.5	10.0	14.0
Water capacity, galls.	3000	4500	6000

Fig. 48.—*Nizam's Guaranteed State Railway:* "H.P.S" *class* 4-6-0, *a standard Indian medium powered type.*

Fig. 49.—*Indian Standard: Class "XA" light branch line "Pacific," built by Vulcan Foundry Ltd.*

Fig. 50.—*Indian Standard: Class "XB" intermediate "Pacific", built by North British Locomotive Co. Ltd.*

Fig. 51.—*Indian Standard: Class "XC" heavy main line "Pacific", built by Vulcan Foundry Ltd.*

Fig. 52.—*Bengal Nagpur Railway: de Glehn* 4-*cylinder compound* 4-6-2 (1928), *built by North British Locomotive Co. Ltd.*

Fig. 53.—*Indian Standard: Class "XD" 2-8-2 heavy freight engine.*

Trials with a few sample locomotives shipped from England in 1924 proved satisfactory, and bulk orders were subsequently placed. As originally built all three classes had spring controlled leading bogies with small wheels of 3 ft. 0 in. diameter, and the trailing trucks were of the Cartazzi type. The intermediate draw-gear was the Goodall freely-articulating type. These features of the Indian Locomotive Standards Committee specification were to result in a good deal of bad riding on track that for various reasons could not be maintained up to the highest standards, and their shortcomings in this respect were unfortunately high-lighted in the most dramatic and tragic manner in the derailment at 60 m.p.h. at Bihta, on the East Indian Railway in 1937, when a 4-6-2 of the "XB" class overturned on straight track, and the subsequent wrecking of the train caused a loss of over 100 lives.

Among non standard locomotives built for India in the period under review must be mentioned the very fine 4-cylinder de Glehn compound 4-6-2s supplied to the Bengal Nagpur Railway by the North British Locomotive Company, in 1928 (Fig. 52). This railway, which was still privately owned at that time, had been one of the earliest users outside Europe of the de Glehn system of compounding, and between the years 1906 and 1913 a total of 17 very handsome "Atlantics", of generous proportions, had been supplied by the North British Locomotive Company. The development, from the "Atlantic" to the "Pacific" type in 1928 was in keeping with developments in the de Glehn compound system in the land of its origin, though these British-built locomotives for India were among the largest and most powerful of the type ever constructed. So far as framing, spring gear, wheel-base and so on this design conformed generally to the Indian standard "XC" class, with 6 ft. 2 in. coupled wheels, and 3 ft. 0 in. bogie wheels; but to derive the greatest advantage from compounding the boiler pressure was fixed at 250 lb. per sq. in., and the heating surfaces were as follows:

Tubes	2228 sq. ft.
Firebox	211 sq. ft.
Total evaporative	2439 sq. ft.
Superheater	637 sq. ft.
Grate area	51 sq. ft.

The cylinders, all with a stroke of 26 in., were $16\frac{1}{2}$ in. diameter high pressure, and 25 in. low pressure. The nominal tractive efforts were:

Working simple	38,000 lb.
Working compound	28,700 lb.

In accordance with the well-known de Glehn principles the engines could be worked simple when required. The change valve was power operated through an auxiliary steam valve controlled from the cab, which, when working simple supplied live steam to the low pressure cylinders. An unusual feature for the year 1928 was the use of balanced slide valves for the low pressure cylinders, as illustrated in the accompanying cross-sectional drawing (Fig. 55). In all 18 of these locomotives were built.

Another notable British-built locomotive for India was the "XE" standard 2-8-2 heavy freight design (Fig. 54), built first by the Vulcan Foundry in 1928.

Fig. 54.—*Indian Standard (East Indian Railway) "XE" 2-8-2 engine for heaviest freight duties* (1928).

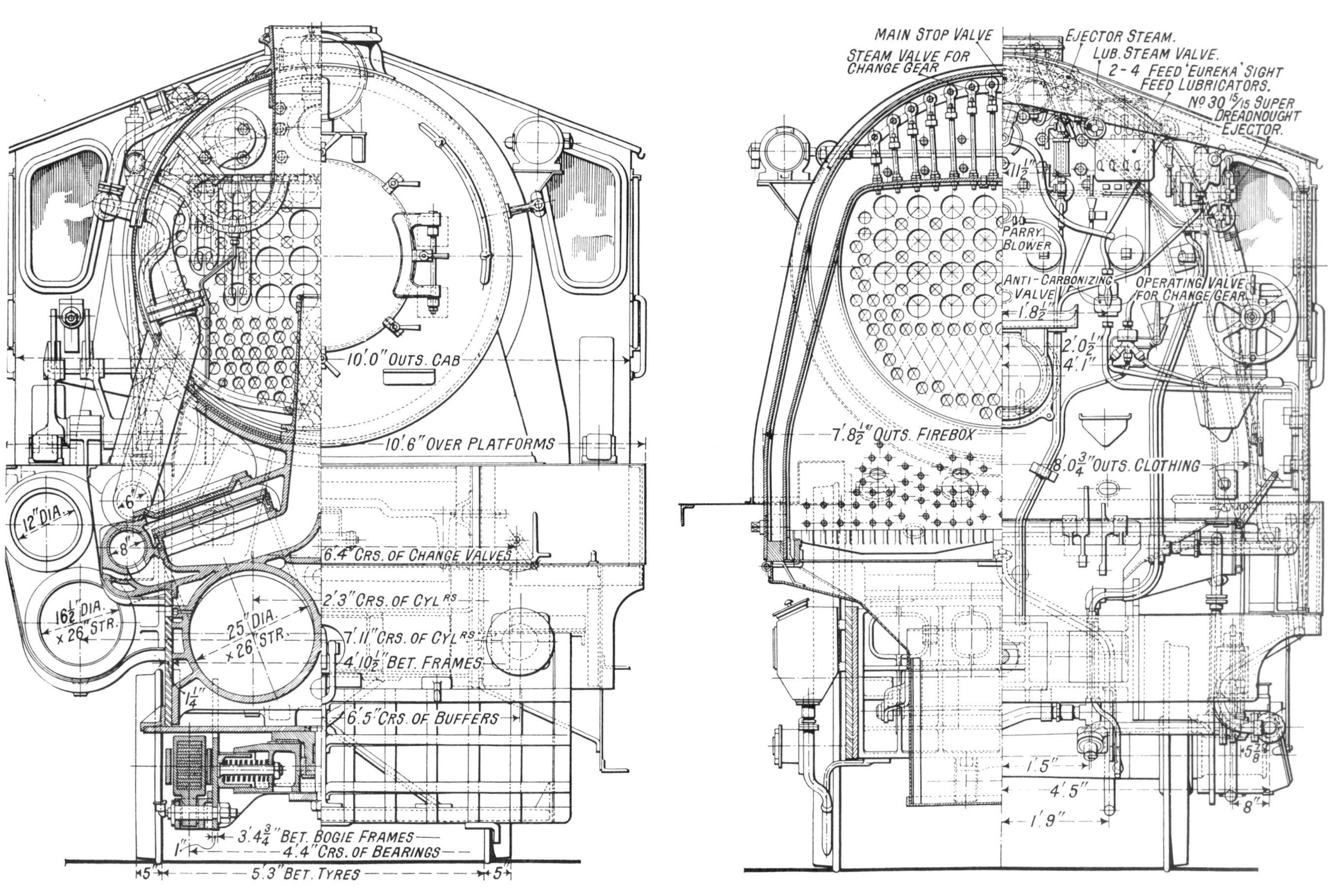

Fig. 55.—*B.N.R. 4-cylinder compound 4-6-2: cross sectional drawings at front end and firebox.*

Fig. 56.—*Buenos Aires Great Southern Railway:* 3-*cylinder* 4-6-2 *for express passenger service, built by Vulcan Foundry Ltd.*

The boiler was considerably larger than that of any of the standard " Pacifics ", having a total evaporative heating surface of 3014 sq. ft., a grate area of 60 sq. ft., and carrying a working pressure of 210 lb. per sq. in. These fine locomotives had a nominal tractive effort of no less than 48,086 lb.; cylinders 23 $^1/_2$ in. diameter by 30 in. stroke, and coupled wheels 5 ft. 1 $^1/_2$ in. diameter. It will be apparent from the accompanying illustration that despite their great power, and the comprehensive nature of the details of equipment a remarkable "clean" and characteristically British outline had been preserved. The smaller standard " XD " 2-8-2 locomotives (Fig. 53) may also be noticed, having 22$^1/_2$ in. by 28 in. cylinders; a boiler pressure of 180 lb. per sq. in. and a nominal tractive effort of 35,260 lb. at 85 per cent. boiler pressure. The boiler has instead the following dimensions:

Heating surfaces	
Tubes	1986 sq. ft.
Firebox	190 sq. ft.
Total evaporative	2176 sq. ft.
Superheater	540 sq ft.
Grate area	45 sq. ft.

In the period under review the Argentine railways, then British owned, took delivery of a number of excellent examples of British locomotive engineering practice, some which were 3-cylinder simples and others compounds. The B.A.G.S. " Pacific " illustrated (Fig. 56), was designed for high speed express work on the Buenos Aires–Mar del Plata, and the Buenos Aires–Bahia Blanca lines, and had the unusually large coupled wheel diameter, for South America, of 6 ft. 6 in. They have three cylinders 19 in. diameter by 26 in. stroke; a boiler pressure of 200 lb. per sq. in. and a nominal tractive effort of 30,685 lb. They proved capable of excellent work hauling loads up to 750 tons, on a schedule that allowed 5 $^1/_2$ hours for a run of 250 miles, with two intermediate stops over a route including much single track. They were oil-fired, and averaged a consumption of 40–42 lb. per mile. These locomotives were built by the Vulcan Foundry.

The Buenos Aires Great Southern Railway operated a heavy suburban traffic, and in the years of expansion after the first world war the facilities for handling this were greatly improved by complete remodelling of the terminus station Plaza Constitution, and by quadrupling the main line to Temperley, 11 $^1/_2$ miles out. The service was entirely steam operated, and in 1925 a new design 2-6-4 suburban passenger tank locomotive was introduced (Fig. 57). Like the express passenger locomotives just described these were 3-cylinder simples, and with a high nominal tractive effort of 29,859 lb. were admirably suited to demands of suburban traffic, requiring rapid acceleration from rest. The leading dimensions of these handsome locomotives were:

Cylinders (3) dia., in.	17 $^1/_2$
stroke, in.	26
Coupled wheels dia., ft. in.	5 – 8
Heating surfaces sq. ft.	
tubes	1188
firebox	135
total evaporative	1323
superheater	302
Grate area, sq. ft.	25
Boiler pressure, lb./sq. in.	200
Coal capacity, tons	4
Water capacity, galls.	1953
Max. axle load	19.25
Total weight in working order, tons	101

Fig. 57.—*Buenos Aires Great Southern Railway:* 3-*cylinder* 2-6-4 *tank engine for suburban work.*

Fig. 58.—*Buenos Aires Great Southern Railway: 3-cylinder oil-burning heavy freight locomotive, built by Beyer, Peacock & Co. Ltd.*

Yet another example of 3-cylinder propulsion on the B.A.G.S. Railway is that of the general purpose, oil-burning 4-8-0, built by Beyer, Peacock and Co. Ltd. (Fig. 58). Chiefly through use of a small coupled wheel diameter of 4 ft. 7 1/2 in., with ample dimensions also in other respects these engines have a high nominal tractive effort of 36,580 lb. at 85 per cent. boiler pressure. The cylinders were 17 1/2 in. diameter by 26in. stroke, as in the Buenos Aires suburban tank engines; the grate area was 29.2 sq. ft., and the fuel capacity 11 1/2 tons of oil.

On the Central Argentine Railway (F.C.C.A.) the use of 2-cylinder cross-compound locomotives had been standard for many years, and it is interesting to recall that as recently as 1940 twenty 2-8-2 locomotives were supplied by Vulcan for work in the Cordoba and Villa Maria districts (Fig. 59). These neat and compact locomotives had the following leading dimensions:

Cylinders:	
H.P. dia., in.	21
L.P. dia., in.	31 1/2
Stroke, in.	26
Dia. of coupled wheels, ft. in.	4 – 7 1/2
Heating surfaces, sq. ft.	
Tubes	1436
Firebox	165
Total evaporative	1601
Superheater	277
Grate area	27.9
Nominal T.E. at 66% boiler pressure, lb.	27270
Maximum axle load, tons	16.5

A very fine 2-8-2 design was supplied by Beyer, Peacock and Co. Ltd. to the Buenos Aires Pacific Railway at the end of 1928, weighing complete with its tender no less than 205 tons (Fig. 60). The rail gauge is 5 ft. 6 in., and with a maximum permitted axle load of 21.65 tons, a very simple and compact design was possible within a total wheelbase that can be considered fairly short for so powerful a locomotive. The two cylinders were 24 1/2 in. diameter by 30 in. stroke; the coupled wheels 5 ft. 7 in. diameter, and the boiler pressure 200 lb. per sq. in. These dimensions provided a nominal tractive effort, at 75 per cent. boiler pressure, of 40,290 lb. Their tractive effort is quoted at 75 per cent. boiler pressure instead of the usual 85 because the valve gear is arranged for a maximum cut-off of 65 per cent. in full gear.

The boiler was short in relation to its diameter of 6 ft. 10 1/2 in. outside. The length of barrel was 21 ft. 11 in., but with a large combustion chamber the distance between tube plates was only 18 ft. 0 in. The total evaporative heating surface was 2760 sq. ft.; the superheater 676 sq. ft.; and the grate area 46 sq. ft. The large tenders had a water capacity of 7700 gall. and a coal space for 15 tons. An interesting item of equipment was the coal pusher on the tender—one of the earliest, if not the earliest use of this device on a British built locomotive.

On the various 3 ft. 6 in. gauge railways in the African Continent, working almost entirely over single lines, the need to operate trains of maximum tonnage at infrequent intervals, rather than resort to the practice of fully developed countries and run a frequent service of relatively light trains, led to the designing of some very large locomotives that had the appearance of great length. Three designs in particular

Fig. 59.—*Central Argentine Railway: 2-cylinder cross-compound 2-8-2, built by Vulcan Foundry Ltd.* (1930).

Fig. 60.—*Buenos Aires Pacific Railway:* 2-8-2, *built by Beyer, Peacock & Co. Ltd.* in 1928. *Tender fitted with coal pusher.*

may be referred to, all for freight service and introduced on to the Gold Coast and Nigerian Railways in 1924–1930. Taken collectively these designs are interesting in that the Nigerian are 3-cylinder simples, and have the Gresley conjugated valve gear. Taking the Gold Coast 4-8-2 design first (Fig. 61), these were based on a maximum permitted axle load of $12\,^1/_2$ tons. A high tractive effort of 30,415 lb. was obtained by use of coupled wheels of only 3 ft. 9 in. diameter. The two cylinders were $19\,^1/_4$ in. diameter by 24 in. stroke, and the boiler pressure was 180 lb. per sq. in. They were built by the Vulcan Foundry, and proved very successful in dealing with heavy ore trains between Kumasi and the port of Takoradi. The boiler proportions are referred to later.

The Nigerian locomotives, with a maximum axle load of $16\,^1/_2$ tons are considerably larger and more powerful. The boiler proportions are compared with those of the Gold Coast 4-8-2 later, but the other basic dimensions are tabulated herewith, for the 2-8-2 of 1925 and the 4-8-2 of 1930 (Fig. 62). The latter class had bar frames. The great length of these

Nigerian Railways
Heavy Freight Locomotives

Year first introduced	1925	1930
Type	2-8-2	4-8-2
Cylinders (3) dia., in.	18	$18\,^3/_4$
stroke, in.	28	24
Coupled wheel dia., ft. in.	4–6	4–6
Total heating surface, sq. ft.	2796	3060
Grate area, sq. ft.	38	45
Boiler pressure, lb./sq. in.	180	200
Total wt. of engine, tons	88.2	90.35
Total wt. of engine and tender, tons	125.85	138.75
Nom. T.E. at 85% boiler pressure, lb.	38,560	39.845

locomotives will be apparent from the accompanying illustration, and the boiler proportions are of interest accordingly. It is worthy of note that whereas the two earlier designs—Gold Coast of 1924, and Nigerian of 1925 included Belpaire fireboxes, the very large Nigerian 4-8-2 of 1930 had a round-topped firebox. All three designs include combustion chambers in the fireboxes to avoid a length of tube that would be excessive in relation to internal diameter, and thus impair the steaming capacity of the boiler.

West African Freight Locomotives
Boiler Dimensions

Railway	Gold Coast	Nigerian	Nigerian
Loco. Type	4-8-2	2-8-2	4-8-2
Nom. T.E. at 85% B.P. lb.	30,415	38,560	39,845
Small tubes, No.	121	119	136
Int. dia., in.	2	$2\,^1/_4$	$2\,^1/_4$
Large tubes No.	18	26	30
Int. dia., in.	$5\,^1/_4$	$5\,^3/_8$	$5\,^1/_2$
Length between tubeplates, ft. in.	$16–5\,^3/_4$	19–6	18–0
Inside dia. of boiler, ft. in.	$5–0\,^3/_4$	$5–5\,^3/_4$	$5–9\,^1/_8$
Heating surfaces, sq. ft.			
Tubes	1418	2080	2235
Firebox	160	210	218
Total evaporative	1608	2290	2453
Superheater	298	506	607
Grate area, sq. ft.	34	38	45

On the South African Railways the era of the really big locomotive was just commencing in the late nineteen twenties, and extended reference will be made later to the many fine examples built in this country. At this stage however a powerful example of what can be termed the penultimate steam era in South Africa is to be seen in the "1800" class of 4-8-2, illustrated herewith. Although usually designated "goods" these fine engines were regularly used on long distance passenger workings. The engine

Fig. 61.—*Gold Coast Railway:* 3*ft.* 6*in. gauge* 4-8-2 *for heavy ore trains. Built by Vulcan Foundry Ltd.*

illustrated (Fig. 63) was one of a series built by Beyer, Peacock and Co., and one noticeable item of equipment is the Hadfield steam reverser. The leading dimensions of these locomotives were:

Cylinders (2) dia. × stroke, in.	$22\ ^1/_2 \times 26$
Coupled wheel dia., ft. in.	4–3
Wheel base, engine, ft. in.	$31\text{–}9\ ^1/_2$
rigid, ft. in.	9–0
total, ft. in.	58–4
Axle load, tons	17
Adhesive weight, tons	68
Wt. of engine in working order, tons	93
Wt. of engine and tender, tons	144
Tractive effort at 85% boiler pressure, lb.	41,670
Boiler pressure, lb./sq. in.	190
Heating surfaces, sq. ft.	
Tubes	2323
Firebox	156
Total evaporative	2479
Superheater	388
Total	2867
Grate area, sq. ft.	39.8
Coal capacity, tons	10
Water capacity, gall.	4250

nearly 24,000 lb. at 85 per cent. boiler pressure, a coupled wheel diameter of only 3 ft. 7 in. was decided upon.

The 2-8-2 locomotives of the "206" class built by Vulcan afford an excellent example of constructional practice. The total weight of engine and tender in working order is no more than $86\frac{1}{2}$ tons, while the principal dimensions are given in the accompanying table. It will be noted that the eight-wheeled tender, which had a capacity for 400 cu. ft. of wood fuel has a rigid wheelbase. The boiler was fed through top feed clackboxes by one Gresham and Craven No. 9 injector, and a Weir pump and heater. The length between tube plates was 15 ft. $5\frac{1}{2}$ in. and the small tubes were 2 in. diameter. In later engines of this same general design, also equipped for wood burning, the 4-8-2 wheel arrangement was adopted, the better to carry the increased weight of an improved front end. Nevertheless the total weight of engine in working order was increased to no more than 60 tons. Apart from that

Fig. 62.—*Nigerian Railway:* 4-8-2 *heavy freight locomotive of* 1930, *built by Vulcan Foundry Ltd.*

The experience of the British locomotive building industry in catering for operating conditions in many parts of the world that are greatly different from those existing on the home railways is shown by the successful designing and production of locomotives to burn wood as their only fuel. Although such machines have been supplied to countries overseas that were in an early stage of development, where railway traffic is not yet heavy, the tasks set to individual locomotives can be be quite severe. The few trains running on the lines can be heavily loaded in themselves. This point will be emphasised more particularly in Chapter 7, where reference is made to the large wood-burning articulated locomotives of the Beyer-Garratt type on the Benguela Railway. Here I am concerned with non-articulated locomotives, and an interesting example is afforded by the metre gauge wood-burning 2-8-2s of the Tanganyika Railway (Fig. 65). This line extends from Dar es Salaam on the East Coast of Africa up to Lake Tanganyika and Lake Victoria Nyanza; it includes heavy gradients, but a serious limitation to locomotive power existed in the low maximum axle load of only 10 tons. Speed was of less importance than haulage capacity, and in designing a locomotive to have a nominal tractive effort of

the basic dimensions of the 4-8-2 variant of the "206" class remained unchanged.

On the Indian railways three different gauges are standard, for various classes of line, the smallest being 2 ft. 6 in. for the lightest of feeder routes. An interesting example of a locomotive for the last-

Tanganyika Railway: 2-8-2 (206 *class*)

Cylinders (2) dia. × stroke, in.	18 × 23
Coupled wheel dia., ft. in.	3–7
Wheelbase engine, ft. in.	26–9
rigid, ft. in.	12–0
total, ft. in.	47–7
Axle load, tons	10
Adhesion Wt., tons	40
Wt. of engine in working order, tons	$57\ ^3/_4$
Wt. of engine and tender in working order, tons	86
Tractive effort at 85% boiler pressure, lb.	23,570
Boiler pressure, lb./sq. in.	160
Heating surfaces, sq ft.	
Tubes	1306
Firebox	139
Total evaporative	1445
Superheater	297
Total	2500
Grate area, sq. ft.	27
Wood capacity, cu. ft.	400
Water capacity, gall.	2500

Fig. 63.—*South African Railways:* " 12 " *class* 4-8-2 *used in general service, and built by Beyer, Peacock & Co. Ltd.*

Fig. 64.—*South African Railways:* " 12*A* " *class* 4-8-2.

Fig. 65.—*Tanganyika Railway: a wood burning* 2-8-2 *of the* " 26 " *class, developed from a* 1925 *design. This line was operated by the East African Railways.*

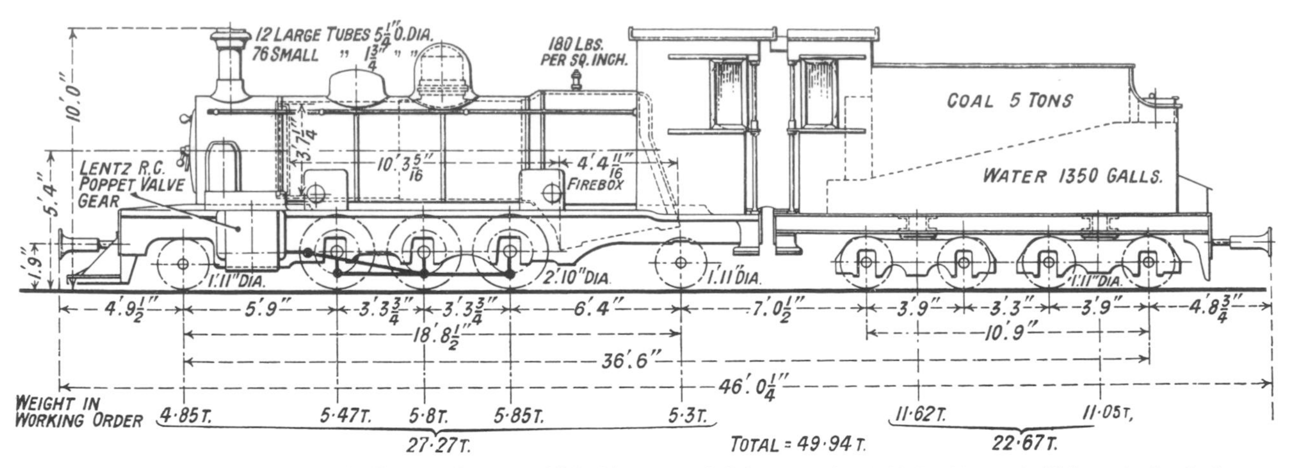

Fig. 66.—*Bengal Nagpur Railway: diagram of 2ft. 6in. gauge* 2-6-2 *locomotive built by Nasmyth Wilson & Co. Ltd.*

mentioned kind of service is afforded by some 2-6-2s for the Bengal Nagpur Railway built by Nasmyth Wilson and Co. Ltd., of which a diagram is shown in Fig. 66. These have a maximum axle load of only 5.85 tons—in fact the 8-wheeled bogie tenders, which have the large coal capacity, for so small a locomotive, of 5 tons have a maximum axle load almost as great as that of the locomotive, namely 5.81 tons. A feature of these engines that created a considerable amount of interest at the time was the use of rotary cam poppet valve gear. In 1929, at the time of their construction very little use had been made in Great Britain of this type of valve gear, and its most striking applications on the continent of Europe were yet to come. The particular arrangement was known as the R.C. type, and was supplied by Lentz Patents Ltd.

The cams were of the "step" kind, and those governing the steam admission provided for three different cut-offs in each direction. The exhaust cams gave two different release and compression points. In addition, there was a circular cam in each steam and exhaust set, the arrangement being such that when the reverser in the cab was in mid-gear position the steam valves were shut, and the exhaust valves open, thus providing a by-pass arrangement. For locomotives designed for branch line duty and relatively slow speed this particular design of poppet valve gear gave good results and economical performance of the locomotive; but it must be mentioned that in heavier duty where finer adjustments of the cut-off were needed, the "steps" represented by the various cams were too great for satisfactory working. There could be times when the cut-off corresponding to one cam would not be sufficient to maintain schedule time, and the next, if used continuously would be more than the steaming capacity of the boiler could maintain. In these Bengal Nagpur 2-6-2s the cut-off positions were 80, 50 and 25 per cent. respectively. The leading dimensions of these engines were:

Cylinders (2) dia. × stroke, in.	12 × 18
Coupled wheel dia., ft. in.	2–10
Heating surfaces, sq. ft.	
Boiler tubes	496
Arch tubes	6
Firebox	58
Superheater	108
Combined total	664
Grate area, sq. ft.	12.5
Boiler pressure, lb./sq. in.	180
Maximum axle load, tons	5.85
Total engine wt. in working order, tons	27.27
Total tender wt. in working order, tons	22.67
Nominal tractive effort at 85% boiler pressure lb.	11,550

Fig. 67.—*Kenya and Uganda Railway:* 2-8-2 *heavy freight locomotive, built by Beyer, Peacock & Co. Ltd.*

Chapter Five Locomotives for Intermediate and Local Duty: 1925–30

Express Passenger Designs

After the grouping of the railways in 1923, while the new "Big Four" companies all gave immediate attention to the designing of new locomotives for the heaviest and fastest duties the extent, particularly of the L.M.S.R. and L.N.E.R. systems, and the diversity of duties and routes to be provided for constituted a major problem when consideration had to be given to replacement of ageing and obsolescent stock. In earlier days the provision of branch line motive power, and of power for secondary main line duties had been met by maintaining in good order locomotives displaced from first class main line duties. But in the period under review some of the main line power becoming obsolescent was precluded from important secondary routes by limitations upon axle-loading. In any case there was a desire to standardise rather than prolong in service older types which included many non-standard features and details of equipment.

An outstanding case was that of the former Great Eastern Railway. Under the last Locomotive Superintendent of that railway, A. J. Hill, the inside-cylinder superheater 4-6-0 design introduced in 1912, had been developed, albeit no more than slightly, to the limit of dead weight per axle permitted by the civil engineer. With the happy combination of a sound mechanical design, and a very high standard of enginemanship on the part of top-link crews some very exacting train services were being successfully worked by engines of this class; but there was very little in reserve, and no margin at all for future acceleration. Fortunately, by the time Sir Nigel Gresley, as Chief Mechanical Engineer of the L.N.E.R., had occasion to review the motive power position on the Great Eastern line the work of the Bridge Stress Committee had shown that axle-loading alone was not the yard-stick by which the effect of a locomotive upon the track and bridges should be judged, and that the total effect could be greatly reduced by improved techniques in balancing. Gresley had already shown himself in favour of 3-cylinder propulsion for all but the smallest locomotives, and while this system was not essential to obtaining a very low total hammer-blow effect upon the track it certainly facilitated the process of balancing, and was favoured on the L.N.E.R. for other factors additionally.

In the "Sandringham" class of 3-cylinder 4-6-0 (Fig. 68), introduced in 1928, the civil engineer was able to accept an adhesion weight of 54 tons, as compared with 44 tons on the standard Great Eastern inside-cylinder 4-6-0s. It is nevertheless well-known that the weight distribution diagram originally proposed by Gresley was not accepted, and a number of difficulties in design had to be overcome before finalisation. But the outcome was the production of a locomotive showing an increase of 19 per cent. in nominal tractive effort over the Great Eastern 4-6-0s, namely from 21,969 to 25,380 lb. at 85 per cent. boiler pressure. Originally a cylinder layout was proposed in which all three cylinders drove on to one axle, as in every other Gresley 3-cylinder locomotive built up to that time; but with this arrangement the weight distribution did not work out satisfactorily, and because of this the drive was divided. Another important consideration in the design of these engines was the overall length, which because of clearance difficulties was not permitted to exceed that of the

Fig. 68.—*L.N.E.R. 3-cylinder* 4-6-0 *"Sandringham" class, for the Great Eastern Line, with short G.E.R. type tender* (1928).

Fig. 69.—*L.N.E.R. One of the later " Sandringhams " for the Great Central Line, with larger tender and named after football clubs.*

Fig. 70.—*L.N.E.R.* 3-*cylinder* 4-4-0 *Class* "D49" *with piston valves and Walschaerts gear, built Darlington works* 1927.

Fig. 71.—*L.N.E.R.* 3-*cylinder* 4-4-0 *Class* " D49 " *with R.C. poppet valve gear " Hunt " class.*

existing 4-6-0 locomotives. The tender, accordingly, was thus of the very short G.E.R. pattern. The first batch of these engines was built by the North British Locomotive Co. Ltd.

The principal dimensions are given, together with those of other locomotives referred to in this chapter, in a separate table; but some features of the design may be specially referred to. The valve gear was arranged in accordance with principles developed in connection with the heavy main line "Pacific" engines, and although the travel in full gear was not more than $5\,^5/_8$ in. the port openings to steam and exhaust were good, and the engines free-running and economical in fuel consumption. Apart from the attenuated look of the tenders the "Sandringham" class locomotives had the characteristic "Gresley" appearance, and the design was eventually multiplied up to a total of 70. The later batches, which were used mainly on the Great Central Line, had much larger tenders of a modern design (Fig. 69). The finest work of these engines was undoubtedly performed by the stud stationed at Leicester shed in the years 1936-9.

In providing new locomotives for intermediate duties in the North Eastern and Scottish areas of the L.N.E.R. Gresley did not have to submit to such serious limitations in axle loading, and again using the 3 cylinder system of propulsion he produced a 4-4-0 of a relatively high nominal tractive effort design to take over duties previously worked by various classes of "Atlantic" and large 4-4-0s, of North Eastern and North British origin. In view of his long and successful experience with locomotives having the wide type of firebox this adoption of the 4-4-0 type, in preference to a modernised 3-cylinder version of the famous Great Northern Atlantic might have seemed surprising. But there is no doubt that Gresley had been impressed with the good work and solid reliability of the former Great Central Railway "Director" class 4-4-0s, 24 new examples of which had been built in 1924 for service in Scotland. The new 3-cylinder 4-4-0s, of Class "D49" (Fig. 70), had quite a few points of similarity to the Great Central locomotives, particularly in the proportions of the boilers. The earliest engines of the class, built at Darlington Works in 1927, had 8 in. diameter piston valves, actuated by Walschaerts gear, and the Gresley conjugated motion for the middle cylinder. Later engines, built in 1932, had the R.C. poppet valve gear (Fig 71). Although capable of very good work these engines never fully superseded the "Atlantics" of North Eastern and North British design, though in the hands of able and enthusiastic crews the piston valve "D49" 4-4-0s did occasionally put up remarkably fine performances. The R.C. poppet valve engines were used mostly on North Eastern services between Leeds, Hull, Scarborough and Newcastle.

In the period under review the L.M.S.R. introduced no entirely new designs for intermediate duties; but an interesting development took place of the well-known Midland Class 2 superheater 4-4-0 (Fig. 72), but with 6 ft. 9 in. coupled wheels, and a modified design of boiler, as follows:

Class "2" 4-4-0 boilers

Design	M.R.	L.M.S.R.
Heating surfaces, sq. ft.		
Tubes	1045	1034
Firebox	127	124
Total evaporative	1172	1158
Superheater	313	246
Combined total	1485	1404
Grate area, sq. ft.	21.1	21.1
Boiler pressure, lb/sq. in.	160	180

The original Midland engines had a slightly higher nominal tractive effort, from the use of $20\,^1/_2$ in. diameter cylinders, against 19 in. But the piston valve diameter was the same in each case, and with larger port areas in relation to cylinder volume, and higher boiler pressure the L.M.S.R. standard design was the better all-round machine. In all 138 were built. They were used extensively in Scotland, on the former Glasgow and South Western line, and to a lesser extent on the L.N.W., L. & Y., Midland, and Somerset and Dorset lines in England.

On the Southern Railway two notable new 4-4-0 designs were prepared during that period, both constituting largely a synthesis of existing standard parts, but which, by skilful arrangement resulted in remarkably successful machines in traffic. Both are of particular interest as demonstrating the fact that one does not necessarily have to start with a clean sheet of paper to produce a good design; and that even in coping with the exigencies of the moment—lack of time, limitation in capital expenditure, and so on—skill and experience can produce quite outstanding results. The first instance was the demand, put to the Chief Mechanical Engineer at short notice, for additional locomotives for the Folkestone service. The best trains had for some time been running to a schedule of 80 min. over the $69\,^1/_2$ miles between Charing Cross and Folkestone Central; but the insertion of one intermediate stop within the same overall timing, and the increase in train loading, by use of corridor coaches instead of the S.E. & C.R. non-corridor stock, put a severe task upon the "L" class 4-4-0s already engaged upon the service, and made desirable the introduction of an improved design when additional 4-4-0s were authorised.

There was no time to prepare an entirely new design, with the attendant work necessary in making new patterns, flanging blocks and other tools; but experience with the former Wainwright 4-4-0 on the principles laid down by Maunsell in 1919 had shown clearly the advantages to be realised from improved front-end design. The result showed, in a classic example, how great an improvement could be effected with the very minimum of capital expenditure. Thus the valve travel was increased only to the maximum permitted by the existing patterns, by adjustment of the eccentrics, and by modification of the lever arms. The boiler was the same as that of the "L" class so far as pressed-out parts were concerned, but modified to take a higher boiler pressure, and with the draughting arrangements modified in accordance with Ashford practice developed on the "N" class 2-6-0 of

Fig. 72.—*L.M.S.R. Standard Class* "2P" 4-4-0 *for intermediate duty* (1928) *developed from the M.R. No.* 2 *class superheated* 4-4-0.

1917. The resulting locomotives, Class " L1 ", had the steam lap increased from $^7/_8$ in. on Class "L" to 1 $^3/_{16}$ in., and valve travel in full gear of 5 $^9/_{16}$ in. The improvement over the " L " class can best be judged by the maximum loads hauled with reliability on the 80-min. Folkestone express, namely 225 tons, with " L " class, and 320 tons with " L1 ". There was practically no difference in weight or nominal tractive effort of the two classes; the difference lay in the valve gear; in the use of smaller cylinders and a higher boiler pressure, and improved draughting.

The second Southern 4-4-0 design, the very famous " Schools " class of 1930 (Fig. 73), which on the basis of nominal tractive effort was the most powerful 4-4-0 ever to run in Europe, and was one of the most successful of all Maunsell designs, originated from a desire of the Traffic Manager for a locomotive of " Lord Nelson " characteristics but of intermediate power rating, and which would have a greater route availability than the 4-6-0 " King Arthur " class. The Chief Mechanical Engineer's department found that the desired result could be obtained very simply by taking the " Lord Nelson " cylinders and valves, but using three cylinders instead of four, while the boiler was a shortened version of the " King Arthur ". Standard patterns and tools were used throughout, and a locomotive equal in nominal tractive effort to the " King Arthurs " produced, but with a total engine weight, in working order, of 67 tons, against 81 tons in the 4-6-0. The " Schools " class, of which a total of 40 was built, incorporated the best features of both 4-6-0 designs—the cylinders, valves and front-end

Fig. 73.—*Southern Railway. R. E. L. Maunsell's 3-cylinder* 4-4-0 " *Schools* " *class, built Eastleigh* 1930.

Southern Railway: Waterloo–Southampton 79.2 miles

Engine No.	Gross load behind tender tons	Actual time	Net time	Net average speed m.p.h.
932	415	82 1/2*	83†	57.3
925	445	88 1/4	87	54.7
931	480	87 1/2	84	56.7
925	485	87 1/4	84 1/2	56.2
926	490	91 1/4	88	54.1
932	510	86 1/2	86 1/2	55.0
926	525	87 1/2	87 1/2	54.3
927	525	88	87 1/2	54.3

* Passing time † Equivalent time to stop

of the " Nelson ", and the free steaming boiler of the " King Arthur ".

In service the " Schools " class engines proved very fast and versatile motive power units. Their use enabled the 80-min. Folkestone trains to be loaded to 400 tons with good timekeeping; they worked trains of this same magnitude on the heavily graded Hastings line; they enabled the Portsmouth service from Waterloo to be accelerated to a 90-min. run (79.3 miles) over a most difficult route, and when this latter route was electrified in 1936 the stud of locomotives previously stationed at Fratton was transferred to Bournemouth and took over the working of the London expresses from the " King Arthurs " at a time when the service was the fastest in its history. Perhaps the greatest factor contributing to the universal success of the " Schools " was the manner in which they responded to varying techniques in driving. Providing they were handled on a light rein when starting away from rest it did not matter if they were worked on a long cut-off with a narrow regulator opening, or linked right up and driven with the regulator full open. They responded equally well, and just as economically to either treatment. Their finest work was performed on the Bournemouth route, and although more detailed reference is made to this in a later chapter devoted entirely to a survey of performance in this period summary details of eight typical runs are given here. Although the Traffic Manager had asked for a locomotive that would haul 400-ton trains at an average speed of 55 m.p.h. it is clear from the tabulated details that the Chief Mechanical Engineer had provided a locomotive capable of maintaining the specified speed, but with loads of 500 tons.

Passenger Tank Locomotives

In the last 12 years prior to grouping a great variety of passenger tank engines had been built, mostly for fast local passenger services, but some, like the 4-6-2 and 4-6-4 designs of the London Brighton and South Coast Railway, for the fastest express passenger work. Almost without exception these pre-grouping passenger tank engines designs incorporated a leading bogie. Nevertheless the increasing popularity of the 2-6-0 type for mixed traffic tender engines led to a new phase in fast passenger tank engine design. This phase was inaugurated in 1917 on the South Eastern and Chatham Railway, with the " K " class 2-6-4, having many parts interchangeable with the " N " class 2-6-0, and from 1929 onwards an interesting series of tank engine designs, all with a leading pony truck was produced on the L.M.S.R., on the Great Western, and on the L.N.E.R. The Derby-built locomotives of the L.M.S.R. had the 2-6-4 wheel arrangement, as favoured on the Southern Railway, while the G.W.R. and the L.N.E.R. adopted the 2-6-2 type. Actually the Great Western design could not be regarded as an entirely new one, since a 2-6-2 tank formed one of the original Churchward standard types of 1901. But the fast passenger tanks of the G.W.R. were originally of the 4-4-2 type, and the 2-6-2 was conceived mainly as a short-haul mixed traffic freight engine.

Before passing on to the new designs the interesting 3-cylinder variant of the Southern " K " class 2-6-4

Intermediate Express Passenger Locomotives

Railway	L.N.E.R.	L.N.E.R.	L.M.S.R.	S.R.	S.R.
Type	4-6-0	4-4-0	4-4-0	4-4-0	4-4-0
Class	" Sandringham "	" D49 "	" 2P "	" L1 "	" Schools "
Cylinders:					
Number	3	3	2	2	3
dia. × stroke, in.	17 1/2 × 26	17 × 26	19 × 26	19 1/2 × 26	16 1/2 × 26
Coupled wheel dia., ft. in.	6-8	6-8	6-9	6-8	6-7
Heating surfaces, sq. ft.					
Tubes	1508	1226.28	1034	1252.5	1604
Firebox	168	171.5	124	154.5	162
Superheater	344	271.8	246	235.0	283
Total	2020	1669.58	1404	1642.0	2049
Grate area, sq. ft.	27.5	26.0	21.1	22.5	28.3
Valve gear	Walschaert Gresley Conjugated	Walschaert Gresley Conjugated	Stephenson —	Stephenson —	Walschaert (3 sets)
Boiler pressure, lb./sq. in.	200	180	180	180	220
Nom. T.E. at 85% boiler pressure, lb.	25380	21556	17729	18910	25130
Max. axle load, tons	18.35	21.25		19.5	21.0
Total engine wt., tons	77.25	66.0	54.05	57.8	67.1
Total wt. engine and tender, tons	129.25	114.3	95.25	98.3	109.5

Fig. 74.—*Southern Railway. The Maunsell* 3-*cylinder* 2-6-4 *tank engine* River Frome, *with conjugated valve gear. Later converted to the* 2-6-0 *tender type.*

must be mentioned. This engine, No. 890 *River Frome* (Fig. 74) had the Holcroft arrangement of conjugated valve gear. The valve chests of the outside cylinders were set inwards, and the valves operated through rockers, as on the standard Great Western 2-cylinder designs. This arrangement provided clearance for the links to the conjugated valve gear, which were attached to the rear end of the outside cylinder valve rockers instead of in front of the valves as in the usual Gresley arrangement. By this method the factor of expansion of the valve spindles, which had to be taken into account on the Gresley locomotives was eliminated. The layout of the conjugated gear was arranged so that the levers were clear of the steam chests and the valves could therefore be removed without dismantling any of the motion. This engine, in common with all the remainder of the " K " class 2-6-4s, was converted to a 2-6-0 tender engine after the derailment of engine No. 800 at Sevenoaks in August 1927, when many lives were lost.

In the early years after grouping two new passenger tank engine designs were introduced on the L.M.S.R. by Sir Henry Fowler, the " 2300 " class 2-6-4 (Fig. 75), and " 15500 " class 2-6-2 (Fig. 77). The 2-6-4 has a characteristic Midland styling although the boiler and firebox, though similar in basic design was not standard

Fig. 75.—*L.M.S.R. Sir Henry Fowler's* 2-6-4 *passenger tank engine, original design,* 1927.

Fig. 76.—*L.M.S.R. A later variation of the Fowler* 2-6-4 *tank* (1933) *with side-windowed cab and doors.*

with any existing type, such as the 3-cylinder compound, and the Class "4" superheater 0-6-0 goods. The comparative details of the boilers of these three well-known designs are as follows:

Class	Compound 4-4-0	0-6-0 Goods	2-6-4 Tank
Heating surfaces, sq. ft.			
Tubes	1169	1044	1083
Firebox	147	123	137
Total evaporative	1316	1157	1220
Superheater	291	253	266
Combined total	1607	1410	1486
Grate area, sq. ft.	28.4	21.1	25
Boiler pressure, lb./sq. in.	200	175	200

The 2-6-4 tank engines were fitted with long-lap, long-travel valves, and proved exceptionally fast and economical locomotives. In working the outer suburban services from Euston, many of which trains ran non-stop to and from Watford Junction, it was normal practice for these engines to be linked up to cut offs as short as 7 or even 5 per cent. Speeds in excess of 80 m.p.h. were common on the inward run from Watford to Euston. They were the first L.M.S.R. locomotives to be built new with long-lap, long-travel valves, and similar valve settings were subsequently used on such locomotives as the "Royal Scots", and the "Patriot" 4-6-0s of 1930. The basic dimensions, in comparison with those of contemporary Great Western and L.N.E.R. locomotives are given in the table on page 63. The second new L.M.S.R. tank engine design was prepared for lighter work, and was in some ways a scaled down version of the larger engine. These, too, were excellent machines, of which a total of 70 was eventually built.

Fig. 77.—*L.M.S.R. Sir Henry Fowler's small* 2-6-2 *passenger and mixed traffic tank engines* (1930).

Fig. 78.—*G.W.R. Standard* 2-6-2 *passenger tank engine* " 6100 " *class* (1931)*: a development of the* " 4100 " *class of* 1929, *with* 225 *instead of* 200 *lb. boiler.*

The original Great Western main line 2-6-2 tank, of 1905, was a heavy mixed traffic unit with maximum axle-load exceeding 19 tons and having a taper boiler with a combined heating surface of 1,670.24 sq. ft. When the type was adopted for fast passenger suburban working (Fig. 78), principally in the London and Birmingham areas, the smaller standard taper boiler, having a combined total heating surface of 1349.1 sq. ft. was used in combination with slightly smaller cylinders. In passing it may be mentioned that during C. B. Collett's time at Swindon some of the standardisation precepts, so carefully built up by Churchward, were discarded, and that in the case of the 2-6-2 tank engines of this general series there were eventually three different sizes of driving wheel diameter as follows:

Class	Boiler Standard	Coupled Wheel diameter ft. in.
3190	No. 2	5 3*
6100	No. 4	5 8
8100	No. 4	5 6

* Originally 5 ft. 8 in. as standardised from 1905

The " 3100 " class was later concentrated on the South Wales main line, and the reduction in wheel diameter from the original standard provided a useful increase in nominal tractive effort.

At the time of grouping, and for some years thereafter the L.N.E.R. was well furnished with tank engines suitable for fast residential traffic, and immediate post-grouping needs had been met by building further engines of the Great Central " A5 " 4-6-2 type. Later in meeting the need for a fast suburban tank engine in Scotland of greater capacity than the ex-North British 4-4-2 tanks Sir Nigel Gresley applied the 3-cylinder principle, with conjugated valve gear to the design of a neat and handsome 2-6-2 for service around Edinburgh and Glasgow (Fig. 79). Built up on well-tried principles these engines rapidly achieved success, and were followed later by a more powerful version, using a boiler pressure of 200 lb. per sq. in. instead of 180 as in the first series. The basic dimensions of these locomotives, in company with other passenger tank engines of the period are given in the table opposite.

Included in this table is the Great Western " 5600 "

Fig. 79.—*L.N.E.R. Gresley* 3-*cylinder passenger tank engine class* " VI ", *built* 1930.

Passenger Tank Locomotives

Railway Type Class	L.M.S.R. 2-6-4 " 2300 "	L.M.S.R. 2-6-2 " 15500 "	G.W.R. 2-6-2 " 5100 "	L.N.E.R. 2-6-2 " V3 "	G.W.R. 0-6-2 " 5600 "
Cylinder:					
Number	2	2	2	3	2
dia. × stroke, in.	19 × 26	17 1/2 × 26	18 × 30	16 × 26	18 × 26
Coupled wheel dia., ft.-in.	5-9	5-3	5-8	5-8	4-7 1/2
Heating surface, sq. ft.:					
Tubes	1082	691	1145.0	1198	1145
Firebox	138	104	121.8	127	121.8
Superheater	246	186	82.3	284	82.3
Total	1466	981	1349.1	1609	1349.1
Grate area, sq. ft.	25	17.5	20.35	22.08	20.35
Boiler press., lb./sq. in.	200	200	225	200	200
Nom. T.E. at 85% boiler press., lb.	23125	21486	27340	24960	25800
Max. axle load, tons	18.15	16.0	17.6	20	18.8
Total wt. in working order, tons	86.25	71.8	78.45	86.8	68.6

class 0-6-2 designed for general service on the local railways in South Wales that were amalgamated with the Great Western at the time of the grouping (Fig. 80). A general service passenger and freight engine was required, and the standard tapered boiler used on the " 5100 " class 2-6-2 was incorporated on a simple straightforward design, which, apart from the boiler lay outside the general pattern of Great Western practice and was a type well-known and long appreciated in South Wales. The standard boiler used on these engines and on the " 5100 " class 2-6-2s was distinguished by having a very small superheater, including a heating surface of no more than 82.3 sq. ft.

Goods and Mixed Traffic Locomotives

The period between 1925 and 1930 saw the production of only one new British design for heavy mineral traffic, the L.M.S.R. " 7F " 0-8-0 (Fig. 81), and this was in some way a modern adaptation of the London and North Western " G2 " designed by Bowen-Cooke. The same boiler was used, but the prevailing technique, where the modernisation of existing designs were concerned, was used, namely to use smaller cylinder diameters, higher boiler pressure, and long-lap, long-travel valves. The new L.M.S.R. 0-8-0s had Walschaerts gear inside.

Another straightforward 0-6-0 for heavy traffic was the L.N.E.R. " J39 " class (Fig. 82), which was a product of Darlington works and incorporated many features of North Eastern Railway practice. The boilers were interchangeable with those of the " D49 " class 4-4-0 express passenger engines already referred to. Except in the design of the largest express passenger locomotives Sir Nigel Gresley allowed his various manufacturing centres, such as Gorton, Darlington and Stratford full latitude in the working out of the new designs once he had settled the broad specifications, and the " J39 ", as worked out at Darlington was in many ways a development of the North Eastern " P3 " class. A small batch of 0-6-0s similar to the "J39", with coupled wheels 4 ft. 8 in. diameter (Class J38) was built specially for short-haul mineral

Fig. 80.—*G.W.R. Heavy* 0-6-2 *tank engine for general passenger and mineral service in South Wales* (1924).

Fig. 81.—*L.M.S.R.* 0-8-0 *mineral engine* (1929) *developed from L.N.W.R.* " G2 " *class.*

Fig. 82.—*L.N.E.R.* 0-6-0 *Class* " J39 ", *built for general service* (1926-1941).

service in Scotland. The " J39 " class, although intended primarily for general goods service, were frequently used on local passenger and short distance excursion trains.

Just at the end of the period covered by this chapter the Great Western brought out a new type of light branch locomotive, the 0-6-0 " 2251 " class, weighing no more than 43.8 tons in working order (Fig. 83). A scaled-down version of the smallest taper boiler was designed for these engines; but with coupled wheels only 5 ft. 2 in. diameter and a working pressure of 200 lb. per sq. in. the nominal tractive effort was 20,155 lb. These engines were used particularly on steeply graded branch lines in Central and North Wales, and their relatively high tractive effort—for so small an engine—gave them a good accelerative power.

During this period however the most interesting trend was in the gradual development of the general utility locomotive, as represented by the L.M.S.R. 2-6-0 of 1926, and by the Southern and Great Western 4-6-0s. Since the construction of the " 43XX " class of outside cylinder 2-6-0s on the Great Western Railway, in 1911, the type had become very popular for fast goods and mixed traffic duties, and prior to the grouping of the railways successful designs had been produced on the Great Northern, London Brighton and South Coast, South Eastern and Chatham, Caledonian, and Glasgow and South Western Railways. By the year 1925 the Southern was becoming an increasingly large user, while on the L.N.E.R. Sir Nigel Gresley was building his 3-cylinder " K3 " class for duties on many parts of the system. The L.M.S.R. 2-6-0 of the " 13000 " class (Fig. 84) was a product of the Horwich drawing office, and had many features that proclaimed its Lancashire and Yorkshire Railway lineage. They were excellent engines, and in addition to their principal duties in England they were allocated to regular passenger work on the Glasgow and South Western, and Highland sections in Scotland.

These Horwich " Moguls " were nevertheless among the last of their type designed for the heaviest general utility service, for which the 4-6-0 type rather than the 2-6-0 thereafter came into favour. The Southern Railway following upon the practice of the London and South Western had been building small-wheeled variants of the standard

Goods and Mixed Traffic Locomotives

Railway	L.M.S.R.	L.M.S.R.	L.N.E.R.	S.R.	G.W.R.	G.W.R.
Type	2-6-0	0-8-0	0-6-0	4-6-0	0-6-0	4-6-0
Class	" 5F "	" 7F "	" J39 "	" S15 "	" 2251 "	" Hall "
Cylinders						
number	2	2	2	2	2	2
dia. × stroke, in.	21 × 26	19 1/2 × 26	20 × 26	20 1/2 × 28	17 1/2 × 24	18 1/2 × 30
Coupled wheel dia., ft. in.	5-6	4-8 1/2	5-2	5-7	5-2	6-0
Heating surfaces, sq. ft.:						
Tubes	1345	1402	1226.28	1716	1069	1686.6
Firebox	160	150	171.5	162	102	154.78
Superheater	307	342	271.8	337	74	262.62
Total	1812	1894	1669.58	2215	1245	2104.0
Grate area, sq. ft.	27.5	23.6	26.0	28.0	17.4	27.07
Boiler pressure, lb./sq. in.	180	200	180	200	200	225
Nom. T.E. at 85% boiler pressure, lb.	26.580	29,746	25.664	29,860	20,155	27,275
Max. axle load, tons	19.6	17,85	19.65	19.9	15.75	18.95
Total engine wt., tons	66.0	60.75	57.85	79.25	43.8	75
Total wt. engine and tender, tons	108.2	101.95	102	135.6	83.2	121.7

express passenger 4-6-0s; but the " H15 " class of 1924, having 6 ft. diameter coupled wheels, seemed to lack the freedom in running necessary for a true general utility machine. The " S15 " illustrated on page 64 (Fig. 85), was a variant of the " King Arthur " class, and did very good work on the heavy fully-fitted night goods trains between London and Southampton and between London and Exeter. But in utilisation the " S15 " was essentially a goods rather than a general utility engine.

The progenitor of the modern " all purposes " 4-6-0 locomotive was the Great Western " Hall " (Fig. 87), very easily developed from the standard 2-cylinder 4-6-0s of the " Saint " class by the simple process of substituting 6 ft. 0 in. for 6ft. 8 in. coupled wheels. The prototype of this very numerous class was actually a " Saint "—No. 2925 *Saint Martin* which was rebuilt in December 1924 for trial purposes. It was not until 1928 however that production commenced of the " Hall " class proper. The ad-

Fig. 83.—*G.W.R. light goods and mixed traffic* 0-6-0, *first built Swindon,* 1930.

Fig. 84.—*L.M.S.R. Horwich design mixed traffic* 2-6-0 (1926).

Fig. 85.—*Southern Railway Maunsell's "S.15" class mixed traffic* 4-6-0 (1927).

Fig. 86.—*G.W.R. Mixed traffic* 4-6-0 *"Grange" class with* 5*ft.* 8*in. coupled wheels* (1936).

(below)
Fig. 87.—*G.W.R. mixed traffic* 4-6-0 *"Hall" class with* 6*ft.* 0*in. coupled wheels* (1928).

vantage these locomotives showed over previous mixed traffic designs was that, equipped with the Swindon standard No. 1 boiler, as on the Churchward 4-6-0s, they could be allocated with reliability to express passenger trains. Despite 6 ft. coupled wheels they proved very speedy machines, and on one occasion worked the high speed " Bristolian " express to time in emergency.

In referring to these engines it is interesting to recall that Churchward's original proposals of 1901 for a complete standard range of locomotives for the Great Western Railway included a large 4-6-0 with smaller diameter driving wheels than those of the high speed express passenger class. The diameter he specified then was 5 ft. $8\frac{1}{2}$ in., and this engine actually materialised in 1936 when the new " 6800 " class, named after " Granges " was introduced to replace the original 2-6-0s of 1911 for fast mixed traffic work (Fig. 86). The " 6800 " class had the same boilers, cylinders and motion as both the " Saint " and " Hall " classes, and the capacity of the three classes can be compared thus:

Class .. .. .. Name	2900 " Saint "	4900 " Hall "	6800 " Grange "
Dia. of coupled wheels ft. in.	6-8 $^1/_2$	6-0	5-8
Nom. T.E. at 85% boiler pressure, lb.	24,395	27,275	28,875

Chapter Six Locomotive Performance 1927–30

The summer services of 1927 marked the first appreciable change from a timetable pattern in Great Britain that had remained virtually the same since the opening of the Great Western shortened routes to the West of England in 1906, and to Birmingham in 1910. On the East Coast and West Coast routes to Scotland it had been traditional to stop at Grantham, York, and Newcastle on the one hand, and at Rugby, Crewe and Carlisle on the other. Some trains called additionally at Darlington and Berwick, on the East Coast route, and at Preston, Lancaster or Carstairs on the West Coast. At least three, and sometimes four different locomotives were utilised on the journeys from London to Edinburgh or to Glasgow, quite apart from occasions when assistant engines were needed over certain sections of the routes. There had been little disposition on the part of the northern lines to follow the Great Western practice of long engine workings, such as Paddington to Plymouth, and the marathon duty worked on the West of England postal specials whereon one locomotive worked over the entire distance from Paddington to Penzance via Bristol, 326 miles.

In the summer of 1927, superficially it might have seemed that both the L.M.S.R. and the L.N.E.R. had taken up the challenge of Great Western supremacy in length of non-stop run, and the historic rivalry between the East Coast and West Coast services to Scotland had embarked upon a new form of competition—albeit not at very high speeds. The relief section of the Flying Scotsman leaving King's Cross at 9.50 a.m. was booked to run non-stop to Newcastle, 268.3 miles, while the 10 a.m. from Euston to Edinburgh and Glasgow made no passenger stops at all, but called intermediately at Carnforth to change engines, and at Symington for the Glasgow and Edinburgh sections to divide. The Newcastle non-stop, at an average speed of no more than 48 1/2 m.p.h. was a very easy task for the Gresley "Pacifics", but on the L.M.S.R. the new 3-cylinder 4-6-0s of the "Royal Scot" class were not ready in time for the summer service. The trailing tare load of the train was 420 tons, but under the L.M.S.R. administration the former L.N.W.R. locomotives were restricted in maximum load limits, and the Euston–Carnforth run at 53 1/2 m.p.h. was always double-headed, usually with a "George the Fifth" class 4-4-0 and a 4-cylinder 4-6-0 of the "Claughton" class. Over the northern stage of the journey the train was worked by a pair of Midland compounds.

The respective publicity departments made the most of the record lengths of non-stop run; for both the King's Cross–Newcastle, and the Euston–Carnforth workings exceeded in mileage the famous run of the Cornish Riviera Express. Both the Anglo-Scottish runs attracted a great amount of public attention, and the trains themselves were well loaded. Passengers appreciated the comfort of long spells without the interruption and occasional discomfort of intermediate stops, moving of luggage and so on; but despite the outward publicity both companies were beginning to feel their way towards a rather broader conception of locomotive working. As the nineteen twenties drew towards their close the financial results from the grouped railways did not show the substantial economies that had been hoped for through amalgamation and standardisation of equipment, and economies in working were being sought by improved methods of operation. One very desirable aim was the reduction of the total locomotive stocks by making individual engines work longer monthly mileages, and take a higher proportion of the total work.

Studies were therefore made of engine diagrams to determine the additional mileages that could be obtained per engine, and in theory this could be achieved by booking locomotives to longer through workings. If a certain fixed time were needed for shed duties between two successive train workings the utilisation of the locomotive would obviously be higher if the duty involved running from Euston to Carlisle and back, instead from Euston to Crewe and back. Other factors were of course involved. Some locomotives by the nature of their design would not be suitable for greatly increased lengths of run, and psychological factors obtruded when locomotives were required to be remanned en route, instead of being confined to regular crews. But the advantages to be gained by longer through workings were such that the gradual trend towards this feature of operation was soon to become a point of major policy with both the L.M.S.R. and the L.N.E.R.

By the autumn of 1927 sufficient of the new "Royal Scot" class 3-cylinder 4-6-0s were available for a further re-casting of the Anglo-Scottish services from Euston to be carried out. The "Royal Scot" train (10 a.m. from Euston) (Fig. 88) was run non-stop between London and Carlisle, 299.1 miles, on both the northbound and the southbound service, and with a gross trailing load of about 440 tons this proved a severe task, particularly through the winter months. The scheduled average speed was 52 m.p.h. On the afternoon West Coast service, then officially known as the "Midday Scot", engines were changed

Fig. 90.—*L.M.S.R. ex-L.N.W.R. " Claughton " class* 4-6-0 *No.* 5969 *on a Manchester-Euston express.*

that had also become a standard feature of Crewe practice for superheater engines, but in the Midland case without trick ports. A critical examination of certain " Royal Scot " class engines, the coal consumption of which had increased by *80 per cent.* since their last general overhaul, revealed that an altogether inordinate amount of internal leakage was taking place past the piston valves and leading directly to a very high proportion of the increase in coal consumption. After some experiments a new design of solid valve head was adopted, having six narrow Ramsbottom rings, and with this relatively simple change in design the defect of a high increase in coal consumption with increased mileage was largely overcome.

This defect having been discovered solid valve heads with six narrow rings were also fitted to a number of the ex-L.N.W.R. 4-cylinder 4-6-0s of the " Claughton " class. Although the all-round performance of these engines did not measure up to the standards demanded in the late nineteen twenties, there were 130 of these fast-running and powerful units in service, and reliance had to be placed upon them for many first class duties just below the level of those allocated to the " Royal Scots ". The steaming was not always reliable, and in 1930 engine No. 6001 was fitted experimentally with the Kylala blastpipe, which was later adapted in France, by M. André Chapelon on the Paris–Orleans Railway, into the well-known and highly successful Kylchap double blastpipe (Kylala–Chapelon). At that time engines of the " Claughton " class (Fig. 90) had largely replaced the Midland compounds on the Leeds–Carlisle main line, over Aisgill summit, and that historic test route for L.M.S.R. locomotives was once again utilised for a series of runs with the dynamometer car, in which engine No. 6001 with Kylala blastpipe was compared with a standard engine of the class. The latter was found to be in poorer mechanical condition than the Kylala engine, and so liable in the test results to give a misleading impression. So after the first set of tests the Kylala blastpipe was removed and the standard arrangement restored. The results are given in the table on page 69, together with those obtained from a " Claughton " in original L.N.W.R. condition in 1924. Although this latter could not be considered a truly representative engine of the class the disparity in coal consumption figures between No. 2221 and No. 6001 with standard blastpipe emphasises how points of detail design can affect locomotive performance. The Kylala blastpipe made the steaming still more free, but at the expense of higher coal consumption, and it was not adopted.

To enhance the capacity of the " Claughton " class locomotives, and provide units intermediate in capacity between the original L.N.W.R. engines and the " Royal Scots ", a small batch was fitted with larger boilers, carrying a pressure of 200 lb. per sq. in., and the nominal tractive effort was increased in proportion to the increase in pressure from 175 to 200 lb. per sq. in. In ten of these rebuilt engines the original front-end was retained, though with the important modification to the piston valves incorporated in the " Royal Scots ", and in the standard " Claughton " engines. Another ten of these rebuilt engines were fitted with the Caprotti valve gear, which is referred to with other arrangements of poppett valves in Chapter 8 of this book. The large-boilered

L.M.S.R. Tests with " Claughton " Class Engines: Leeds–Carlisle
300-*ton Trains*

Date	Engine No.	Piston valve type	Blastpipe	Mileage out of shops†	Load trailing tons tare	Average speed m.p.h.	Coal per mile lb.	Coal per d.h.p. hour lb.	Evaporation lb. of water per lb. of coal
2-12-24	2221*	Schmidt	Standard	23,825	301.6 305.6	47.4 48.0	49.6	5.36	5.87
3-12-24	2221*	Schmidt	Standard	23,825	302.6 325.6	47.25 50.1	54.0	5.77	6.0
12-5-30	6001	Solid 6-ring	Kylala	11,111	305 312	46.9 49.4	38.4	4.08	7.91
13-5-30	6001	Solid 6-ring	Kylala	11,111	307 310	46.4 49.4	3.81	3.79	7.57
14-5-30	5973	Solid 6-ring	Standard	35,894	304 316	46.2 46.0	45.5	4.64	7.53
15-5-30	5973	Solid 6-ring	Standard	35,894	316 311	47.8 49.6	47.2	4.69	7.53
3-6-30	6001	Solid 6-ring	Standard	11,111	311 313	47.1 51.0	35.7	3.61	8.37
4-6-30	6001	Solid 6-ring	Standard	11,111	311 320	47.4 49.4	35.5	3.64	8.16

*Original numbering: later No. 5927 † At start of tests

" Claughtons ", particularly those with piston valves, did very good work, and although they incorporated no change in valve gear from the original L.N.W.R. layout, with short lap, short travel valves, their basic coal consumption in heavy express passenger service between Euston and Manchester, and Euston and Holyhead was between 3.0 and 3.5 lb. per d.h.p. hour. Mention of Holyhead is a reminder that this was one of the earliest extensions of locomotive working on the L.M.S.R., and instead of the former practice on the Irish Mails of changing engines at Crewe, the locomotives were rostered to work through over the 263 miles between Euston and Holyhead, with the same crews on a " double home " basis. This was done with the original standard " Claughton " engines at first, and later with the large boilered variety.

All previous records in length of through engine working were broken in 1928, when from May 1 the Flying Scotsman of the L.N.E.R. was booked to run non-stop daily, in both directions between King's Cross and Edinburgh Waverley, a distance of 392.7 miles (Fig. 91). Although no curtailment of overall time was made in the schedule laid down, and an average speed of no more than 47.5 m.p.h. was involved, the event attracted a great amount of public attention, and an eye-witness of the scene at King's Cross wrote afterwards: " Departure platform No. 10 was thronged with what is believed to be the largest gathering of the public ever known on any such occasion ". The engine was the historic Gresley "Pacific" No. 4472 *Flying Scotsman*, and the load 386 tons tare. In itself a very easy task was set to engine and crews; but regular non-stop running over such a distance inevitably increases the need for the utmost reliability of the locomotive concerned, and a quality of fuel, and skill in firing that will avoid the formation of clinker, or inefficiency in steaming towards the

Fig. 91.—*L.N.E.R. The up " Flying Scotsman " non-stop from Edinburgh to King's Cross near Grantham, hauled by " A3 " class 4-6-2 No.* 2795 Call Boy, *with corridor tender.*

Fig. 92.—*L.N.E.R. Rear-view of engine No.* 4472 Flying Scotsman *fitted with corridor tender.*

end of the journey. In certain quarters the relatively low overall speed was criticised, seeing that no acceleration of service between London and Edinburgh had resulted from the elimination of the stops at Grantham, York, Newcastle and Berwick. But the London and North Eastern Railway was proceeding gradually in its development of the techniques of long non-stop running, and in ten years time from the first introduction of the " non-stop " the time between London and Edinburgh had been cut by 1 1/4 hours, giving an average speed of 56, instead of 47 1/2 m.p.h.

The most interesting development in connection with the running of the London–Edinburgh " non-stop ", in 1928, was the introduction of Sir Nigel Gresley's corridor tenders (Fig. 92). The limit, on a fast express run of both time and distance for one crew had been reached in the London–Carlisle non-stop runs on the L.M.S.R.: 5 hr. 55 min. for a 299-mile run on the up Royal Scot express. The non-stop " Flying Scotsman " when first introduced in May 1928 had a journey time of 8 1/4 hours and arrangements had to be made for one crew to relieve another, at the half-way point. The designing of special tenders with a corridor connection through which the relief engine-men could pass from the train to the footplate was an ingenious idea. The passage way had a width of 18 in. and a height of 5 ft., and circular windows were provided, one at each end, for lighting the corridor. Far from restricting the fuel capacity, the corridor tenders carried 9 tons of coal, against 8 tons in the original " Pacific " tenders. As previously the tenders had rigid axles, and weighed 62.4 tons in fully loaded condition. Appropriately the new service was inaugurated, from the London end, by engine No. 4472 *Flying Scotsman*. On the first non-stop train from Edinburgh to King's Cross the engine was No. 2580 *Shotover*.

The new service saw a further integration of locomotive workings over the East Coast Route as between the former Great Northern, North Eastern and North British Railways, and from the outset the crews taking turns on the " non-stop " included four from Edinburgh (Haymarket). Regular "double-home" workings were also instituted on many trains between King's Cross and Newcastle, with engines for the most part confined to regular crews. Under this arrangement North Eastern Area men from both Gateshead and Heaton sheds worked regularly to London, though on the East Coast route it was not until the year 1938 that the practice of long through workings was carried further, and engines rostered

Fig. 93.—*L.N.E.R. ex-G.N.R. superheated Atlantic engine, as used on the Leeds Pullman workings until* 1937.

to cyclic duties which involved their being handled by many different crews in the course of a single "diagram". Despite the developments in the Anglo-Scottish services from 1928 onwards the fastest running on the East Coast main line was still to be seen on the Leeds expresses, until the important accelerations in the late spring of 1932.

In the last paragraph reference was being made particularly to the working of the Gresley "Pacific" engines. But no mention of locomotive performance on the East Coast route in this period would be complete without some reference to the work of the large-boilered Great Northern "Atlantics" used on the Pullman trains between King's Cross and Leeds (Fig. 93). It is true that the origin of these locomotives dates back to 1902; but no more than a passing reference was made to them in Ahrons' classic work, and the dimensions quoted related to their original non-superheated condition. A process of modernisation was carried out by Sir Nigel Gresley, and this reached its ultimate form in 1919, by the fitting of 32-element superheaters. Some engines of the class retained their original 18 3/4 cylinders and slide valves, while a number were rebuilt with 20 in. diameter cylinders, and piston valves. All the modernised engines did very good work; but the piston valve variety came to play a most important part in the working of the traffic—so much so that although this modernisation came a little before the beginning of the period covered by this volume their outstanding work belongs wholly to it, and their dimension are quoted herewith:

L.N.E.R. (G.N.) Atlantics: Class "C1"

Cylinders (2) dia. × stroke, in.	20 × 24
Piston valve dia., in.	8
Motion:	
Maximum valve travel, in	4 3/8
Steam lap, in.	1 1/4
Exhaust clearance, in.	5/16
Lead, in full forward gear, in.	5/32
Cut-off in full gear, %	70
Heating surfaces, sq. ft.:	
Tubes	1824
Firebox	141
Superheater	568
Combined total	2533
Grate area, sq. ft.	31
Coupled wheels dia., ft. in.	6-8
Boiler pressure, lb./sq. in.	170
Adhesion weight, tons	40
Total wt. of engine, tons	69.6
Total wt. of engine and tender, tons	112.7
Tractive effort at 85% boiler pressure, lb.	17,340

With the last-mentioned figure in mind, and no less the limited adhesion weight of 40 tons, it might seem incredible that one of those engines could take on the haulage of a 500-ton Anglo-Scottish express at Grantham, at a moment's notice, and work that train over the 82 3/4 miles to York in 86 or 87 min. Yet this feat was achieved not once, but on three separate occasions, with different engines and different crews. These were in conditions of emergency, when the locomotives concerned were not prepared for maximum efforts of haulage. But until the year 1938 they were the standard motive power on the high-speed Pullman trains, booked to cover the 185.8 miles between King's Cross and Leeds in 194 min. northbound, and 193 min. southbound, with a minimum load of about 280 tons tare.

Features of these locomotives in their fully modernised condition that contributed to their remarkable weight-hauling and high-speed abilities were the very fine boiler, and high degree superheating. The 32-element superheater actually provided more heating surface in the superheater elements than was available on the original Gresley "Pacific" locomotives with boilers carrying a pressure of 180 lb. per sq. in. The valve gear was traditional with short lap and short travel; but the piston valves were very large in relation to the cylinder volume, and the arrangements of ports coupled with the high degree of fluidity of such highly superheated steam permitted large volumes of steam to enter at each stroke, and a generous amount of exhaust clearance enabled a free exhaust to occur at normal running cut-offs, which were about 40 to 45 per cent. These engines indeed provided a classic case where the tractive effort formula was of no assistance in assessment of the haulage capacity of the locomotive.

The following table gives a summary of six runs on the Leeds–King's Cross Pullman trains in the years 1932–37. The distance from start to stop is 185.7 miles. The initial length of 29.7 miles to Doncaster includes several hindrances to sustained fast running, and therefore the performance over the 156 miles from Doncaster to King's Cross is also quoted:

Run No.	1	2	3	4	5	6
Engine No.	4436	4456	3280	4444	3284	4444
Load tons, gross trailing	285	290	295	300	325	335
Total time, min. sec.	193-00	189-35	209 55	195 25	191 15	195 15
Net time, min.	188	176	186 1/2	190	181 1/4	190
Net time Doncaster to King's Cross, min.	149 1/2	140	147	152 1/2	144 1/2	152
Net av. speed Doncaster to King's Cross, m.p.h.	62.7	66.9	63.7	61.4	64.8	61.6
Max. speed, m.p.h.	85	93	84	82	88	85 1/2

L.N.E.R. Grantham York

G.N. Atlantic No.			4415		4404	
Gross trailing load, tons			545		585	
Miles		Sch.	m.	s.	m.	s.
0.0	Grantham	0	0	00	0	00
14.6	Newark	15	17	13	17	23
33.1	Retford	35	34	16	36	15
47.7	*Black Carr Jc.*		46	56	48	38
50.5	Doncaster	53	49	34	51	19
54.7	*Shaftholme Jc.*		53	51	55	36
67.5	*Brayton Jc.*		66	04	67	41
68.9	Selby	73	67	35	69	17
80.7	*Chaloners Whin Jc.*		83	33	83	09
			—		sig. check	
82.7	York	90	86	56	87	40
Newark–Black Carr						
Av. speed, m.p.h.			66.8		63.6	
Equivalent d.h.p.			953		922	
Shaftholme–Brayton						
Av. speed, m.p.h.			62.8		63.6	
Equivalent d.h.p.			835		922	

Runs 2 and 5 in particular included some exceptionally fine running for engines of these proportions.

Two emergency occasions when engines of this class were called upon, almost at a moment's notice to take over haulage of very heavy East Coast expresses provide remarkable examples of engine performance. Summary details are given on page 71. From a series of observations on these engines the following ratios of equivalent drawbar pull to nominal tractive effort have been calculated.

Speed, m.p.h.	40	55	60	65
Ratio of D.B. pull to nom. T.E.	60.0	38.0	34.0	31.0

On the Great Western Railway the period under review was marked by the working of the " King " class locomotives (Fig. 94), with very heavy loads, on the West of England service, and the development of sustained high speed running on the " Cheltenham Flyer " with the " Castle " class locomotives. Taking the West of England trains first, by the end of the period under review the historic London–Plymouth non-stop run of the Cornish Riviera Express had been abandoned; but so far as power output was concerned the run of 173.7 miles in 175 min. from Paddington to Exeter involved an even harder effort. Details of a typical run can be summarised as follows:

London–Westbury section:
- Distance, miles 95.6
- Time start to pass, min. sec. 92-5
- Load, trailing tons tare 498
- Cut-off varied between 15% and 18%
- Regulator between $^7/_8$ and full open on level and uphill stretches
- Maximum speed, m.p.h. .. .. $82\,^1/_2$

Westbury–Taunton section:
- Distance, miles 47.6
- Time, pass to pass actual, min. sec. 48-7
- Time, pass to pass net, min. 46
- Load, trailing tons tare 434
- Cut-off varied between 15% and 18%
- Maximum speed, m.p.h. 72

The final stretch between Taunton and Exeter with a load of 372 tons was subject to severe checks on account of engineering works on the line; but although delays caused a total loss of $9\,^3/_4$ min. the arrival at Exeter was only $2\,^1/_4$ min. late. The net time showed an average speed of 62 m.p.h. from start to stop, which in relation to the economical working of the engine must be considered as a very good result. The gross trailing loads over the three sections were 530, 460, and 395 tons respectively. Details of the engine working, recorded by an observer on the footplate indicate that the steaming was almost perfect throughout. There was a slight drop in pressure to 245 lb. per sq. in. at Whiteball summit, but otherwise the pressure was 250 lb. per sq. in. throughout. Such a performance was made possible by the high standards of day-to-day maintenance always characteristic of Great Western practice; high quality coal, expertly fired, and a driver who clearly knew the road and its peculiar features to the last detail. It was in fact an ideal performance, which enabled the Churchward principles in design to be exploited to great advantage. Unfortunately the ideal conditions which largely contributed to the making of such runs were not infrequently missing in later years and the traditional Swindon practice had to be modified accordingly, as will be recounted in later chapters of this book.

The management of the Great Western Railway was justifiably proud of the records held by its express train services; but when the northern lines eclipsed its record length of non-stop run of 225.7 miles from Paddington to Plymouth by runs from London to Carlisle and from London to Edinburgh an attempt

Fig. 94.—*G.W.R. " King " class* 4-6-0 *No.* 6025 *on the Cornish Riviera Express.*

Fig. 95.—*G.W.R. "Castle" class* 4-6-0 *on down South Wales express near Cholsey.*

was made to make secure the existing British record for the fastest scheduled start-to-stop run. The 2.30 p.m. express from Cheltenham Spa to Paddington was ideal for such an exploitation, for it was booked non-stop over the 77.3 miles from Swindon to Paddington. This slightly downhill and excellently aligned route was utilised at a quiet time of the day. From the summer of 1923 the train had been scheduled to make the run in 75 min. In 1929 this was cut to 70 min., and in September 1931 to 67 min. On three days in succession following this last mentioned acceleration the run was made in $59\frac{1}{2}$, $58\frac{1}{2}$ and $58\frac{1}{4}$ min. respectively, the last mentioned, with its average speed of 79.6 m.p.h. from start to stop, constituting a world's record. But the Great Western authorities were not content with the slender margin by which this record had been secured, and on June 5, 1932, a specially staged attempt was made to set up a new record,

Fig. 96.—*Southern Railway: Urie "*N.15*" class* 4-6-0 *No.* 747 *with Maunsell modifications, on Bournemouth and Weymouth express near Worting Junction.*

Fig. 97.—*Southern Railway: " Lord Nelson " class* 4-6-0 *No.* 859 *on up Continental express in Folkestone Warren.*

not only for the Cheltenham Flyer itself, but for the down train leaving Paddington at 5 p.m. The following are details of what was then a world's record run:

G.W.R. Swindon–Paddington
Load: 195 gross trailing tons
Engine: No. 5006 *Tregenna Castle*

Dist. miles		Sch. min.	Actual m. s.	Speeds m.p.h.
0.0	SWINDON	0	0 00	—
5.7	Shrivenham		6 15	54.7
10.8	Uffington		9 51	85.3
16.9	Wantage Road		14 05	86.3
24.2	DIDCOT	21	18 55	89.2
28.8	Cholsey		21 59	90.2
35.8	Pangbourne		26 33	92.2
41.3	READING	34	30 11	90.6
53.1	Maidenhead		38 08	89.0
58.8	SLOUGH	47	42 10	84.7
68.2	Southall	54 1/2	48 51	84.3
76.0	Westbourne Park	61	54 40	80.6
77.3	PADDINGTON	65	56 47	

On this journey the regulator was gradually opened to the full by the time Shrivenham was passed, and cut-offs were varied between 17 and 18 per cent. from Shrivenham to within 2 miles of Paddington. Part of the very fast overall time it must be admitted was due to an exceptionally rapid run in from Westbourne Park. On the same day engine No. 5005 *Manorbier Castle* hauled the down train, loaded to 210 tons, from Paddington to Swindon in 60 min. 1 sec. On this journey the average speed between mileposts 6 and 76 was 81 m.p.h., against the very slight gradient, and the cut-offs required, with full regulator varied between 19 and 21 per cent. The " Castle " class locomotives frequently demonstrated their ability to run to 90 m.p.h. on virtually level track. Such performances, taken with the heavy load haulage capacity of the " King " class locomotives previously referred to, represented the ultimate development of the Churchward principles in locomotive design, at any rate so far as express passenger service was concerned on the Great Western Railway.

Chapter Seven   Articulated Locomotives

Introduction

At the beginning of the period under review the different types of articulated locomotive being manufactured by the British locomotive building industry lay in four broad categories:

1. The Mallet.
2. The Fairlie, or modified Fairlie.
3. The Kitson Meyer.
4. The Beyer-Garratt.

Generally speaking the Fairlie group and the Kitson-Meyer tended to disappear, and whereas the Mallet in its modified form, using high-pressure steam in all cylinders was developed to locomotives of very large size in the United States of America it was mainly the Beyer-Garratt type in which the British articulated locomotive developed from 1925 onwards. Prior to that time Mallet articulated locomotives in their true form, that is, as compounds, had been built in considerable numbers for service overseas by the North British Locomotive Company and a number of these very imposing locomotives are illustrated herewith. But whereas the American loading gauge and the relative straightness of many of the most severely graded main lines in the U.S.A. favoured the development of the Mallet in its 4-cylinder simple form, the operating conditions on the majority of the railways for which the British locomotive industry catered, favoured the development of the Garratt.

A study of the proportions of some of the largest Mallet articulated compound locomotives built up to the year 1925 shows that the boiler proportions were approaching the limit for satisfactory steaming; indeed some of the locomotives having very long boiler barrels would appear to have gone almost beyond that limit. A boiler with a very long barrel is not an ideal proposition where heavy continuous steaming is required, and the inherent physical characteristics of the Beyer-Garratt permitted of the mounting of boilers of ideal proportions, while still coming within the limits of a restricted loading gauge, sub-standard gauge track, and a sharply-curved line. The Garratt boiler, with its barrel slung between the two engine units, could be of very large diameter, and the requisite heating surface could be obtained with a relatively short distance between the tube plates—in other words, it was ideal for hard continuous steaming. In this respect the Garratt had also the advantage over the modified Fairlie; for whereas the latter also had a boiler mounted between the two engine units, the existence of a long rigid frame from end to end of the locomotive meant that this had to be strong and deep in the centre and this imposed a restriction on boiler size which was not so marked in the case of the Garratt.

The origin of the large articulated locomotives of the period covered roughly by the years 1910 to 1960 lay primarily in the need to have a large locomotive with a more flexible wheel base, and a lower individual axle loading than could be provided with locomotives of conventional type. Both the Mallet and the Garratt, in their respective spheres of activity, provided the answer and an alternative to regular double heading. Both were normally medium speed machines. The curves and gradients, that are so frequently to be found on routes with a severely restricted axle loading also precluded the running of trains at normal British standards of express speeds; but in certain countries where the alignment and physical conditions were more favourable, locomotives of the Beyer-Garratt type have been designed for express passenger duty, and notable examples have been used on the Sao Paulo Railway, and in Algeria. Locomotives for the latter railway were not built in Great Britain, though nevertheless under the Beyer-Garratt patents.

In recent years the latest and largest Garratts have been built to meet the needs of railways where the question of line capacity was becoming acute, and where it was found desirable to run trains of heavier formation to avoid increasing the number of trains on the line, and at the same time to avoid double heading. Nevertheless the Garratt, like the largest Mallet articulated compounds is virtually two locomotives in one; and when the nominal tractive effort of the complete unit exceeds 60,000 or 70,000 lb. special arrangements have to be made for firing on those railways which did not use mechanical stokers. On some of the largest Garratts that were hand-fired a crew of four men was carried, usually consisting of driver, fireman and two extra hands to get supplies of coal forward from the bunker. This, of course, was a more economical proposition than using two locomotives each with two skilled men on the footplate. The Beyer-Garratt, which has been supplied in all shapes and sizes to railways all over the world, except in North America, has been one of the greatest and most successful products of the British locomotive building industry, and in this chapter illustrations and descriptions are given of a variety of locomotives in this category, which have been built by Beyer-Peacock and Co. Ltd. at their Gorton Works.

Fig. 98.—*South African Railways:* 2-6-6-0 *Mallet compound locomotive, built by North British Locomotive Co. Ltd.*

The Mallet

The characteristics of a typical Mallet compound of the period just prior to 1925 are well illustrated by consideration of the South African 2-6-6-0 No. 1660 illustrated herewith (Fig. 98). The gauge is 3 ft. 6 in. and the maximum axle-load permitted on the route for which the locomotives were required was 14 tons. Yet the duty needed a machine of 30,000 lb. tractive effort at 50 per cent. boiler pressure. By use of the Mallet principles, it was possible within the limitations laid down to design a locomotive having a combined total heating surface of 2,311 sq. ft., a grate area of 40 sq. ft. and to provide an adhesion weight of 79 tons, and yet with a maximum rigid wheelbase of only 8 ft. 4 in. As usual with the true Mallet the high pressure cylinders, 16 1/2 in. by 24 in., drove the rear group of coupled wheels, and the low pressure cylinders drove the leading group. The low pressure cylinders were 26 in. diameter by 24 in. stroke, and the coupled wheels were 3 ft. 6 1/2 in. diameter.

Brief mention was made in Ahrons' work of the giant South African Mallet of the "MH" class, which was of the 2-6-6-2 type, and with its tender weighed 179.8 tons in working order (Fig. 99). This was, of course, a main line freight unit with a maximum permitted axle load of 18.2 tons, and was at one time the largest locomotive in the world operating on the 3 ft. 6 in. gauge. The nominal tractive effort at 50 per cent. boiler pressure, was 43,330 lb. In this great engine particular interest naturally centres in the boiler, which from the photograph appears immensely long in relation to the general proportions of the locomotive itself. The diameter of the first ring was 5 ft. 11 in., and the distance between the tube plates 22 ft. The cylinders were 20 in. diameter high pressure,and 31 1/2 in. low pressure, both with a stroke of 26 in.,and the coupled wheel diameter was 4 ft.

Fig. 99.—*South African Railways: The giant Mallet compound "MH" class.*

The heating surfaces were:

S.A.R. 2-6-6-2 *Class "M.H."*
Heating surfaces, in sq. ft.

Tubes	2961
Firebox	250
Total evaporative	3211
Superheater	616
Combined total	3827
Grate area	53
Boiler pressure, lb./sq. in.	200

The Burma Railways also made considerable use of the Mallet type of locomotive, particularly on the Lashio line in the north of the country where there is a length of 11 miles continuously at 1 in 25. Furthermore this section includes numerous reverse curves, uncompensated for gradient, the effect of which is equivalent to a gradient of 1 in 21. On the remaining 138 miles of the Lashio line the ruling gradient is 1 in 40. An interesting design of Mallet articulated compound locomotive was built by the North British Locomotive Company, and illustrated herewith (Fig. 100). The maximum axle load permitted was 10 tons yet obviously a very powerful engine was needed over such exceedingly severe gradients. The Mallet compounds illustrated were of the 0 6 6 0 type, with 3 ft. 3 in. wheels and a maximum rigid wheelbase of 8 ft. 3 in. Within a total weight of 59.4 tons it was not possible to provide a boiler with more than 1552 sq. ft. of heating surface, and a grate area of 33 sq. ft. These engines were not superheated, and worked at a pressure of 180 lb. per sq. in. The cylinder diameters were $15\frac{1}{2}$ in. high pressure and $24\frac{1}{4}$ in. low pressure, with a common stroke of 20 in. At 50 per cent. boiler pressure the nominal tractive effort was 22,176 lb. These engines were found capable of hauling a load of 145 tons on the 1 in 25 gradient, and of 260 tons on the 1 in 40. The coal consumption averaged 67.9 lb. per 100-ton train miles.

A final example of a Mallet articulated compound is to be seen in the 0-6-6-0 for the Uganda Railway illustrated herewith (Fig. 101). The basic dimensions of this engine are exactly the same as those of the Burma locomotive just described, and the same detailed design was followed with no more than minor modifications. The more liberal loading gauge on the Uganda Railway permitted of a taller chimney and larger steam dome. The latter railway used the Westinghouse brake, and the main reservoir for this system was carried on the top of the boiler, thus rather detracting from what would otherwise have been a handsome locomotive. The two designs taken together—Burma and Uganda represented powerful engines for the metre gauge at the time of their construction, just before World War I.

Fig. 100.—*Burma Railways:* 0-6-6-0 *Mallet compound for the Lashio Line.*

In early days of the use of the Mallet compound type of locomotive the 0-6-6-0 wheel arrangement, as embodied in the Burma and Uganda engines just described, and in the first British-built Mallet for the Pekin-Kalgan Railway illustrated in Chapter 4, was also used on the South African Railways. These engines were introduced particularly for assisting heavy freight trains bound for Johannesburg on the difficult gradients leading from Natal, and from the south on to the high central plateau of the Transvaal.

The Kitson-Meyer

Whereas the Mallet, so far as British design and construction was concerned, became completely

Fig. 101.—*Uganda Railway:* 0-6-6-0 *Mallet compound, similar in design to the Burma type, but with Westinghouse brake.*

Fig. 102.—*Kalka-Simla line, North Western Railway of India: Kitson-Meyer articulated locomotive.*

superseded by the Beyer-Garratt for heavy, exceptional duties the Kitson-Meyer type which was first introduced in 1894, was used for new designs down to the year 1929. As distinct from the Mallet, in which the rear engine unit is carried on the main frames and in which the front unit pivots in a bearing provided in a cradle beneath the forward part of the boiler, in the Kitson-Meyer the superstructure, which carries the boiler, the fuel and the water bunkers, was carried on two bogies; and the whole of the weight was passed to the bogies through the pivots, which were placed as near as possible to the centre of the adhesive wheelbase. As a result, what could be termed the executive part of the locomotive turns readily according to the line being traversed.

The engine illustrated in Fig. 102 was built in 1928 by Kitson and Co. Ltd., for the Kalka-Simla Railway in India, for use on 2 ft. 6 in. gauge. The line is an exceedingly difficult one, with gradients of 1 in 25 and curves of 120 ft. radius. High speed was, of course, out of the question, and a remarkably high nominal tractive effort of 26,025 lb. was achieved, at 85 per cent. boiler pressure, by use of cylinders 13½ in. diameter by 14 in. stroke, in combination with coupled wheels of only 2 ft. 6 in. diameter. The cross-sectional drawing reproduced herewith (Fig. 103), shows how a relatively large boiler and firebox was accommodated on the 2 ft. 6 in. gauge. Fortunately the loading gauge permitted of an overall width of 7 ft. 6 in.

In the case of any articulated locomotive a point of interest, and always a potential source of weakness in the case of a new design, is the means of achieving the flexibility in the steam connections to the pivoting units. In the Kitson-Meyer type, in the course of experience of building no fewer than 71 different types of locomotives the makers have used a ball and socket design with uniform success. The steam pipes for either engine unit were carried rigidly on the superstructure to a position central with the bogie pivot, where they were received in the ball and socket joint and thence continued to the cylinders. The exhaust steam from the cylinders was brought back to the smokebox in a similar way, a ball and socket joint being installed close to the twisting point. The reversing and handbrake gears, operated from the footplate, were provided with universal joints in the connections.

The leading dimensions of this interesting locomotive design were:

Cylinders (4) dia., in.	13½
Cylinders stroke, in.	14
Coupled wheel dia., ft. in.	2-6
Wheelbase, coupled, ft. in.	6-0
Wheelbase, total, ft .in.	44-10
Heating surfaces, sq. ft.:	
Tubes	904.7
Firebox	110
Superheater	212
Combined total	1226.7
Grate area, sq. ft.	27
Boiler pressure, lb./sq. in.	180
Adhesion weight (front group), tons	24.6
Adhesion weight (hind group), tons	23.95
Total wt. in working order, tons	68.55

A very powerful example of the Kitson-Meyer type was built in 1935 by Robert Stephenson and Co. for Colombian National Railways, having the 2-8-8-2 wheel arrangements. The rail gauge is 3 ft. 0 in., and the locomotive was required to haul trains of 330 tons up a gradient of 1 in 22 at 9–10 m.p.h. The maximum axle load permitted was 14½ tons, and it was possible to design a relatively large locomotive for the duty. The boiler had a total evaporative heating surface of 2567 sq. ft. and a grate area of 51 sq. ft., and with four cylinders 17¾ in. diameter by 20 in. stroke, coupled wheels 3 ft. 1½ in. diameter, and a boiler pressure of 205 lb. per sq. in. the nominal tractive effort at 85 per cent. boiler pressure was no less than 58,564 lb.—an altogether exceptional figure for the 3 ft. 0 in. gauge. This remarkable engine had an overall length of 66 ft. 4¾ in. and a total weight in working order of 130½ tons.

The Modified Fairlie

The original Fairlie type locomotives, made famous in Great Britain by the celebrated "double-engines" of the narrow gauge Festiniog Railway, had two boilers fed from a central firebox, and the largest ever built in this country were three supplied by the Vulcan Foundry Ltd. to the Mexican Railway, in 1911, which had a total weight of no less than 138 tons, and an axle load of 23 tons. The nominal tractive

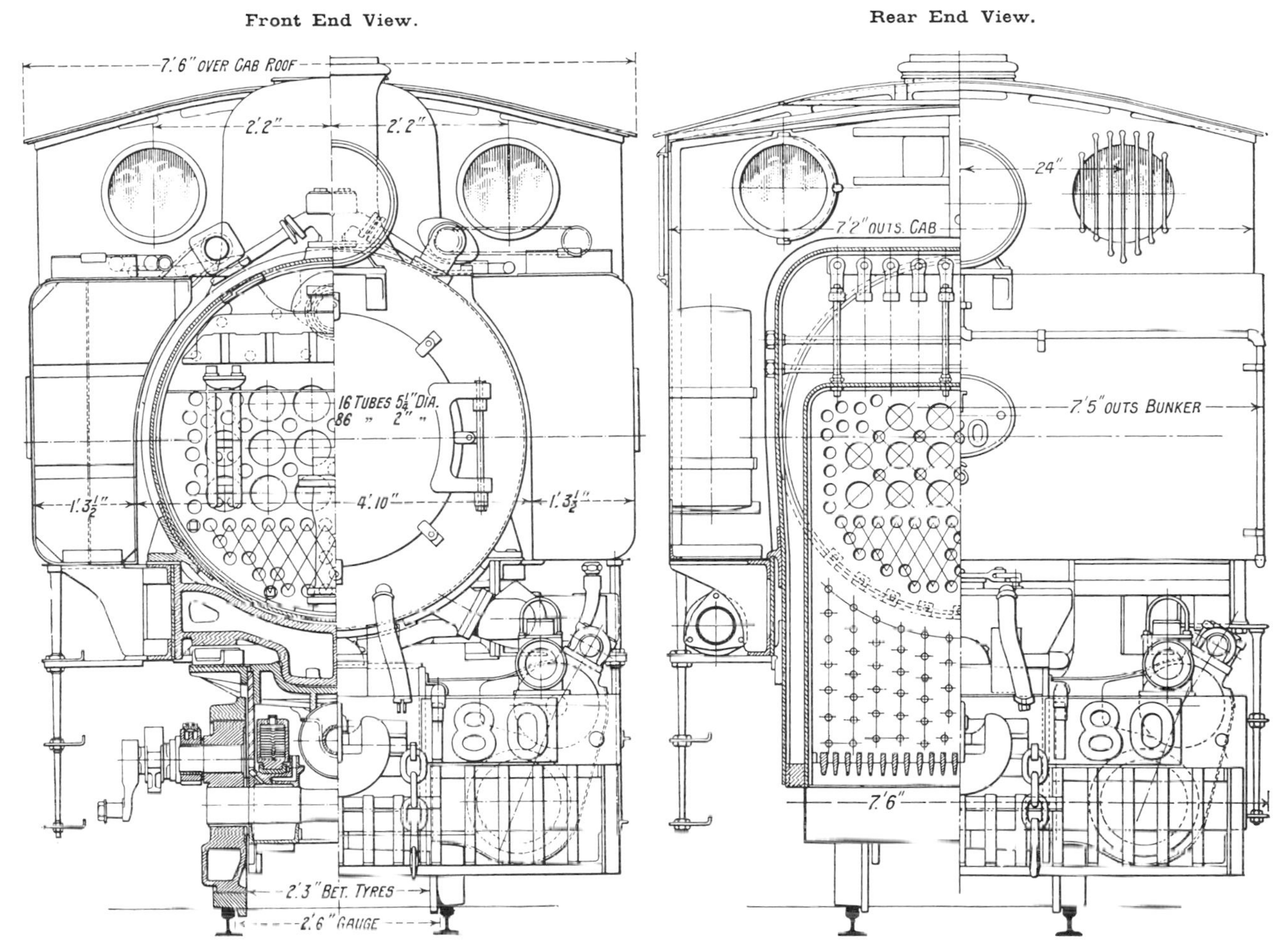

Fig. 103.—*Kitson-Meyer locomotive: cross-sectional drawings showing ingenious design for 2ft. 6in. gauge.*

effort at 75 per cent. boiler pressure was no less than 52,176 lb. But for a large and powerful engine developing so high a tractive effort the true Fairlie arrangement is hardly ideal from the firing point of view, and a totally different form of articulated locomotive was subsequently patented under the name of " modified Fairlie ". In actual appearance it was more like the Garratt, as will be appreciated from the accompanying illustration (Fig. 104), showing an engine of the 2-6-2 + 2-6-2 type built for the South African Railways in 1926 by the North British Locomotive Co. Ltd.

The " modified Fairlie " has some points of resem-

Fig. 104.—*South African Railways: " Modified Fairlie " type locomotive, built by North British Locomotive Co. Ltd. in* 1926.

Fig. 105.—*Burma Railways: Metre-gauge* 2-8-0 + 0-8-2 *Beyer-Garratt locomotive.*

blance to the Kitson-Meyer in that the boiler, tanks, and coal bunker are mounted on a rigid frame, supported on pivots engaging with bearing members on the two engine units. But while the Kitson-Meyer has the normal locomotive arrangement of the tender portion, entirely at one end, in the " modified Fairlie " the water tanks are arranged both fore and aft of the boiler, as in the Garratt. The South African locomotive illustrated, Class " FD ", was designed for a route with a maximum permitted axle load of 12 $^1/_2$ tons. The following were the leading dimensions:

Cylinders (4) dia., in.	15
Cylinders stroke, in.	24
Coupled wheel dia., ft. in.	3-9 $^1/_2$
Coupled wheelbase, ft. in.	8-6
Total evaporative heating surface, sq. ft.	1745
Superheater, sq. ft.	362
Boiler pressure, lb./sq. in.	180
Grate area, sq. ft.	40.87
Nominal tractive effort at 85% boiler pressure, lb.	36,312
Total wt. in working order, tons	104
Rail gauge, ft. in.	3-6

The Beyer-Garratt

As briefly mentioned by Ahrons the inception of this most successful type of articulated locomotive was due to H. W. Garratt, Inspecting Engineer for the New South Wales Government Railways. His patent application was made in 1907 and, eventually, Messrs. Beyer, Peacock & Co. Ltd. collaborated closely with him. Garratt's idea was the simplest of all ways of producing an articulated locomotive, in having the boiler on a separate cradle frame suspended between the two engine units. Unlike the Kitson-Meyer and the Modified Fairlie the tanks and coal bunkers are carried on the engine units, leaving the boiler-carrying cradle as short as possible. The fact that the underside of the firebox is completely free from obstruction permits the fitting of large self-cleaning ashpans. On conventional locomotives of the non-articulated type a designer was usually faced with the alternative of having a wide shallow firebox or a deep and long narrow one; but with the Garratt type of engine one can provide a firebox that is both wide and deep. It lends itself to the satisfactory combustion of a variety of fuels, and some notably successful designs have been produced for burning very low grade coal with an ash content as high as 40 per cent.

A total of eighteen different designs of Beyer-Garratt locomotive have been chosen for illustration and their basic dimensions are tabulated herewith. This table has been arranged in order of gauges. It is nevertheless most appropriate to begin with the 2-8-0 + 0-8-2 of the Burma Railways, because it was the performance of these, delivered in 1924, that first established the ascendancy of the Beyer-Garratt type over the Mallet. In tests carried out on the Lashio line the Garratt proved capable of hauling a load of 220 tons on the 1 in 25 gradient, against 145 tons by the Mallett, and the coal consumption per 100 ton miles showed an economy of 18.5 per cent. in favour of the Garratt (Fig. 105). It is true that the 2-8-0 + 0-8-2 was a more modern and considerably more powerful engine; but the main point was that the additional power was obtained at more economical fuel rates.

Following these comparative tests and the regular working of the engine in succeeding months, a further

Fig. 106.—*Leopoldina Railway: Metre-gauge* 4-6-2 + 2-6-4 *for low grade fuel: wood or briquettes.*

test was made to ascertain whether any advantage would be obtained with a compound Garratt. A compound engine was supplied by Beyer Peacock & Co. Ltd. in 1926, but no appreciable benefit was to be obtained from this change, and all future orders for engines of this class were for the standard arrangement of the Beyer-Garratt, with all four cylinders taking high pressure steam. The more recent history of Beyer-Garratt operation on the Burma Railways, and of the same engines in the neighbouring Bengal Assam Railway is referred to in the later chapter dealing with the part steam locomotives played in World War II.

Another interesting example of a metre-gauge Beyer-Garratt design is that of the Leopoldina Railway, reference 2 in the table on page 86. This was the largest British owned railway in Brazil. These engines were first supplied in 1930 for working the heavy night passenger service over the 200-mile section between Campos and Victoria, and in the course of the run the line passes through difficult mountain country, with a ruling gradient of 1 in 33 and numerous curves of about 4 chains radius. Although the new 4-6-2 + 2-6-4 locomotives did not rank among the largest of metre-gauge Beyer-Garratts they proved an immediate success and many more were subsequently put to work. They were particularly interesting in being designed expressly for low-grade fuel—either wood or briquettes—and their fuel bunkers were arranged for the convenient stacking of blocks or logs.

The Kenya and Uganda Railways, now the East African Railways, provide a remarkable instance of the application of the Beyer-Garratt type of locomotive to a railway operating through difficult country, on rails weighing no more than 50 lb. per yard, and where it was necessary to work heavy loads on individual trains in order to minimise line occupation. The accompanying gradient profile and curvature notes (Fig. 107) will enable the physical conditions existing in the 879 miles between Mombasa and Kampala to be appreciated. Two types of Beyer-Garratt locomotive have been chosen for illustration, the "EC2" (Fig. 108), developed from the original "EC1" first introduced in 1926, and the gigantic "EC3", of 1939 (Fig. 109). The "EC1" and "EC2", of which there were 36 in service up to the outbreak of World War I established a fine record of availability and freedom from trouble. In the year 1937, for example, the entire stud averaged 43,928 miles per annum, with many individual engines exceeding a total of 60,000 miles. Having regard to the slow speed of operation on this railway, and the fact that the average of 43,928 for the whole stud included the time spent in shops for overhaul these were remarkable results.

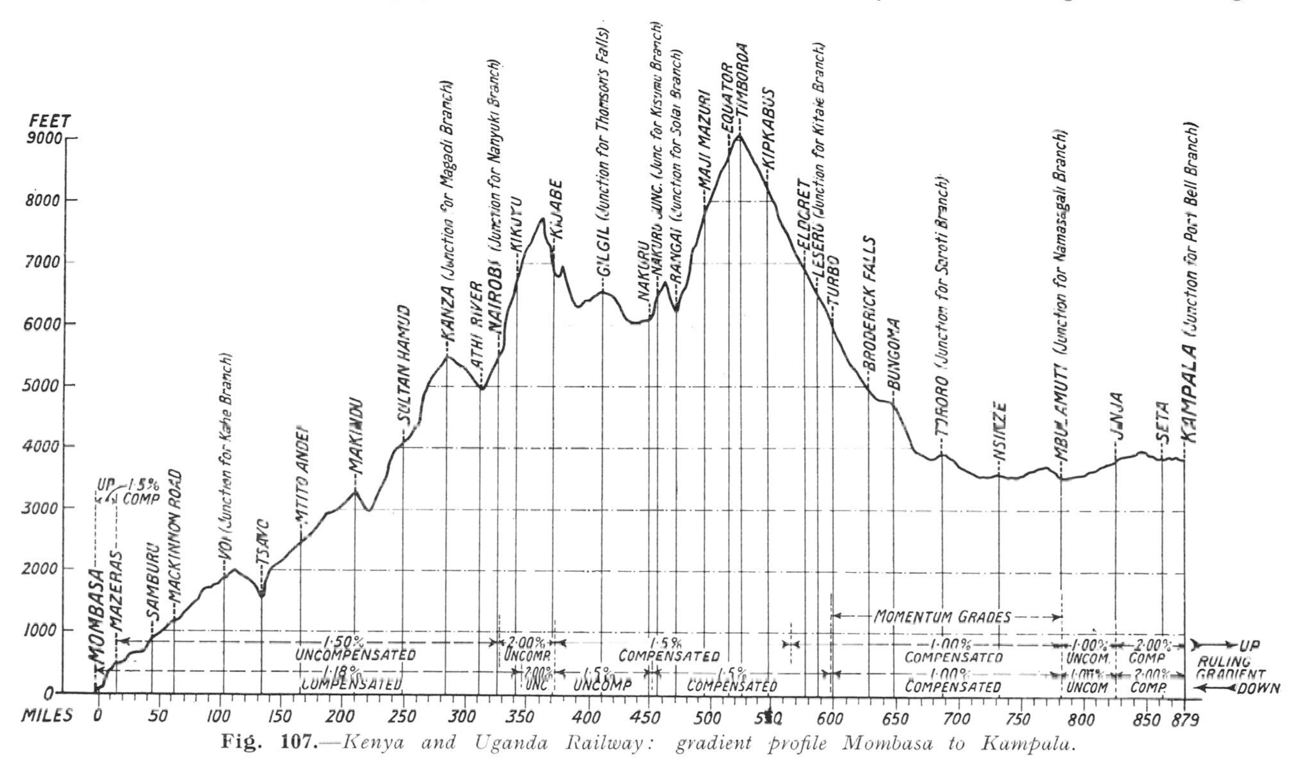

Fig. 107.—*Kenya and Uganda Railway: gradient profile Mombasa to Kampala.*

The "EC3" class, introduced in 1939 (ref. 4) established certain new records for steam traction. They were the first ever to have the wheel arrangement 4-8-4 + 4-8-4; they were the largest and heaviest engines yet to work on a 50 lb. rail, and yet their maximum axle-load was less than 12 tons. It was interesting to find that in these engines the coupled wheel diameter was increased from 3 ft. 7 in. in Class "EC2" to 4 ft. 6 in. and this feature, which contributed to better running and reduced maintenance costs for the running gear, also reduced the nominal tractive effort in a comparative sense, so that the increase over the Class "EC2", 46,100 lb.

Fig. 108.—*K.U.R.* 4-8-2 + 2-8-4 *locomotive, class* "*EC2*".

Fig. 109.—*K.U.R.* 4-8-4 + 4-8-4 *locomotive, class* "*EC3*", *first introduced* 1939.

Fig. 110.—*Benguela Railway: wood burning* 4-8-2 + 2-8-4 *locomotive.*

Fig. 111.—*Nigerian Railway:* 4-6-2 + 2-6-4 *locomotive for working over track laid with* 45 *lb. rails.*

against 40,260, is not so great as the increase in boiler proportions and increase in boiler pressure might suggest. The caboose system of operation was used on the K.U.R., and with the " EC3 " engines through working over the 1,100 miles from Nairobi to Kampala and back was introduced. Locomotive operation on this railway is referred to again in later chapters of this book.

In turning to locomotives working on the 3 ft. 6 in. gauge, illustrations and details are included relating to railways all of which have difficult operating conditions. The Benguela Railway uses wood fuel on the coastal section, and the 4-8-2 + 2-8-4 (ref. 5) (Fig. 110) works on the wood of eucalyptus trees grown alongside the railway specially as a locomotive fuel. They were required to haul a load of 450 tons on a 1 in 40 gradient. The Nigerian 4-8-2 + 2-8-4 (ref. 7) (Fig. 111) is interesting for its very low axle-load for working over 45 lb. rails, and also for the fact of the 28 locomotives of the Beyer-Garratt type in use on this railway, up to the year 1944 being driven, fired and repaired entirely by Africans, under European supervision.

On the Sudan Government Railway it is the existence of desert conditions rather than exceptional curvature and gradients that provides one of the most severe tests of locomotive operation. A large portion of the line was laid with 50–52 lb. rails, and with sandy conditions, a complete lack of water, and a sun temperature reaching over 160 deg. Fahr. there were many points of detail design to be given the most careful attention. Although these Beyer-Garratt engines (Fig. 112) have very large water capacity in their tanks it was necessary to haul an auxiliary water tank on certain duties, where no water was available intermediately for a distance of 150 miles. These engines have successfully worked freight trains of 1600 tons over the desert sections from Atbara to Wad Medani, and loads of 900 tons were taken on the 1 in 100 gradients of the line from Port Sudan through the Red Sea Hills.

The working of the Beyer-Garratt locomotives on the Rhodesia and South African Railways has become something of a legend in 3 ft. 6 in. gauge railway operation, and the four designs chosen for illustration and dimensional reference are no more than a very few among the numerous and successful classes that have been developed on these two railways. On the Rhodesian Railways prior to the inception of diesel traction more than half the total tractive effort of the entire system was in the form of Beyer-Garratt locomotives, while in the period from 1933 to the end of World War II the tonnage of freight conveyed on the railway *trebled*. The 4-6-4 + 4-6-4 illustrated (Fig. 113), has been a particularly successful general purpose locomotive, used on the mail trains, and also on freight, while the 4-8-2 + 2-8-4 (Fig. 114) was used principally on the northern main line on freight working. The following logs show typical performance of the 4-6-4 + 4-6-4 type in both passenger and goods service.

Fig. 112.—*Sudan Government Railway:* 4-6-4 + 4-6-6 *locomotive for working heavy freight trains in desert conditions.*

Fig. 113.—*Rhodesia Railways: General purpose main line* 4-6-4 + 4-6-4 *locomotive.*

Fig. 114.—*Rhodesia Railways: Heavy freight* 4-8-2 + 2-8-4 *locomotive for the northern main line.*

Specimen Logs of Passenger and Goods Trains
Beyer-Garratt Engine 271

	Feet above sea level	Miles between stations	No. 7 Down "Rhodesia Express" 28/1/41		Miles per hour	No. 53 Down Goods 20/1/41		Miles per hour
			Arr.	Dep.		Arr.	Dep.	
Gwelo	4,650		—	15.5		—	3.35	
		24			35.1			20.6
Hunters Road	4,218		15.46	15.47		4.45	4.48	
		17			30.9			21.7
Que Que* ..	3,979		16.20	16.27		5.35	9.50	
		26			30.0			16.4
Battlefields ..	3,664		17.19	17.20		11.25	11.28	
		11			31.4			20.6
Umsweswe Siding ..	3,524		17.41	17.42		12.00	12.10	
		11			28.7			22.0
Gatooma ..	3,814		18.5	18.8		12.40	12.43	
		20			31.6			19.1
Hartley ..	3,900		18.46	18.56		13.46	13.56	
		8			34.3			24.0
Gadzema Siding ..	3,800		19.10	19.12		14.16	14.17	
		21			30.7			20.7
Makwiro ..	4.307		19.53	20.3		15.18	15.28	
		22			26.9			19.1
Norton	4,499		20.52	21.4		16.37	16.40	
		26 1/2			28.4			16.7
Salisbury ..	4,825		22.00	—		18.15	—	
		Total 186 1/2			Average 30.5			Average 19.3

Intermediate sidings at which no stop made omitted.
Load of No. 7 Down 18 bogies, 687 tons, 72 axles.
Load of No. 53 Down 18 bogies, 7 shorts, 1,052 tons, 86 axles.
NOTE.—The running time of the "Rhodesia Express" from Gwelo to Salisbury is 6 hr. 25 min. for a load of 550 tons. In the log shown, 18 min. were made up with a load of 687 tons.
Load tons—2,000 lb.
*Crew-changing station for goods trains.

The South African Railways was one of the first administrations in the world to experiment with locomotives of the Beyer-Garratt type, as early as the year 1921. Superiority over the Mallet being soon established, the development of the Garratt proceeded rapidly from the "GB" class, having the 2-6-2 + 2-6-2 wheel arrangement, and a nominal tractive effort of 20,610 lb. to the gigantic "GL", of 1929 (ref. 10) (Fig. 115), with a nominal tractive effort of 89,130 lb. This latter is, of course, a main line unit with a maximum axle-load of 18 1/2 tons; but it is nevertheless an astonishing machine to operate on the 3 ft. 6 in. gauge. The "GM" class (ref. 11) (Fig. 116) was designed for more general service, on routes laid with 60 lb. rails, and was another very fine example of locomotive design, to the general Beyer-Garratt principles. Their introduction on the section between Johannesburg and Zeerust, in 1938 increased the capacity of that line by 50 per cent. Whereas the previous locomotives had been able to handle loads of 450 to 500 tons on the 1 in 40 gradients, the new "GM" class handled 700 to 750 tons on the same schedules.

The well-known 2-6-0 + 0-6-2 type of the London Midland and Scottish Railway (ref. 12) (Fig. 117) was designed to eliminate double heading on the heavy coal trains of the Midland Division, between Toton Yard and Cricklewood. It consisted of two standard 2-6-0 engine units supplied by one very large boiler in the Beyer-Garratt tradition. Again it was a case of marshalling very heavy trains to minimise line occupation and to ensure reasonably punctual working on a route heavily used by a variety of passenger trains. Loads of up to 1,400 tons was taken by these engines, of which a total of 33 was eventually in service. The average speed of the south-bound loaded coal trains was increased from 18 to 21 m.p.h. by use of these engines, together with a saving of some 15 per cent. in fuel. The second example of a Beyer-Garratt, on the 4 ft. 8 1/2 in. gauge, the Iranian 4-8-2 + 2-8-4 (ref. 13) (Fig. 118), makes an interesting contrast to the British engine. Here were to be found the typical conditions

Fig. 115.—*South African Railways: "GL" class* 4-8-2 + 2-8-4 *locomotive, with tractive effort of* 89,130 *lb. on* 3*ft.* 6*in. gauge.*

Fig. 116.—*South African Railways: "GM" class general service Beyer-Garratt locomotive* 4-8-2 + 2-8-4 *type.*

Fig. 117.—*L.M.S.R.* 2-6-0 + 0-6-2 *Beyer-Garratt locomotive for the Toton-Cricklewood coal traffic with self trimming coal bunker on the rear unit.*

Fig. 118.—*Iranian Railways:* 4-8-2 + 2-8-4 *locomotive,* 61840 *lb. tractive effort, oil-fired, for service on mountain gradients.*

Fig. 119.—*U.S.S.R. The most powerful Beyer-Garratt locomotive ever built:* 4-8-2 + 2-8-4, *with a tractive effort of* 89,200 *lb., only slightly greater than the S.A.R. "GL" class (see Fig.* 115).

Beyer-Garratt Locomotives

Ref. No.	Rail Gauge ft.-in.	Railway	Wheel Arrangement	Max. Axle Load tons	Cylinders (4) dia. stroke in.	Coupled Wheel dia. ft.-in.	Total Heating surface sq. ft.	Grate Area sq. ft.	Boiler press. lb./sq. in.	Nom. T.E. at 85% b.p. lb.	Illustration Fig. No.
1	Metre	Burma	2-8-0+0-8-2	10.5	15½-20	3-3	2055	43.7	200	41890	105
2	Metre	Leopoldina	4-6-2+2-6-4	10.5	15½-22	4-2	2051	34	185	33240	107
3	Metre	Kenya	4-8-2+2-8-4	10.5	16½-22	3-7	2437	43.6	170	40260	108
4	Metre	Kenya	4-8-4+4-8-4	11.75	16-26	4-6	2692	48.5	220	46100	109
5	3-6	Benguela	4-8-2+2-8-4	13.	18½-24	4-0	3014	51.5	180	52360	110
6	3-6	Sudan	4-6-4+4-6-4	12.5	16¾-26	4-9	2400	43.2	200	43520	111
7	3-6	Nigeria	4-6-2+2-6-4	9.75	13½-26	4-0	1850	31.4	200	33560	112
8	3-6	Rhodesia	4-6-4+4-6-4	13.25	17½-26	4-9	2830	49.5	180	42750	113
9	3-6	Rhodesia	4-8-2+2-8-4	13.35	18½-24	4-0	2816	49.5	180	52360	114
10	3-6	South Africa	4-8-2+2-8-4	18.5	22-26	4-0	4185	74.5	200	89130	115
11	3-6	South Africa	4-8-2+2-8-4	15	20½-26	4-6	3820	63.4	200	68800	116
12	4-8½	L.M.S.R.	2-6-0+0-6-2	21	18½-26	5-3	2637	44.5	190	45620	117
13	4-8½	Iranian	4-8-2+2-8-4	14.75	19.29-25.98	4-5.2	4486	68.25	200	61840	118
14	5-0	U.S.S.R.	4-8-2+2-8-4	19.75	22 7/16-28	4-11	4540	85.5	220	89200	119
15	5-3	Sao Paulo	4-6-2+2-6-4	18.5	20-26	5-6	3622	49.2	200	53570	120
16	5-6	B.A.G.S.	4-8-2+2-8-4	12.7	17½-26	4-7½	2649	44.2	200	48790	121
17	5-6	Bengal Nagpur	4-8-2+2-8-4	17	20½-26	4-8	4114	70	210	69650	122
18	5-6	Ceylon	2-6-2+2-6-2	13.5	16-22	3-7	2222	44.9	185	41200	123

Fig. 120.—*Sao Paulo Railway: Express passenger Beyer-Garratt* 4-6-2 + 2-6-4 *type, for working on* 5*ft.* 3*in. gauge.*

Fig. 121.—*Buenos Aires Great Southern Railway: oil fired* 4-8-2 + 2-8-4 *for heavy freight service.*

Fig. 122.—*Bengal Nagpur Railway:* 4-8-2 + 2-8-4 *locomotive for haulage of* 1,500-*ton coal trains.*

Fig. 123.—*Ceylon Government Railway:* 2-6-2 + 2-6-2 *showing a* 1947 *development of a* 1928 *design.*

of heavy overseas duty: a ruling gradient of 1 in 36, oil firing, and the existence of many tunnels in mountain country.

The Russian locomotive (ref. 14) (Fig. 119), although the only one of its class, deserves special mention, because at the time it was the largest and heaviest steam locomotive to be built outside the U.S.A. The great height of the Russian loading gauge permitted of tall boiler mountings that to some extent give a false impression of the vast proportions of this locomotive. The boiler is enormous, with a total evaporative heating surface of 3570 sq. ft. and a grate area of 85.5 sq. ft. Owing to the severe conditions of the Russian winter numerous devices were provided throughout the locomotive to prevent freezing. For instance all the steam pipes were arranged to drain, while all the water pipes were provided with heat either from steam or radiation, and special attention was given to the lagging. The initial tests on this locomotive, in 1932, were carried out in 74 deg. Fahr. *of frost*, during which the engine successfully hauled a load of 2,700 tons on a gradient of 1 in 111, at about 10 m.p.h.

The Sao Paulo locomotive (ref. 15), represents the only Beyer-Garratt design prepared specially for express passenger service. It was originally designed with the 2-6-2 + 2-6-2 wheel arrangement, and gave excellent service in the haulage of 500-ton passenger trains; but to provide for extra water capacity the wheel arrangement was subsequently changed to 4-6-2 + 2-6-4 (Fig. 120). The necessary parts for conversion were made in England, and the reconstruction carried out in the railway workshops at Lapa. The second South American broad gauge example (ref. 16) (Fig. 121) represents one of the few instances of a Beyer-Garratt locomotive intended for a fairly level railway, and the Buenos Aires Great Southern ordered engines of this type in 1927 for working freight trains of exceptional weight on the Bahia–Blanca to Neuguen line. These engines were oil-fired.

The Bengal Nagpur Railway, at one time privately owned, operated one of the heaviest traffics in coal, and ores to be found in India. Much of its mileage was laid to the heaviest main line standards with 75, 85 and 90 lb. rails, rock ballasted, and maintained in first class condition. The 4-8-2 + 2-8-4 locomotives (ref. 17) (Fig. 122) represent the result of much experience with the Beyer-Garratt type and were put into service in 1940. They are primarily coal engines, and in haulage of regular loads of 1,500 tons they need the services of four men on the footplate: driver, two firemen, and a coal trimmer. Although they were the latest engines of the Beyer-Garratt type on the B.N.R. at the time of their introduction, they were slightly less powerful and less heavy than the 4-8-0 + 0-8-4 series of 1929, which had a nominal tractive effort of 69,660 lb. These latter engines were built to a permissible maximum axle load of 20.25 tons, for working on sections laid with 90 lb. rails, and the fact that in these high ranges of power a locomotive of almost identical capacity could subsequently be designed for use on lighter rails, with a restriction to a maximum axle load of 17 tons is an example of the design and manufacturing techniques that had been developed in this particularly successful branch of British locomotive engineering.

The last example of Beyer-Garratt design to be referred to in detail is the 2-6-2 + 2-6-2 for the Ceylon Government Railway (ref. 18) (Fig. 123). The engine illustrated represents the perpetuation in 1946, with modern improvements, of a design first worked out in 1928. These examples of the Beyer-Garratt articulated locomotive, which have been chosen with care, do collectively illustrate the diversity in conditions for which the British locomotive building industry has had to provide. One has only to instance the operating requirements of the Lashio Line in Burma; caboose working on the K.U.R.; the wood-burning 4-8-2 + 2-8-4 of the Benguela; desert conditions in the Sudan, and the exceptional climatic conditions in Russia to emphasise this diversity. But the greatest "success-story" so far as Beyer-Garratt locomotives are concerned is undoubtedly in Southern Africa, where the continued development and extensive use of the type on both the Rhodesia and on the South Africa Railways is in itself a testimony to the effectiveness of this notable product of the British locomotive building industry.

Fig. 124.—*East African Railways. Mail train leaving Nairobi for Mombasa hauled by a 60th class Beyer-Garratt Locomotive No.* 6020.

Chapter Eight — The Slump Years and their Effects 1930–35

From the year 1930 onwards while Great Britain in common with the rest of the world was enduring the effects of the greatest period of trade depression in history the home railways were fighting to combat the drain on their resources from rising operating costs and dwindling traffics. Already however some of the advantages of grouping were becoming apparent in the way some companies were able to reduce their total stocks of locomotives, and the policy gradually being evolved had a very considerable effect on future design. On the L.M.S.R. for example, in 1927, the total stock consisted of 10,128 locomotives, of 305 different classes, and at the end of 1931 a significant reduction had already been made, to a total of 9,032, of 261 different classes. Improved repair methods greatly reduced the time that locomotives were out of traffic, so that in turn fewer locomotives were needed to work the traffic. Furthermore, improved methods of servicing at the running sheds enabled the turn-round time between successive duties to be reduced. Reference has already been made, in Chapter Six, to the principle of working longer mileages per engine, so that the percentage of shed time to running time was reduced.

Some years before the crisis of the world depression hit the British railways in all its severity the L.M.S.R. had introduced a very elaborate and comprehensive system of costing for every locomotive on the line, recording systematically the cost of every item of repair, and its running costs in fuel and lubricating oil. The cost of installing such a system might have seemed at first to outweigh the advantages to be gained. But all concerned soon came to appreciate that a magnificent tool of management was now available, which very quickly and convincingly highlighted the efficient, as well as the inefficient classes then on the active list. The repair costs and the fuel consumption records showed the classes which were able to stand up to the new standards of performance, in putting up long daily mileages, and economical rates of fuel consumption.

It soon became clear that relatively few locomotives in the existing L.M.S.R. stock measured up to the desired standards, and thus were suitable for perpetuating in new construction; and it was at this stage that Sir Josiah Stamp as he was then, the President of the Executive, decided that the new Chief Mechanical Engineer, required in succession to E. J. H. Lemon, should be obtained from outside the L.M.S.R. company's staff. Mr. W. A. Stanier, as he was then, formerly Principal Assistant to the Chief Mechanical Engineer of the Great Western Railway, was appointed, and given what was without doubt the most comprehensive remit ever entrusted to a British railway locomotive engineer. With the performance records of every engine on the line available for study Stanier had the task of implementing an enormous programme of " scrap and build ", to modernise the entire locomotive stock. The new locomotives had not only to meet the best existing standards of performance, but to surpass them, and to meet the ever-increasing requirements of long mileages, quick turn-round times, multiple manning, cyclic rosters, and above all, low fuel consumption.

At a paper read before the Institution of Locomotive Engineers in January 1946, Mr. E. S. Cox quoted some figures that had emerged from the individual costing of locomotives, and those that related to well-known express passenger designs are of interest, in view of the bearing they had upon future practice.

3-Years Average Coal Consumption and Repair Costs for Selected Classes 1927–30

Former Rly.	Type	Class	Coal per mile, lb	Comparative Repair Cost Index*
Midland	4-4-0	" 2P "	45.9	100
Midland	4-4-0	Compound	46.5	136
L.N.W.R.	4-6-0	" Prince of Wales "	51.1	157
L.N.W.R.	4-4-0	" George V "	56.4	149
Caledonian	4-6-0	Pickersgill " 60 "	66.3	117
Caledonian	4-4-0	Pickersgill " 115 "	59.1	110

* Taking Midland Class 2 4-4-0 as 100

To some extent coal consumption is a measure of the severity, or otherwise of the duties undertaken by the locomotives in question, though in later years when both the Midland compounds and the L.N.W.R. " Prince of Wales " class became relegated to lesser tasks the overall coal consumption of the " Prince of Wales " class remained about 51–52 lb. per mile, while that of the Midland compounds rose sharply to much the same figure.

So far as boiler design was concerned Stanier's early work could be described as " pure Churchward ". The tapered barrel, so highly developed at Swindon between 1903 and 1910 was used in all new locomotives, and with it also was associated the characteristic Great Western type of Belpaire firebox. It was perhaps inevitable that in the launching of so huge a programme there should have been mistakes at the start. Churchward had always used a moderate degree of superheat, to secure high thermal efficiency by minimising the amount of heat wasted in the steam at exhaust. This practice answered well enough on the Great Western Railway where the quality of coal, and

Fig. 125.—*L.M.S.R. Stanier* 2-6-0, *with taper boiler, built Crewe,* 1933.

the traditions of firing together provided almost perfect steaming of all the modern classes of locomotive. But in the diverse conditions prevailing on the L.M.S.R., where qualities of coal varied greatly from area to area, and firing methods varied equally from the meticulously painstaking, to somewhat carefree tactics, there was widespread trouble from indifferent steaming with many of the early Stanier locomotives, and they were compared unfavourably with engines of comparable power classification that had preceeded them, as for example the new 3-cylinder 4-6-0s of the "Jubilee" class (Fig. 127) with the parallel boilered 3-cylinder 4-6-0s of the "Patriot" class (Fig. 126).

Fig. 126.—*L.M.S.R. 3-cylinder express passenger* 4-6-0 "*Patriot*" *class* (1930), *built in replacement of L.N.W.R.* "*Claughton*" *class.*

Fig. 127.—*L.M.S.R. 3-cylinder* 4-6-0 *No.* 5552 Silver Jubilee; *Stanier design first introduced* 1934.

Stanier 4-6-0 Boilers

Engine Class	6 ft. 0 in. Class "5"			6 ft. 9 in. Class "5X"	
Batch Period	5000–5069 1934–35	5070–5451 1935-37	5452 onwards* 1938 onwards	Original	Modified
Superheater:					
No. of elements	14	24	28	14	24
Heating surface, sq. ft.	227.5	307.0	359.0	227.5	300
Evaporative heating surfaces, sq. ft.:					
Tubes	1460	1460	1479	1462.5	1460
Firebox	156	171.3	171	162.4	181
Combined total, sq. ft.	1843.5	1938.3	2009	1852.4	1941
Grate area, sq. ft.	27.8	28.65	28.65	29.5	31.0

* Including engines numbered below 5000

Fig. 128.—*L.M.S.R. Stanier Class "5" mixed traffic* 4-6-0*: original type with domeless boiler and low degree superheat* (1934).

In broad principle it was only in boiler design that Stanier's work on the L.M.S.R. followed Great Western practice. Walschaerts valve gear outside the frames was used for the new 2-cylinder mixed traffic 4-6-0s, while on the larger engines, three sets of valve gear were used on the "Jubilee" 4-6-0s, and four sets on the "Princess Royal" class Pacifics. Long lap, long travel valves were already an established feature of L.M.S.R. practice. Experience with the earlier Stanier locomotives indicated that a higher degree of superheater was desirable to meet all conditions found on different parts of the line. The new mixed traffic 4-6-0s, later to be known so well as the "Black Fives" (Figs. 128, 129) were intended for use almost

Fig. 129.—*L.M.S.R. Stanier Class "5" mixed traffic* 4-6-0*: later variety, with domed boiler, and larger superheater.*

Fig. 130.—*L.M.S.R. The "Jubilee" class as originally built, with domeless boilers and low degree superheat.*

anywhere on the system, from London and Swansea in the south, to Wick and Thurso. The engines provided the ideal answer to the demand for general purpose machines that could be worked over long mileages, could respond to the handling of a multiplicity of engine crews and which would be low on both fuel and maintenance costs. Although the earliest batches had boilers with only a moderate degree of superheat they at once proved very fast engines that could undertake express passenger workings if necessary.

The leading dimensions of these engines and of the 3-cylinder express passenger 4-6-0s of the "Jubilee" class (Figs. 130, 131) are shown in the table on this page. This table includes particulars of the original and later boilers, thus showing the development that took place as a result of experience on the line. As

Stanier 4-6-0*s*
General Dimensions

Class	"5"	"5X"
Coupled wheel dia., ft.in.	6-0	6-9
Cylinders:		
number	2	3
dia., in.	18 1/2	17
stroke, in.	28	26
Boiler pressure, lb./sq. in.	225	225
Nominal tractive effort at 85% boiler pressure, lb.	25,455	26,610
Motion (Walschaerts)		
Dia. of piston valves, in.	10	9
Max. travel of valves, in.	6 1/2	6 3/16
Steam lap, in.	1 1/2	1 3/8
Exhaust clearance, in.	1/16	Nil
Lead, in.	1/4	5/16
Cut-off in full gear, %	75	76
Total weight of engine in working order, tons	72.1	79.5

Fig. 131.—*L.M.S.R. 3-cylinder "Jubilee" class* 4-6-0, *later variety, with domed boiler, larger superheater, and in final L.M.S.R. style of painting.*

finally developed both classes were very successful, and of the mixed traffic class no fewer than 842 were eventually in service. After nationalisation of the railways in 1948 their use was extended to routes that were formerly part of the L.N.E.R. system. On the Midland Division engines of the " Jubilee " class were regularly rostered to some long and complicated diagrams, with multiple manning, and their use enabled the service to be operated with fewer locomotives than previously.

economy and success, but the turn-round times at each end were shortened to give maximum utilisation. Locomotives were no longer allocated to the corresponding return working, such as the down Royal Scot express one day, and the up Royal Scot on the next; the next return working was taken, so that a locomotive arriving in Glasgow around 9.30 a.m. with the Night Scot sleeping car express, worked south again on the up Midday Scot, leaving Glasgow at 1.30 p.m. Some details of performance in these arduous

Fig. 132.—*L.M.S.R. One of the first two Stanier " Pacifics ", No.* 6201 Princess Elizabeth, *with original boiler and small tender.*

Fig. 133.—*L.M.S.R. Standard " Pacific " engine " Princess Royal " class, as built* 1935, *with high degree superheater.*

The introduction of the large Pacific engines of the " Princess Royal " class (Fig. 132) was followed by through engine working between Euston and Glasgow, 401 miles. Only the first two engines of this class had boilers with a moderate degree of superheat. This was soon found to be inadequate, though the poor steaming originally experienced with these engines could be attributed, in part, to inadequate cross-sectional area through the superheater. The accompanying table sets out the dimensions of these locomotives, and when the later examples (Figs. 133, 135) were in service not only were the London–Glasgow 401-mile runs regularly made with good conditions are given in a later chapter. With exactly the same relevant dimensions, the nominal tractive effort of the first Stanier Pacifics was the same as that of the Great Western " King " class 4-6-0s.

Design features included in the new standard range of locomotives on the L.M.S.R., and applying to the 3-cylinder 2-6-4 suburban tank engines (Figs. 137–139) as well as to the 4-6-2 and 4-6-0 main line locomotives were a smokebox of generous proportions, and carefully designed steam passages to minimise any wire-drawing effects, and a very successful design of coupled wheel axle box. These were of robust construction, being steel castings with pressed-in brasses and the

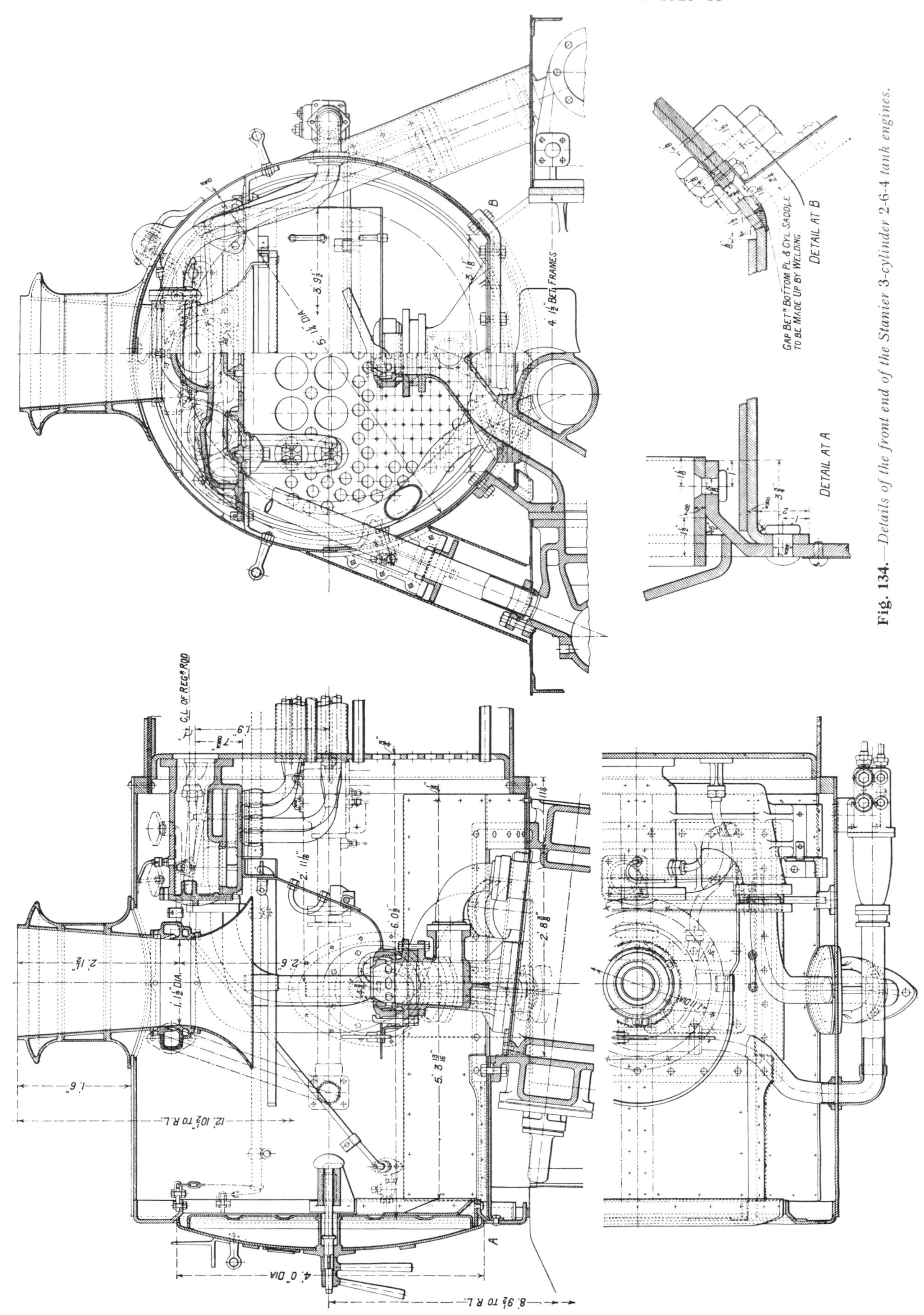

Fig. 134.—*Details of the front end of the Stanier 3-cylinder 2-6-4 tank engines.*

Fig. 135.—*L.M.S.R. The* 4-6-2 *engine* Princess Elizabeth *with modified boiler, as used in the spectacular Euston–Glasgow runs of* 1936.

usual white metal crowns. The accompanying drawing (Fig. 136) illustrates details of the boxes as fitted to the 2-6-4 tank engines. The arrangement of oil delivery will be noted. This type of axle box proved very effective in minimising troubles from heated journals on long runs at a continuously high

L.M.S.R. " Princess Royal " class boilers

Variety	6200 6201 as originally built	6203-6212 and 6200 modified	6201 modified
Dia. of barrel outside			
Max., ft. in.	6-3	6-3	6-3
Min., ft. in.	5-9	5-9	5-9
Distance between tube-plates, ft. in.	20-9	19-3	20-9
Tubes, small			
Number	170	123	119
Outside dia., in.	2 1/4	2 3/8	2 3/8
Superheater flues			
Number	16	32	32
Outside dia., in.	5 1/8	5 1/8	5 1/8
Heating surfaces, sq. ft.			
Tubes	2523	2097	2429
Firebox	190	217	190
Superheater	370	653	594
Total	3083	2967	3213

L.M.S.R. The " Princess Royal " class 4-6-2*s*
Dimensions Common Throughout

Cylinders (4) dia., in.	16 1/4
Cylinders, stroke, in.	28
Coupled wheel dia., ft. in.	6-6
Boiler press., lb./sq. in.	250
Tractive effort at 85% boiler pressure, lb.	40,300
Grate area, sq. ft.	45
Motion:	
Dia. of piston valves, in.	8
Max. travel of valves, in.	7 1/4
Steam lap, in.	1 3/4
Exhaust clearance	Nil
Lead, in.	1/4
Cut-off in full gear, %	73 1/2
Total wt. of engine only in working order, tons	104.5

output of power. Axle boxes of similar design were later fitted to all locomotives of the " Royal Scot " class. A further drawing shows a cross-sectional arrangement of the smokebox of the 2-6-4 tank engine (Fig. 134). This design, like the earliest varieties of 4-6-2 and 4-6-0 tender engines had domeless boilers and regulator in the smokebox.

The earlier part of the period under review, following experimental work in the late nineteen-twenties, saw extended trials of various forms of poppet valve gear on three out of the four grouped railways of Great Britain. At first emphasis was laid upon the mechanical advantage to be derived from use of poppet valves in themselves, without any reference to improvement in steam distribution. Indeed one of the earliest applications was on some 3-cylinder 4-4-0s of the L.N.E.R. Class " D49 " which were fitted with the Lentz oscillating cam type of gear, actuated by the ordinary radial valve gear, in which

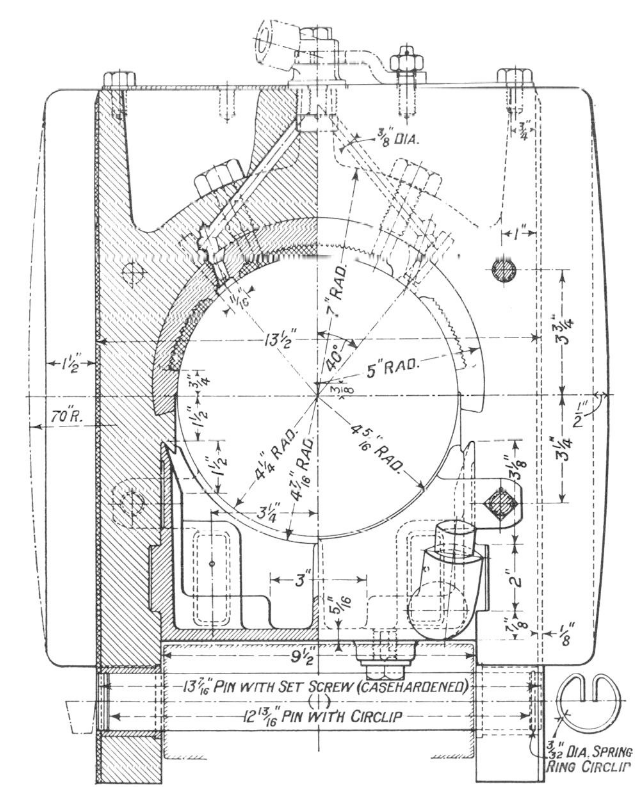

Fig. 136.—*Driving axlebox, L.M.S.R. as introduced on new standard locomotives from* 1934.

Fig. 137.—*L.M.S.R. Stanier* 3-*cylinder* 2-6-4 *tank engine* (1934) *for the London, Tilbury and Southend line.*

Fig. 138.—*L.M.S.R. Stanier* 2-*cylinder* 2-6-4 *tank design for general service* (1935).

Fig. 139.—*L.M.S.R. A later Stanier standard* 2-6-4 *tank engine.*

Fig. 140.—*L.M.S.R. Stanier's first* 2-6-2 *tank engine design of* 1935, *with domeless boiler and top feed.*

Fig. 141.—*L.M.S.R. The first version of the Stanier* " 8F " 2-8-0, *with domeless boiler and top feed* (1935).

Fig. 142.—*L.M.S.R. The later standard version of the* " 8F " *with domed boiler and separate top feed.*

the valve events were inter-related. The invention of the rotating cam type of poppet valve gear, by Arturo Caprotti, in Milan, in 1921, opened the way for some interesting British developments, that realised important economies of the operation of certain locomotives of obsolescent design. The British Caprotti gear had been applied to certain locomotives of the former Great Central " Lord Faringdon " class (B3) (Fig. 145), by Sir Nigel Gresley, while on the L.M.S.R. ten locomotives of the enlarged " Claughton " class (Fig. 144) were fitted. The Great Central engines could not be regarded as a successful design, as originally built, and the use of poppet valve gear made a very considerable improvement.

The ex-L.N.W.R. " Claughton " class locomotives had been greatly improved by substituting solid valve heads and six narrow rings for the original design of piston valves, and the ten engines of the class with larger boilers showed an even greater improvement in general performance. Comparative dynamometer car test runs carried out between Euston and Manchester indicated no advantage to the Caprotti valve gear locomotive, as compared with the modified piston valve locomotive with enlarged boiler. While the test results were unfavourable to the Caprotti engine, over six years of continuous service in express passenger traffic the 10 poppet valve locomotives showed a saving of $6\,^1/_2$ per cent. in coal consumption over the 10 otherwise identical engines with piston valves.

L.M.S.R. Large Boilered Claughton Trials Euston–Manchester

Engine	Caprotti valve gear	Walschaert's Narrow ring piston valves
Average wt. of train, tons ..	418	417
Average running speed, m.p.h.	52.9	52.7
Coal consumption:		
lb. per mile	39.9	38.2
lb. per ton mile (inc. engine)	0.078	0.075
lb. per d.h.p. hr.	3.53	3.25
lb. per sq ft. of grate area per hr.	69.1	65.8

In 1932 trials were made on three British railways with the R.C. poppet valve gear. This differed from the Caprotti in having a series of step cams, which in the case of the steam inlet cams correspond to a definite rate of cut-off. The Caprotti gear, like a screw reverser on a piston valve locomotive, is capable of infinitely variable adjustment. The essential features of the R.C. poppet valve equipment consisted of rotary cam shafts, one for each cylinder, each fitted with a series of step cams, as previously mentioned, four sets of which are fitted to each cam shaft. Each set of cams was made in the form of a solid sleeve, fitted on the shaft and secured by a key. Each step corresponds in the case of the steam inlet cams to a definite rate of cut-off, and in the case of those controlling the exhaust valves, to definite points of release and compression. In full gear no lead was given, and the amount of lead was progressively increased up to the shortest rate of admission. In 1932 R.C. poppet valve gear was fitted to three well-established classes of British locomotives:

(a) The L.M.S.R. " Horwich " 2-6-0.
(b) The L.N.E.R. 3-cylinder " D49 " class 4-4-0.
(c) The G.W.R. " Saint " class 4-6-0.

The L.M.S.R. application was made to five engines of the 2-6-0 class (Fig. 147), and some detailed drawings of the equipment fitted are shown in the accompanying diagrams (Fig. 146). The L.N.E.R. application was made to a complete batch of 40 new locomotives named after " Hunts ", this following trials of both oscillating cam, and rotating car gear on engines of the original " D49 " class, the standard examples of which had piston valves actuated by Walschaerts gear. The " Hunt " class " D49/3 ", had cams providing five positions of cut-off in forward gear, ranging from 15 to 75 per cent. This meant that adjustment of power output had to be made in rather large steps, and in the case of running at or near maximum capacity made the locomotives somewhat inflexible. These 40 engines of the " Hunt ", or D49/3 class were however used mainly in secondary express passenger service in the North Eastern area working from Leeds, Hull and Newcastle, and the trains hauled were mostly of relatively light formation that could be worked comfortably on 15 per cent. cut-off, with good economy. Their most important turn was the morning

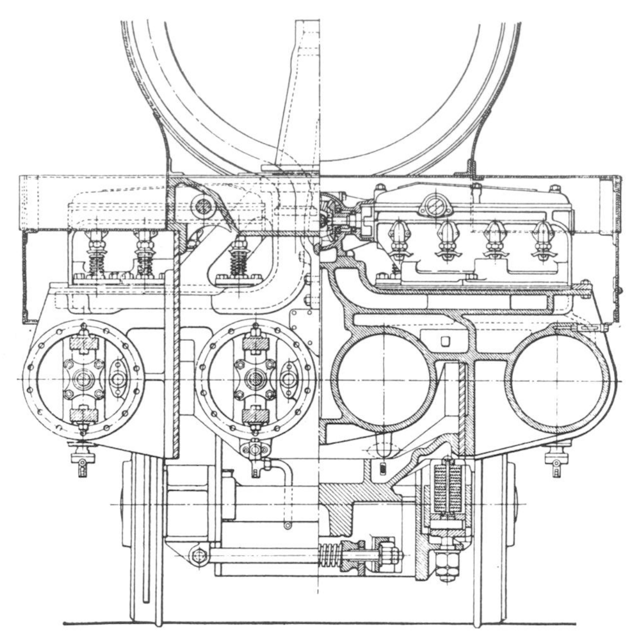

Fig. 143.—*L.M.S.R. Caprotti valve gear, as originally fitted to the L.N.W.R. " Claughton " class* 4-6-0 *No.* 5908 *when retaining the original boiler.*

Fig. 144.—*L.M.S.R. Large-boilered " Claughton "* 4-6-0 *fitted with Caprotti valve gear.*

Fig. 145.—*L.N.E.R. ex-G.C.R. 4-cylinder 4-6-0 No.* 6167 (*" Lord Faringdon " class*) *with Caprotti valve gear.*

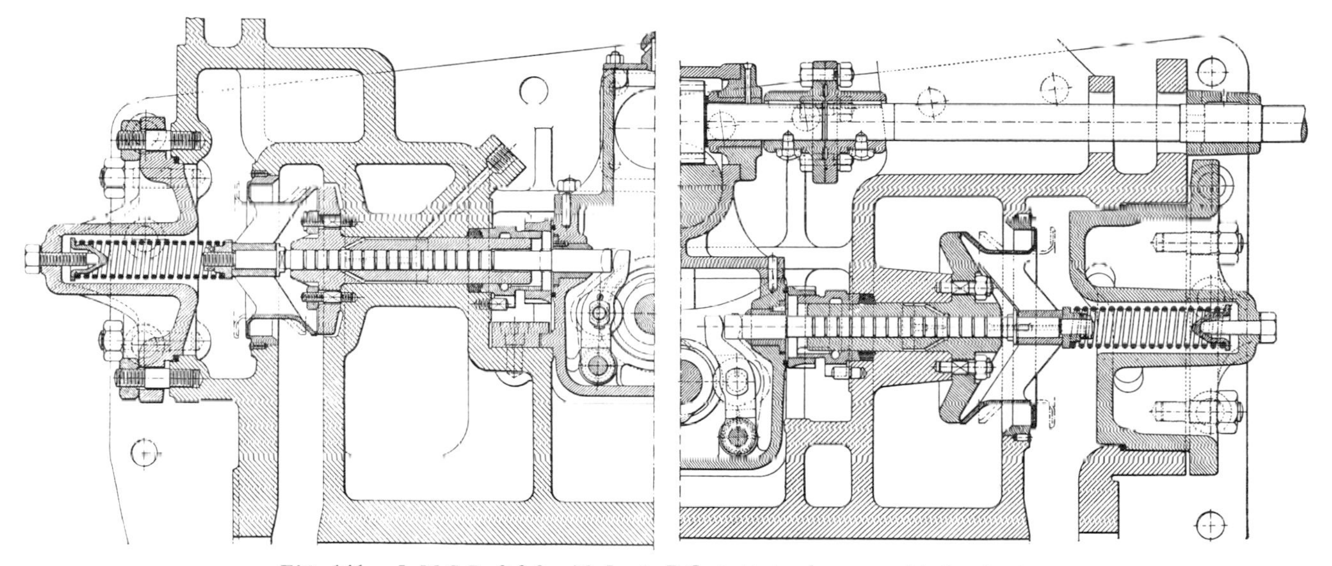

Fig. 146.—*L.M.S.R.* 2-6-0 *with Lentz R.C. poppet valve gear: details of valves.*

Fig. 147.—*L.M.S.R. Horwich* 2-6-0 *with Lentz R.C. poppet valve gear* (1931).

A small hole 27 was introduced in the preliminary throttle valve 6; this was to permit a certain amount of steam being continuously fed to the booster cylinders in order to keep them warm. To idle the booster the idling valve 28, shown in the reverse lever pilot valve 2, was turned so that the passage connected with the pipe 1 and allowed the steam to flow direct to pipe 2 leading to the preliminary throttle valve, and shutting it off from pipe 3 leading to clutch cylinder 7. The exhaust from the booster was taken through pipe 10 to the steam separator 29 and thence through pipe 11 into an annular space in blast pipe 30 and to atmosphere in the ordinary way. Valve 31 in the steam separator 29, which was balanced by counter-weight 32, was operated by the weight of trapped moisture in the chamber, the moisture being allowed to drain through the opening 33. The by-pass valve 40 was provided as an auxiliary exhaust from the booster main throttle valve operating cylinder. The booster cylinder sight feed lubricator 41 was automatically started by steam from pipe 4 passing along pipe 14 from the preliminary throttle valve and was provided with an anti-syphoning device 42, which broke the syphoning action when steam was cut off by the preliminary throttle valve.

Tests with the first engine to be equipped showed that the design specification had been more than maintained. In starting on level track the following results were obtained:

	Without booster	With booster
Load started, tons	496	746
Max. drawbar pull, tons	9	12 $^1/_4$

On a 1 in 70 gradient, with a 300-ton train from a start at the foot of the gradient without the booster a speed of 18 m.p.h. was attained in 8 min. With the booster in operation a speed of 25 m.p.h. was attained in 5 $^3/_4$ min.

By the year 1932 the attention of British locomotive engineers was being drawn to the remarkable results of rebuilding certain Pacific locomotives on the Paris, Orleans Railways, on principles laid down by M. André Chapelon. The advantages to be derived from long-lap, long-travel valves were by that time generally appreciated, and great improvements in performance had been secured by their adoption. But losses in efficiency were also incurred at other parts of the steam circuit, in long and tortuous passages, in sharp bends, and in poor design of ports and steam chests. Chapelon's work can be collectively and loosely described as " internal steamlining ", and among British engineers there was no one quicker to appreciate the full significance of what was going on in France than Mr. H. N. Gresley—as he still was in 1933. In planning a large new locomotive that would dispense with the need for double-heading north of Edinburgh the work of Chapelon was most carefully studied. The need, north of Edinburgh, was becoming urgent. The limit of load for both high and low-pressure Pacifics was 480 tons northbound, and 420 tons southbound. With the general introduction of third-class sleeping cars on the night trains many of these were loading regularly to over 500 tons.

The working of the 3-cylinder 2-6-0 engines of Class " K3 " on many parts of the L.N.E.R. system had given complete confidence in the standard design of pony truck at speeds of up to 80 m.p.h., while the Cartazzi type of trailing truck was standard on the Pacifics. There was thus every justification in the adoption of the 2-8-2 type for the new locomotives, except perhaps in the very curving nature of the route.

At first two 2-8-2 locomotives were built, one with R.C. poppet valve gear (Fig. 150), and one with the standard Gresley arrangement of Walschaerts gear, and derived motion for the central cylinder (Fig. 151). Apart from the very large proportions of the locomotives in general it was the " front-end " that contained so many novel features. Some time prior to the designing of the 2-8-2 engines Gresley had introduced the principle of casting the cylinders, valve chests, and smokebox saddles of his 3-cylinder locomotives in a single " monobloc " steel casting, and at Gorton Works, under Mr. R. A. Thom, the practice had been perfected to the extent that the intricate task was undertaken of casting the three large cylinders of the

Fig. 150.—*L.N.E.R. Sir Nigel Gresley's first* 2-8-2 *express locomotive, the* Cock o' the North *built Doncaster* 1934, *with poppet valves.*

Fig. 151.—*L.N.E.R. The second* 2-8-2, *the* Earl Marischal, *with standard layout of piston valves, and conjugated valve gear.*

" P2 " engines (21 in. diameter by 26 in. stroke) with their associated valves in a single unit. The accompanying drawing (Fig. 153) shows details of this fine piece of foundry work. This drawing relates to the first engine of the " P2 " class, No. 2001, *Cock o' the North* with R.C. poppet valve gear.

The other feature of these engines that was novel in Great Britain was the double blastpipe and chimney of the Kylchap type, which is shown in the cross-sectional drawing of the smokebox (Fig. 152). The principle of the double blastpipe used on certain later locomotives of British design was to increase the area of the blastpipe orifice, and so provide a free exhaust with a minimum of back pressure without affecting the draughting effect of the column of exhaust steam issuing from the blastpipe, which is in proportion to the surface area of the jet in contact with the smokebox gases. The diameters of the blastpipe tops on the *Cock o' the North* were 5 $^{3}/_{16}$ in. providing together a cross-sectional area of 42 sq. in. A single jet to give the same area would be 7 $^{5}/_{16}$ in.; but the circumference of this jet would be only 23 in. against 32 in. of the two 5 $^{3}/_{16}$ in. diameter jets on the *Cock o' the North*. The effectiveness of the double blastpipe in promoting more rapid steaming without recourse to sharpening the blast itself will thus be appreciated.

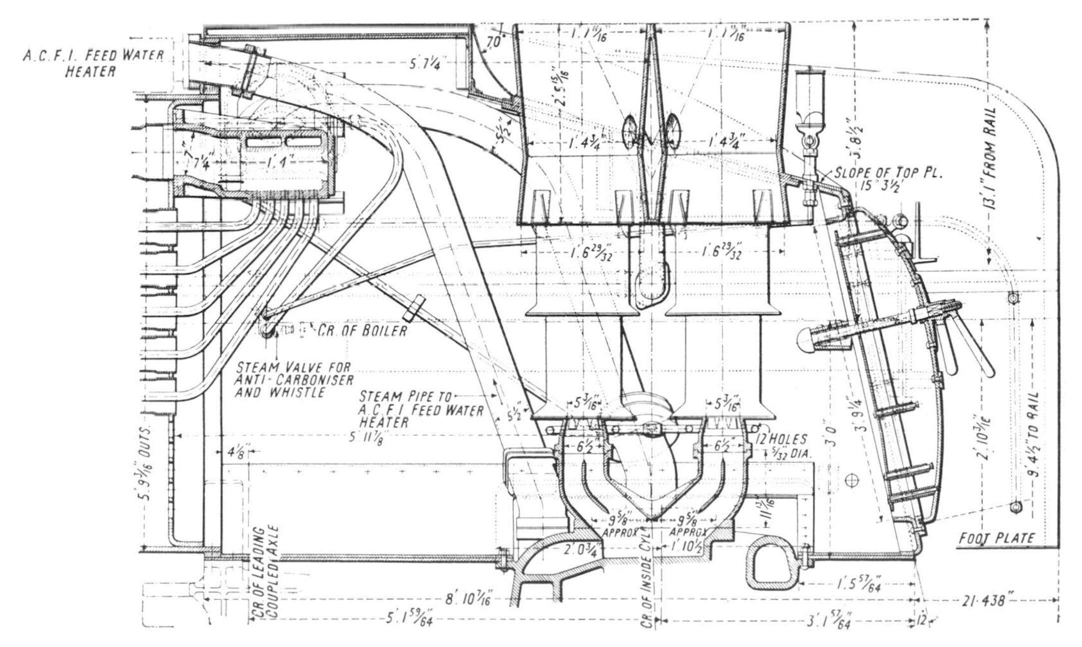

Fig. 152.—*L.N.E.R. Cross-section of smokebox:* 2-8-2 *engine* Cock o' the North.

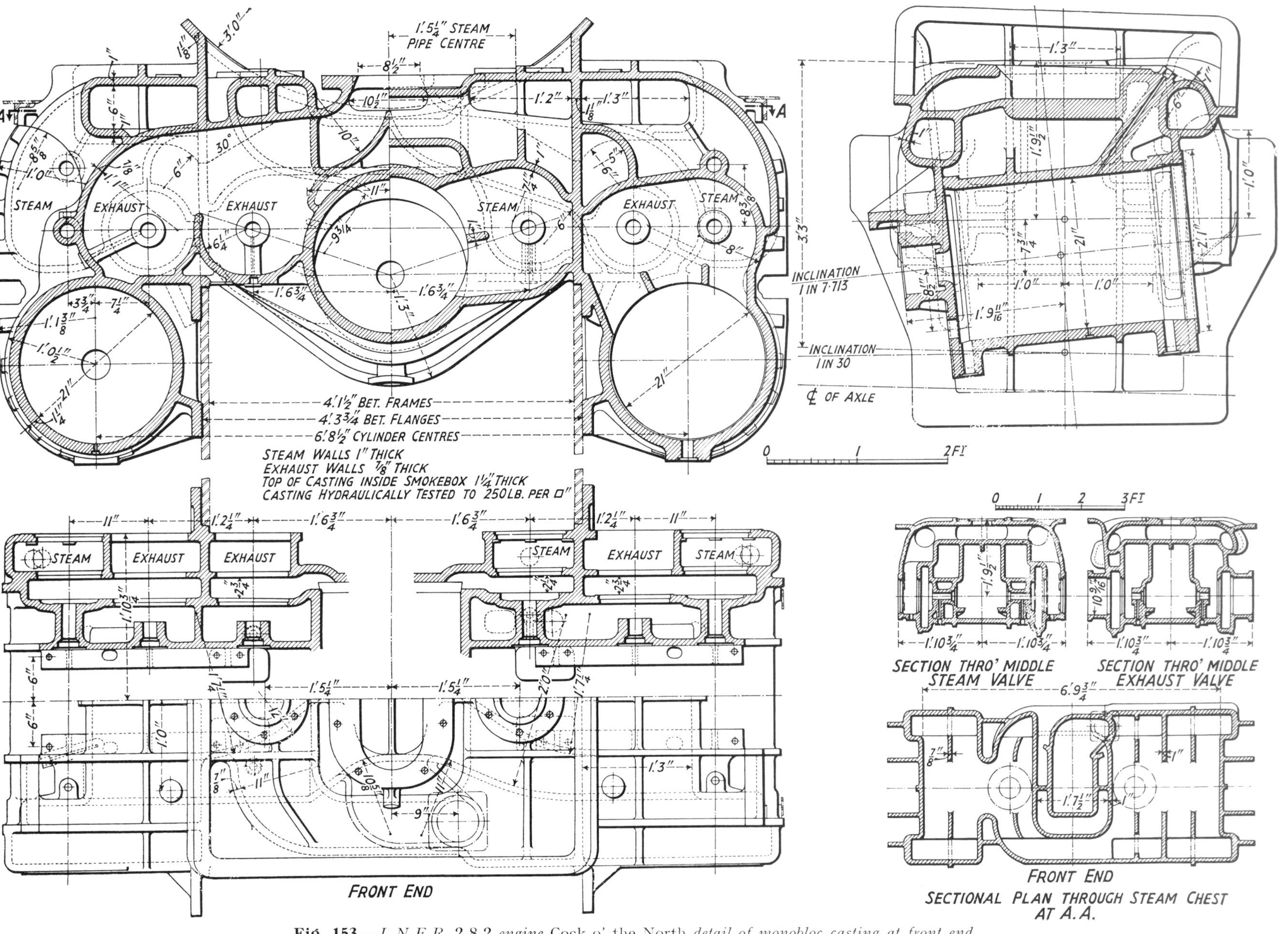

Fig. 153.—*L.N.E.R.* 2-8-2 *engine* Cock o' the North *detail of monobloc casting at front end.*

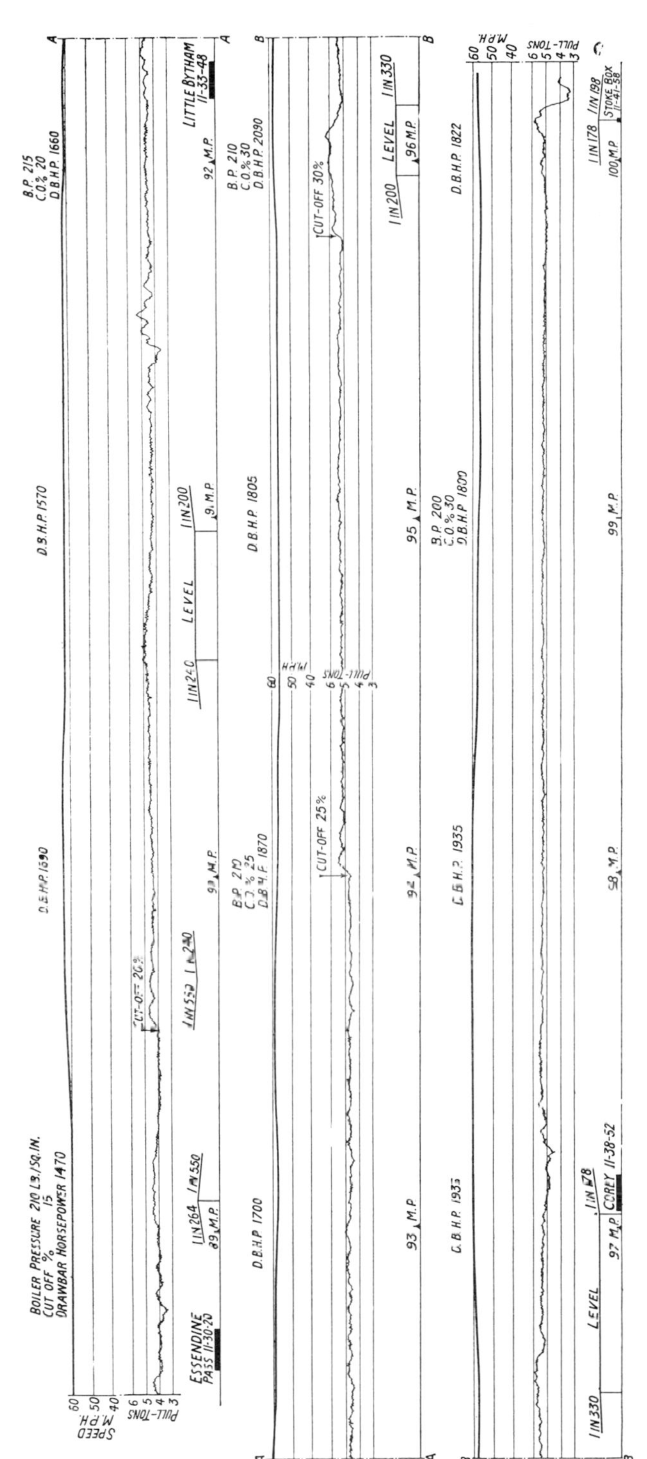

Fig. 154.—*L.N.E.R. Dynamometer car record of test run with the* Cock o' the North.

The engine was fitted with an enlarged version of the standard Gresley Pacific boiler, with a grate area of 50 sq. ft. With 6 ft. 2 in. diameter coupled wheels the *Cock o' the North* was the most powerful passenger locomotive in Great Britain, having a nominal tractive effort, at 85 per cent. boiler pressure, of 43,462 lb.

Although designed for special duties in Scotland the first engine of the class was put through some dynamometer car trials on the Great Northern main line, and ran up to 85 m.p.h. with ease. Some specific details of test performances are given in Fig. 154, which shows high power output on the Stoke bank; but in 1935 the author rode on both Nos. 2001 and 2002 in regular service in Scotland and some impressions may be set down here. Both engines were completely masters of the heaviest work, though it was the piston valve engine that undoubtedly created the most favourable impression. The poppet valve engine was worked in three different positions of cut-off, namely 18, 25 and 35 per cent., whereas the piston valve engine had the gear adjusted with that same precision—of 2 or 3 per cent. at a time—that one could observe regularly on the Gresley Pacifics. The loads hauled by the 2-8-2 engines varied between 515 and 545 tons and these were hauled up the steep gradients with supreme competence. In later years it was sometimes stated that these engines, and the four further engines of the class that were built in 1936 were excessively heavy in maintenance on account of their long fixed wheelbase on the curves of the Aberdeen route. Whatever may have developed later, in their early days the riding was always very smooth and comfortable, with an easy buoyancy and a total absence of anything in the way of jolts and lurches that might have suggested the engines were taking unkindly to the curves.

In the summer of 1936, when the two engines 2001 and 2002, were first in regular service on the Aberdeen route, they were each confined to the same working diagram every day. Engine No. 2001 was stationed at Edinburgh Haymarket shed, and made two return trips to Dundee, while No. 2002 was at Dundee, Tay Bridge shed, and made two return trips to Aberdeen. Each engine was worked only by two sets of men, and there is no doubt that the success of their running was dependent upon the skill and attention of these regular crews. At the same time, from experience on the footplate with both sets of men, on both engines, there were variations in handling as between the different crews that made clearly detectable changes in the engine performance, particularly as regards the coal consumption of engine No. 2002. One of the drivers allocated to this engine seemed reluctant to use a wide open regulator, and his consequent use of longer cut-offs in running heavy trains had a slightly adverse effect on the economy.

Chapter Nine Unconventional Locomotives 1929–35

While improvements in detail design were being made to the conventional steam locomotive, in the form of improved valve gears, better front-ends, and features calculated to provide greater reliability in service some striking attempts were made to break away completely from the conventional. The first of these movements dates back to 1924, when H. N. Gresley became so impressed with the increases in overall efficiency obtained by the use of really high steam pressures in land and marine boilers that he began consideration towards a railway steam locomotive in which it was hoped that similar economies might be achieved. The boiler pressure he decided upon was 450 lb. per sq. in. and in September 1924 he approached Mr. Harold Yarrow, of Glasgow, with a view to that engineer's firm preparing a design of boiler of the water-tube type suitable for use on a locomotive generally of L.N.E.R. Pacific proportions (Fig. 155). The design involved a number of unusual problems, and in fact took more than three years to complete (Fig. 157); but the order was placed for the boiler early in the year 1928, see Fig. 158.

It was built up with one steam drum 3 ft. inside diameter by 27 ft. 11 $^{5}/_{8}$ in. long, and two water drums on either side of the firebox each 18 in. diameter and 11 ft. 0 $^{5}/_{8}$ in. long. There were two other drums under the forward part of the boiler each 19 in. diameter and 13 ft. 5 $^{3}/_{4}$ in. long. The forward drums were connected to the steam drums by 444 tubes, 2 in. diameter, and 74 tubes 2 $^{1}/_{2}$ in. diameter. The drums on each side of the firebox were connected to the steam drum by 238 tubes 2 $^{1}/_{2}$ in. diameter. All the drums were solid forged and machined all over. They were manufactured by John Brown and Co. Ltd. of Sheffield. The use of a pressure of 450 lb. per sq. in. meant a complete departure from all standard railway boiler fittings, and special forms of regulator, safety valves, water gauges and injectors had to be provided. To obtain the fullest advantage from high pressure steam, with the largest possible range of expansion the locomotive was designed as a 4-cylinder compound, with high pressure cylinders 12 in. diameter, and low pressure 20 in., all having a stroke of 26 in.

The main regulator admitted high pressure steam to the high pressure steam chests, but to facilitate starting an additional supply of steam could be admitted through a small 1 in. diameter regulator to the low pressure steam chests. That regulator had to be closed directly the locomotive was under way. A special form of superheater was fitted in the main flue on the boiler side of the regulator, so that the elements were always under full steam pressure. The elements were connected to two forward headers located immediately in front of the water tubes. The combustion gases after passing between the tubes were taken down each side of the boiler through two flues. There was an air space between these flues and the outer casing, and the air supply to the ashpan for combustion purposes was taken through that air space, the object being to pre-heat the air and also to prevent the outer casing of the boiler from becoming overheated. The intake was at the front of the smokebox through a large rectangular hole in the centre and two smaller rectangular holes, one on each side. The air supply to the ashpan was controlled by means of a damper. If necessary cold air could be admitted to the ashpan through the front damper, though this latter was intended primarily to facilitate ash removal. It was

Fig. 155.—*L.N.E.R. The experimental high pressure* 4-6-4 *locomotive, with Yarrow water tube boiler.*

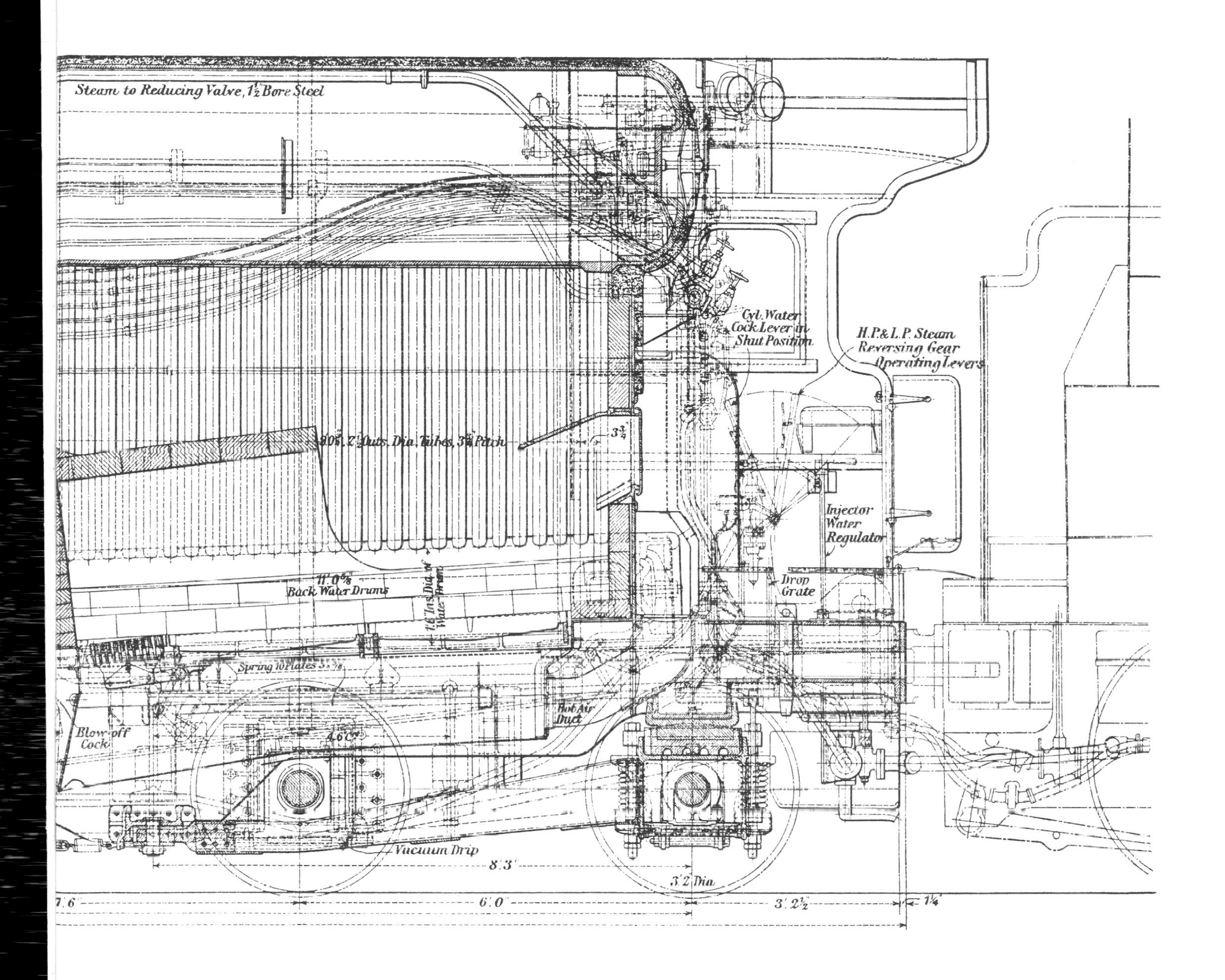

Steam to Reducing Valve, 1½ Bore Steel
Cyl. Water Cock Lever in Shut Position
H.P. & L.P. Steam Reversing Gear Operating Levers
Injector Water Regulator
Drop Grate
11' 0⅝" Back Water Drums
Spring 10 Plates
Blow-off Cock
Hot Air Duct
Vacuum Drip
8' 3"
3' 2" Dia
6' 0"
3' 2½"
1¼"

Fig. 159.—*L.N.E.R. Engine* 10000; *general view.*

Fig. 160.—*L.N.E.R. Engine* 10000 *entering King's Cross station with an up Scottish express.*

After some preliminary working it was found that the work of the locomotive was not correctly divided between high and low pressure cylinders—a common fault in many compounds. On engine No. 10000 this was largely corrected by lining up the high pressure cylinders to 10 in. diameter. The object of building this experimental locomotive was not to secure increased power, or speed as compared with the standard Pacific engines, but if possible to secure substantial fuel economies. The early results were very promising, and in a classic paper read before the Institution of Mechanical Engineers in January 1931,

Fig. 161.—*L.M.S.R. The Fowler* 900 *lb. experimental compound* 4-6-0 *No.* 6399 Fury.

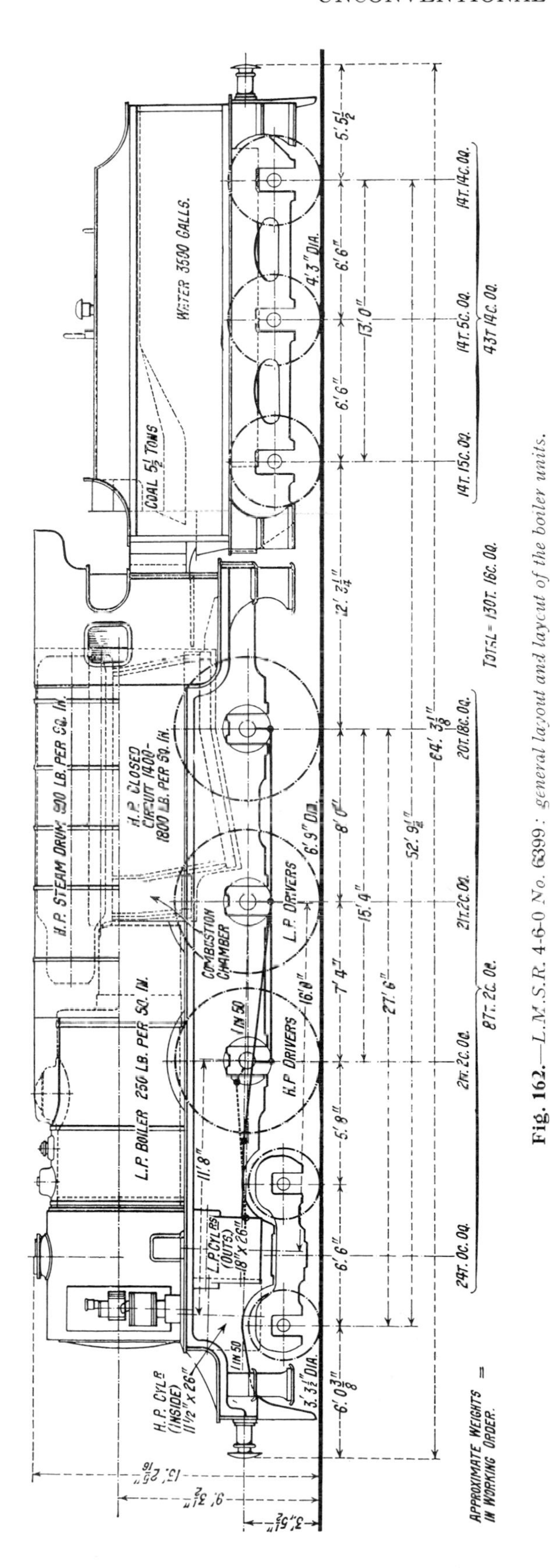

Fig. 162.—*L.M.S.R. 4-6-0 No. 6399: general layout and layout of the boiler units.*

Gresley was able to report that engine No. 10000 " has worked trains of over 500 tons weight for long distances at express speed with consistent reliability and success, and although it has not been possible so far to carry out any extensive trials there is every indication that it will prove more economical in fuel consumption than express engines of the latest normal type ".

During the summer of 1930 the engine was working in the ordinary Pacific link from Gateshead shed including a long-mileage double-manned diagram that included the evening Glasgow–Leeds express with a section booked at over 60 m.p.h. start to stop over the 44 miles from Darlington to York. In the same year, on one occasion, the engine worked the non-stop Flying Scotsman from King's Cross to Edinburgh, and returned on the corresponding southbound train next day (Fig. 160). The author witnessed her punctual arrival at King's Cross, although, of course, the booked average speed was not high. Unfortunately for this bold experiment, which can be ranked beside that of Sir Cecil Paget's sleeve-valve 2-6-2 on the Midland Railway, the early promise was not fulfilled, and No. 10000 became such a troublesome engine to maintain that she was scrapped in 1938.

Another striking novelty in British locomotive design that appeared almost at the same time as the L.N.E.R. " 10000 " was the ultra-high pressure 3-cylinder compound 4-6-0 No. 6399 of the L.M.S.R. (Fig. 161). In this engine, which was built by the North British Locomotive Co. Ltd., the very high pressure of 900 lb. per sq. in. was used. As in the case of the L.N.E.R. engine, the L.M.S.R. 4-6-0 was designed for direct comparison with the " Royal Scot " 4-6-0s and the proportions were arranged so that the nominal tractive effort was almost exactly the same as follows:

Cylinders:	
High pressure	
(1) diameter, in.	11 ½
stroke, in.	26
Low pressure	
(2) diameter, in.	18
stroke, in.	26
Coupled wheels, dia., ft., in.	6-9
Coupled wheelbase, ft., in.	15-4
Boiler press., lb./sq. in.	
High pressure	900
Low pressure	250
Adhesive weight (estimated), tons, cwt.	63-2
Weight of engine in working order (estimated), tons, cwt.	87-2
Weight of engine and tender in working order (estimated), tons, cwt.	130-16
Tractive effort, lb.	33,200

The accompanying line drawing (Fig. 162) shows the general disposition of the various units of the special steam raising installation, which with great designing ingenuity were arranged on the chassis of a " Royal Scot " class 4-6-0. The principal feature of interest was, of course, the boiler, the design of which was largely due to The Superheater Co. Ltd. From the diagram it will be seen that the boiler consisted of three portions, generating steam at 1,400 to 1,800 lb. 900 lb., and 250 lb. per sq. in. The 1,400 to 1,800 lb.

Fig. 163.—*L.M.S.R.* 4-6-0 *No.* 6399 *assembling the boiler in the works of the North British Locomotive Co. Ltd.*

section consisted of a water-tube firebox forming a closed circuit. The vertical tubes, the locations of which are indicated on the outline drawing, were connected at their lower ends to a foundation ring, and to the ring forming the base of the combustion chamber. At their upper ends they were expanded into two horizontal cylindrical equalising drums from which coils passed to the interior of a large steam drum arranged partly between and partly above the drums of the closed circuit. The steam circulating in these coils evaporated the water in the large steam drum, and generated steam at a pressure of 900 lb. per sq. in. The material employed for this larger drum, which was not in contact with the fire, was nickel steel. It was a machined forging and was supplied by John Brown and Co. Ltd. of Sheffield, who specialised in the manufacture of high-pressure boilers and drums. The steam there generated, after passing through a superheater was used in the high pressure cylinder (Fig. 162).

The forward portion of the boiler corresponded generally to the orthodox locomotive boiler barrel, and consisted of a nickel steel barrel with front and back tube plates of mild steel, through which the tube flues carried the hot gases to the smokebox. Steam was generated in this portion at 250 lb. per sq. in.

For feeding the high pressure drum, a pump was provided, which took its supply from the low-pressure boiler. Water was supplied to the low-pressure boiler by a live steam injector located on the driver's side

Fig. 164.—*L.M.S.R.* 4-6-0 *No.* 6399*: a right-hand side view.*

and an exhaust steam injector on the fireman's side. As steam was raised much more quickly in the high-pressure drum than in the low-pressure boiler, arrangements were made to by-pass, by means of an intercepting valve, any excess steam from the high-pressure drum into the low-pressure boiler, thus avoiding waste through blowing-off.

The regulator handle worked both the high-pressure and the low-pressure regulators simultaneously. On opening the regulator, steam was admitted into the high-pressure cylinder, after passing through the high-pressure superheater situated in the lower boiler tubes. Exhausting from the high-pressure cylinder, the steam entered a mixing chamber, where it was met by steam at 250 lb. pressure, which had previously passed through a low-pressure superheater situated in the upper boiler tubes. From the mixing chamber steam entered the two outside cylinders.

The engine was named *Fury* (Figs. 164, 165) which had previously been borne by a " Royal Scot " and which became redundant when the name of the particular locomotive was changed to that of a regiment of the British Army. The name *Fury* proved to be one of ill-omen. The engine never entered revenue earning service. Its trials were conducted in Scotland, and on one of these the failure of one of the small connecting tubes caused a serious blow-back as the engine was passing through Carstairs station. One of the testing staff was killed, and the experimental work was halted, never to be resumed. The engine languished for some years in the paint shops at Derby, and then in 1935 the special boiler and cylinders were scrapped, and the frames used for a new " Royal Scot " class engine with taper boiler.

Ever since the successful application of the steam turbine to marine propulsion railway locomotive engineers have sought means of realising the increased thermal efficiency theoretically possible over that of the reciprocating steam engine. In other respects also, the turbine would appear to have attractions for rail propulsion, since the absence of heavy reciprocating parts would largely eliminate problems of balancing, and the resulting locomotives would be easier on the permanent way. To secure the greatest practical temperature range, and consequently, the highest thermal efficiency, earlier turbine driven locomotives had been equipped with condensers. The Reid-Ramsay turbo-electric locomotive will be recalled: also the Beyer-Ljungström machine (Fig. 166) which worked in regular express passenger service for a short time between St. Pancras and Manchester, in 1928. The trouble with these experimental machines, like those tried out in various countries on the continent of Europe, was that the complications arising from the use of condensers and all their attendant equipment involved much additional maintenance work, and so reduced the availability of the locomotive in service that any reductions in fuel consumption due to increased thermal efficiency were far outweighed by the disadvantages.

Fig. 165.—*L.M.S.R.* 4-6-0 *No.* 6399: *a rare view showing the engine in steam.*

At the same time turbine propulsion in itself holds out so many inherent advantages that the idea was not fully abandoned and for this reason a Swedish experimental locomotive, built in 1932, attracted considerable attention among locomotive engineers in this country. In that year the Grangesberg-Oxelösund Railway put into service a 2-8-0 main line locomotive using a non-condensing turbine. The

Fig. 166.—*Beyer-Ljungström condensing turbine locomotive run experimentally on the L.M.S.R. in* 1928.

railway in question had a number of 2-8-0 reciprocating steam locomotives of the same general power classification, using the same boiler, and the only difference between the turbine and the rest lay in the drive, which in the former was effected through triple reduction gearing and a jack shaft. The rated tractive effort of locomotives of this class was 47,040 lb., with a rated power output at the rail of 1,270 h.p. at 27 m.p.h. The maximum operating speed was 43.5 m.p.h. In view of the use of turbine propulsion on the experimental unit of the class it is interesting to note the relatively low boiler pressure of 185 lb. per sq. in. The extreme simplicity of the turbine drive appealed to British engineers who were forever seeking increased monthly mileages from their locomotives and immunity from failure, and at the invitation of the late Dr. H. L. Guy, later Sir Henry Guy, and then Chief Turbine Engineer of the Metropolitan-Vickers Company, W. A. Stanier, by then Chief Mechanical Engineer of the L.M.S.R. visisted Sweden to see the locomotive at work. Dynamometer car tests indicated savings of 7 $^1/_4$ per cent. in coal consumption and 15 per cent. in water, in favour of the turbine, and a serious consideration began to be given on the L.M.S.R. to the building of an experimental locomotive on similar lines. The Swedish locomotive was used in freight service, and the dynamometer car trials previously mentioned were carried out with loads of 1,500 tons.

Some very broad indications as to the advantages of turbine drive have already been outlined. One could perhaps group the respective arguments, for and against, under three main headings: (1) non-condensing turbine in place of cylinders, valves, and valve gear, (2) elimination of reciprocating parts, and (3) totally enclosed rotary gear drive, with efficient lubrication.

Under heading (1), reference has already been made to the improved theoretical efficiency possible in a turbine, but the elimination of the valves and valve gear was a point by which it was hoped to reduce intermediate maintenance work. Wear of the cylinder, valve, and steam chest surfaces, together with deterioration in steam tightness of valves, glands and packings can, through steam leakage result in a considerable falling off in efficiency as mileage increases. Dynamometer car tests made on the L.M.S.R. with an individual engine of the " Royal Scot " class showed an increase of 8 per cent. in coal consumption over a mileage of 28,000. In consequence piston and valve examinations were given every six to eight months, incurring maintenance costs, and loss of

Fig. 167.—*L.M.S.R. Stanier non-condensing turbine* 4-6-2 *locomotive No.* 6202, *built Crewe* 1935.

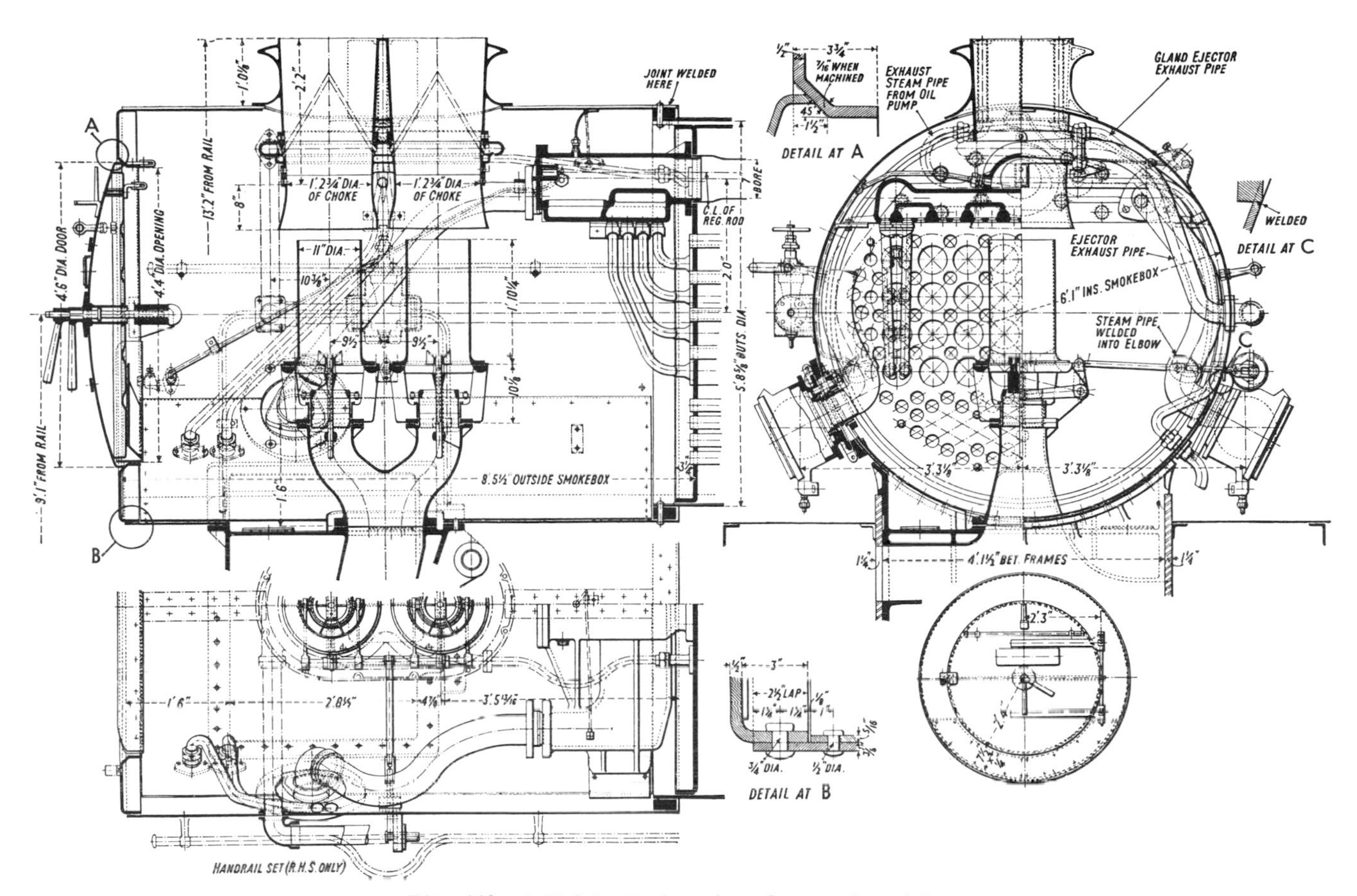

Fig. 168—*L.M.S.R. Turbomotive; layout of smokebox.*

availability of the locomotive. With the turbine drive it was hoped to avoid such periodic examinations.

The elimination of reciprocating parts results in the elimination of hammer blow. With reciprocating steam locomotives the civil engineer of the L.M.S.R. permitted a static axle load of 22 ½ tons maximum; but with turbine drive and a complete absence of hammer blow a static axle load of 24 tons maximum was permitted, thus providing for a most valuable increase in adhesion weight.

With regard to (3), by the use of a totally enclosed gear drive it was hoped to secure a substantial reduction in frictional losses, with a consequent reduction in wear and tear; it was also hoped to eliminate repairs to the running gear, and further to reduce the incidence of heating due to grit penetrating the working faces. All three factors, with their numerous ramifications gave promise of a more powerful locomotive within the limits of weight laid down by the civil engineer's department, greater availability and a reduced coal and water consumption over reciprocating locomotives doing the same work. With all this in mind Stanier began active preparations for the design of an express passenger Pacific, generally

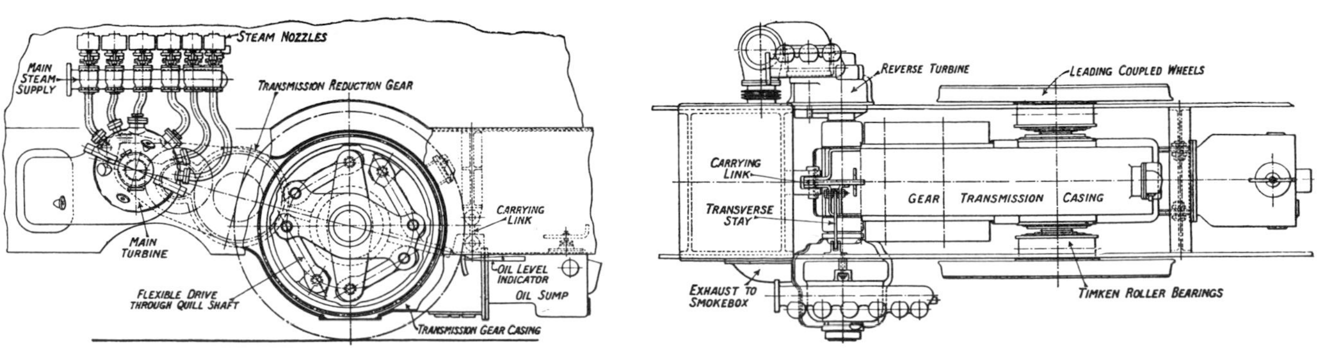

Fig. 169.—*L.M.S.R. Turbomotive: layout of main and reverse turbines.*

similar to the " Princess Royal " class, but to be driven by a non-condensing turbine of the " Ljungstrom " type. The collaboration of the Metropolitan-Vickers Company and Ljungstrom was obtained, and the power characteristics of the proposed new locomotive were worked out by Dr. Guy and his staff. In passing it may be mentioned that in developing the " Princess Royal " class 4-6-2 it was originally intended to build three units for trial purposes; but with the turbine proposition already in mind one set of frames and other materials were held back, and only two 4-6-2s were put on the road in 1933.

The expectations of the new locomotive in comparison with those of the " Princess Royal " class were as follows:

	Reciprocating	Turbine
Boiler pressure, lb./sq. in. ..	250	250
Steam temperature, deg. Fahr.	650	650
Evaporation, lb./hr.	30,000	30,000
Output at rail, h.p.		
at 30 m.p.h.	1570	2050
at 40 ,, ...	1700	2270
at 50 ,, ...	1770	2350
at 60 ,, ...	1800	2400
at 70 ,, ...	1770	2350

It is interesting to learn that the turbine locomotive was designed for maximum efficiency at 62 m.p.h., thus suggesting heavy hauls at relatively moderate speed, rather than the working of lightweight high speed services. The tractive effort at starting, was estimated at 40,000 lb. as with the " Princess Royal " class, but at 70 m.p.h. the turbine was expected to sustain a drawbar pull of $5\,^1/_4$ tons.

Dr. Guy was responsible for the design of the turbines and the gear drive, while the design of the chassis and boiler was developed by the Derby drawing office in association with Guy and his staff (Fig. 168, 169, 170). There were two turbines, one for forward running and one for reverse, mounted very neatly on the running plates on the left hand and right hand side of the locomotive respectively. There were sixteen stages in the forward turbine, and with a view to using the locomotive (and others of her type that might be built subsequently) on the through Anglo-Scottish workings already allocated to the " Princess Royal " engines, the internal arrangement of the blading was designed to maintain high efficiency over a wide range of speed. A high standard of performance was needed equally when slogging up Shap and Beattock at 30 m.p.h. and in making averages of 60 to 65 m.p.h. over long stretches south of Crewe. Steam was passed via one or more of six hand-controlled valves to the nozzle group in the turbine casing. The valves were opened in succession and the power output of the turbine was regulated by the number of valves open. The turbine was permanently connected through double-helical triple-reduction gearing to the leading coupled axle. The reverse turbine on the other hand, was normally not connected with the main drive; its reduction gear was connected when required through a sliding splined shaft and dog clutch mechanism (Fig. 169). Originally there was a steam servo motor to effect engagement, but this was later discarded in favour of a simple hand control from the cab. Before the war Guy was at one time engaged in designing a form of interlocking control which would have made the operation of the forward and reverse turbines largely foolproof. But the onset of war brought this development to a stop.

As originally built No. 6202 had a domeless boiler with proportions generally similar to those of the modified " Princess Royal " type with 32-element superheater, as in the 6203–6212 batch; but the layout of small tubes was different. A double blastpipe and chimney was fitted from the outset, and with a continuous exhaust, as compared with the pulsations of a reciprocating engine, some modifications to the draughting were necessary. The relevant boiler

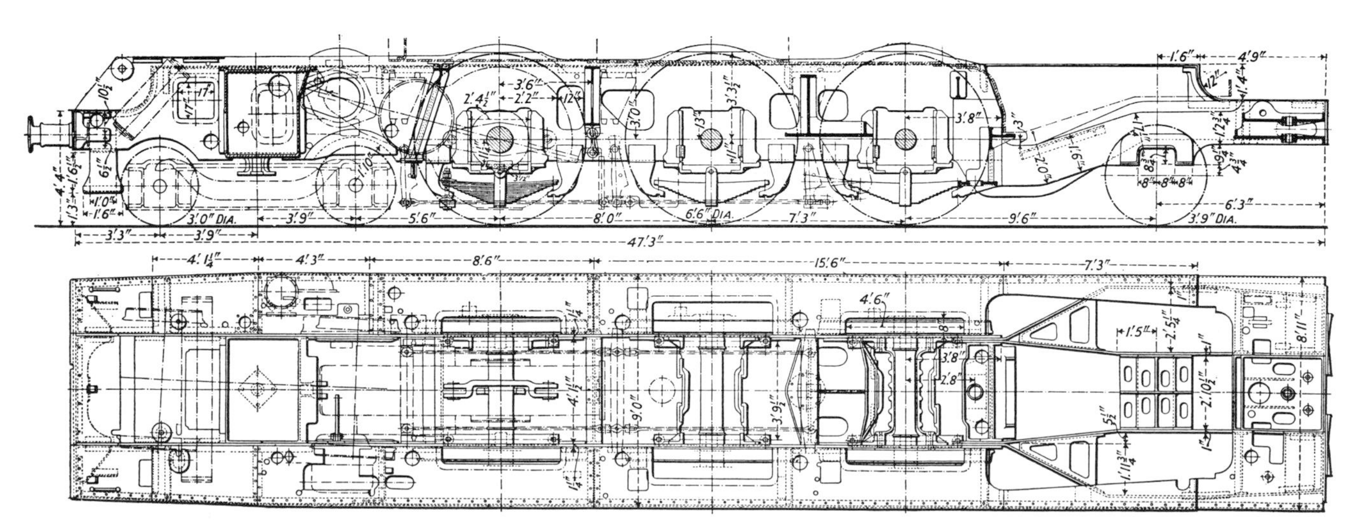

Fig. 170.—*L.M.S.R. Turbomotive: layout of framing and drives.*

proportions were as follows:

Engine Nos.	6200; 6203–6212	6201 (1935 trials)	6202 original	6202 later
Small tubes, Number	123	119	112	81
outside dia., in. ..	2 3/8	2 3/8	2 1/4	2 1/4
Superheater flues, Num.	32	32	32	40
outside dia., in. ..	5 1/8	5 1/8	5 1/8	5 1/8
Heating surface superheater elements, sq. ft.	653	594	653	577

Engine No. 6202 as shown above was later fitted with a boiler having a 40-element superheater that provided a free area through the tube flues of 69.3 per cent. of the total, as compared with 53.7 per cent. in the original domeless boiler. This actually resulted in a slight reduction in superheating surface, though at a still later date a 40-element superheater with triple elements as in the " Duchess " class was fitted, and this increased the superheater heating surface to 832 sq. ft.

The locomotive was put into service in June 1935 and the most superficial observations showed that it was capable of the hardest work set to any L.M.S.R. locomotive of the day. It was used principally between Euston and Liverpool. The engine rode very smoothly and was highly appreciated by the footplate staff. At first there was some trouble with exhaust steam beating down and obscuring the view from the cab glasses, but this was obviated by the fitting of deflector shields.

With an entirely new type of locomotive and one that remained the only example of its kind running in Great Britain, it was not altogether surprising that troubles were experienced in service. The remarkable thing is that they were so few—that is, until war conditions put an entirely different aspect upon locomotive operating in general. In a most frank and comprehensive paper presented to the Institution of Locomotive Engineers in 1946, Mr. R. C. Bond, now General Manager, British Railways Workshops, gave a complete record of the availability of the locomotive, together with an account of all the major failures that occurred. Until the outbreak of war in 1939, the locomotive had averaged 54,205 miles per annum compared with about 80,000 miles per annum for the " Princess Royal " class; for a novel and experimental machine this must be considered as a remarkably fine record, especially in that during the year 1936 a total of 73,268 miles was run. In the early days of the war, however, since this locomotive needed a good deal of specialised attention it was taken out of traffic altogether, and stored at Crewe. When the war effort began to work up to its full intensity, and every engine that could turn a wheel was needed, No. 6202 was put into service again, and it was unfortunately under the stress of wartime conditions that the more serious failures took place. The overall picture, as presented in 1946, by Mr. Bond showed that since construction, but deducting the time when the engine was stored, the average annual mileage was only 28,500. In war conditions the annual mileages of the " Princess Royal " and " Duchess " classes had dropped from the pre-war 80,000 for both classes to 53,000 and 73,000 respectively, so that both were in a different category altogether to that of No. 6202.

While some of the troubles experienced with No. 6202 were definitely attributable to war conditions, others were part of the price one would expect to pay for experience with a type of propulsion new on a British main line express passenger locomotive. Lack of previous experience with turbine drive tended to make the inspection staff a little over-cautious, but while this is understandable and a good fault, it sometimes resulted in the locomotive being " stopped " when it actually proved to be unnecessary, and so contributed to the relatively poor availability record. Another important factor, sometimes, was the time spent in getting replacement parts. Where the turbines were concerned this meant obtaining them from the Metropolitan-Vickers Co. and with the latter very heavily engaged in vital war production it was difficult to get isolated production or repair jobs for No. 6202 fitted in. The delays following upon failures

L.M.S.R. Dynamometer Car Test Runs
10 a.m. Euston–Glasgow

Date	4.5.36	6.5.36	11.5.36	13.5.36	19.10.36	26.10.36	22.6.37	24.6.37
Engine No.	6202	6202	6212	6212	6210	6202	6202	6202
Boiler	A	A	—	—	—	B	C	C
Load, tons tare								
Euston–Crewe	564	556	539	569	552	535	484	489
Crewe–Symington	470	474	477	507	482	475	484	489
Symington–Glasgow	331	304	477	337	313	305	315	320
Length of trip (miles)	401.4	401.4	401.4	401.4	401.4	401.4	401.4	401.4
Time actual running, min.	452.16	449	464.1	459.7	464.5	442.75	448.5	436.5
Average speed, m.p.h.	53.3	53.7	52.0	52.5	52.0	54.5	53.8	55.3
Average d.h.p.	831.5	787.6	771	770	863.5	828.6	851.2	858.2
Coal consumption:								
lb. per mile	44.6	41.7	45.0	45.5	47.6	44.3	44.3	42.4
lb. per d.h.p. hr.	2.86	3.14	3.04	3.10	2.86	2.91	2.81	2.74
lb. per sq. ft. of grate area per hr.	51.3	49.8	52.1	53.1	55.0	53.7	53.1	52.2
Water, galls. per mile	35.8	31.5	37.7	37.9	39.6	35.4	40.3	37.7
Evaporation								
lb. of water per lb. of coal	8.0	7.56	8.38	8.34	8.32	7.99	9.11	8.87
lb. of water per hr. (running time)	19,030	16,900	19,580	19,900	20,600	19,300	21,670	20,750

were therefore much greater than they would have been in normal times, and far more than if a stud of turbine locomotives had been at work, and the railway shops fully equipped to deal promptly with their requirements. For these reasons the availability record of No. 6202 cannot be compared on any fair basis with that of the reciprocating Pacifics of the L.M.S.R. All the more serious failures took place during the war period, including the rather alarming breakage of the forward turbine spindle at 60 m.p.h. These failures may, in part, be attributed to increasing mileage since construction and as part of the experience necessary to be bought before a turbine locomotive of this kind could be considered as a thoroughly reliable all-round motive power unit.

At its best the road performance of the locomotive was so good as to suggest that any amount of time and persistence was worth expending in order to achieve the reliability so essential in the heaviest traffic. At three different periods, in 1936 and 1937, a series of dynamometer car trials was conducted on through Euston—Glasgow workings, matching the engine against standard " Pacifics " of the " Princess Royal " class. These records are particularly interesting in that they represent the work of No. 6202 with three different boilers, and at varying periods since last general overhaul. For the most part the loads conveyed were very heavy. The tabulated summaries on page 115 show the average running speeds made throughout, but to present the tasks given both to No. 6202 and to the competing reciprocating engines, the intermediate timings are given below:

(a) 10 a.m. EUSTON TO GLASGOW

	Dist.	Sch.	Av. speed
Euston–Rugby	82.6	87	56.9
Rugby–Crewe	75.5	80	56.6
Crewe–Carlisle	141.0	159	53.3
Carlisle–Symington	66.9	81	59.6
Symington–Glasgow		reduced load	

(b) 10 a.m. GLASGOW TO EUSTON

Glasgow–Symington		reduced load	
Symington–Carlisle	66.9	71	56.5
Carlisle–Euston	299.1	334	53.9

(c) 2 p.m. EUSTON TO GLASGOW

Euston–Crewe	158.1	163	58.2
Crewe–Lancaster	72	79	54.7
Lancaster–Penrith	51.2	59	52.1
Penrith–Carlisle	17.9	19	56.5
Carlisle–Glasgow	102.3	116	52.9

Except between Symington and Glasgow where the load of the " Royal Scot " express was generally lighter through the detaching of the Edinburgh portions the gross loads behind the tenders were generally above, rather than below 500 tons. In view of this the record of coal consumption, not only of the turbine but also of the " Princess Royal " class engines, is an exceedingly fine one. For easy reference, in the tables which have been set out in detail, the different boilers fitted to engine 6202 are referred to as A (domeless), B and C (both with domes) thus:

Ref. letter	Number of Superheater elements	Superheater heating surface, sq. ft.
A	32	653
B	40	577
C	40*	852

* with triple flow elements

L.M.S.R. Dynamometer Car Test Runs
2 p.m. Euston – Glasgow

	Oct. 20, 1936	Oct. 27, 1936
Date	Oct. 20, 1936	Oct. 27, 1936
Engine No.	6210	6202
	—	B
Load tons tare:		
Euston–Crewe	480	454
Crewe–Carlisle	512	486
Carlisle–Glasgow	496	486
Length of trip, miles	401.4	401.4
Time—Actual running, min.	450.9	454.9
Average speed, m.p.h.	53.4	53.2
Maximum speed, m.p.h.	82.2	85.0
Average d.h.p.	840.8	943.5
Coal consumption:		
per train mile, lb.	46.8	49.6
per d.h.p. hr./lb.	2.96	2.78
per sq. ft. of grate area per hr./lb.	55.7	58.5
Water—gallons per mile	38.9	38.5
Evaporation:		
lb. of water per lb. of coal	8.28	7.71
lb. per hour (running time)	20,750	20,280

L.M.S.R. Engine 6202

Speed	Load ton tare	Gradient (rising)	Drawbar pull tons	Drawbar h.p.	Total approx. I h.p.
34	489	1 in 75	8.26	1,678	2,224
46 1/4	489	1 in 131	5.44	1,501	2,046
58	489	1 in 99	4.43	1,531	2,177
72 1/2	483	1 in 333	3.50	1,517	2,336

On two occasions, once with the "Princess Royal" class engine No. 6210 and once with No. 6202, the 400-mile run between London and Glasgow was performed four times in three days, by use of the 10.45 p.m. up sleeping car express thus:

(1) 10 a.m. Euston to Glasgow
(2) 10.45 p.m. Glasgow to Euston
(3) 2 p.m. Euston to Glasgow
(4) 10 a.m. Glasgow to Euston

The work performed by both engines particularly on the difficult 2 p.m. down " Mid-day Scot ", showed that they had stood up to this severe test remarkably well. In this connection the tests on the 2 p.m. down (see tables, above) were perhaps the most severe of all. Apart from these two, the normal loads of the trains concerned were taken, with the addition only of the dynamometer car; but on the 2 p.m., which at that time had its load reduced to about 300–350 tons for its fast non-stop run from Carlisle to Glasgow, the load was specially augmented, so that tare loads little short of 500 tons were conveyed.

Some of the individual performances involved were remarkable, especially the very fast ascents of the Shap and Beattock inclines by engine No. 6202 with boilers B and C. At the same time some of the running of this engine was unnecessarily hard over certain sections. The six valves did not permit of the fine adjustments of control that an experienced driver can exercise with the ordinary screw reverser on a reciprocating steam locomotive. On the final series of tests with No. 6202 in June 1937, it would seem that the optimum performance was obtained at a standard that appears definitely above the maximum achievements of the " Princess Royal " class. What this actually meant in running conditions can be appreciated from certain selected performances from the dynamometer car test runs made on the " Royal Scot " in June 1937:

Carnforth–Oxenholme: 12.8 *miles*

Engine No.	Load tare	Time m. s.	Av. speed m.p.h.
6212	477	13 55	56.6
6210	512	13 50	56.9
6202(A)	474	12 55	59.5
6202(B)	475	14 55	54.5
6202(C)	489	12 10	63.1

Oxenholme–Tebay: 13.1 *miles*

Engine No.	Load tare	Time m. s.	Av. speed m.p.h.
6212	507	17 20	45.3
6210	482	14 10	55.5*
6202(A)	470	16 20	48.2
6202(B)	486	14 55	55.7*
6202(C)	484	16 00	49.1

* Mid-day Scot

Tebay–Shap Summit: 5.5 *miles*

Engine No.	Load tare	Time m. s.	Av. speed m.p.h.
6212	507	9 30	34.7
6210	512	8 40	38.1
6202(A)	470	8 50	37.4
6202(B)	486	7 10	46.1
6202(C)	484	7 35	43.5

Beattock Station–Beattock Summit: 10 *miles*

Engine No.	Load tare	Time m. s.	Av. speed m.p.h.
6212	477	18 20	32.7
6210	482	18 40	32.2
6202(A)	474	16 50	35.7
6202(B)	486	15 15	39.4
6202(C)	489	16 10	37.1

On the basis of these tests and of the general performance of the locomotive during its first four years of service the results could be regarded as very encouraging, and despite the subsequent experience with it during the war years in 1946 it was certainly felt by many engineers that the experiment was worth pursuing still further. A point of considerable importance was that the uniform blast avoided any grooving of the firebox plates, and the boiler and firebox consequently had a better life than that of the reciprocating engines.

In the circumstances that developed after that time considerable expenditure would have to have been incurred in renewing the main turbine, and other parts, and it was decided to rebuild her conventionally. Thus ended one of the most notable British attempts to break away from the conventional reciprocating steam locomotive, and in so doing to try and obtain higher thermal efficiency.

Chapter Ten Locomotives for Overseas 1930–40 Non-articulated types

An outstanding feature of the work of the British locomotive building industry during this period was the construction of a series of progressively larger engines, of advanced design and great tractive power, for the 3 ft. 6 in. gauge South African Railways. Although the rail gauge is narrow compared with that of Great Britain the loading gauge width is generous, and permits of a width over cab of 10 ft. From an early date, 1904 indeed, the 4-8-2 type had been used for general service on the main lines, and in 1930 the administration had more than 100 locomotives of this wheel arrangement at work. In 1930 the very powerful "15CA" class was introduced, as shown in the accompanying photograph (Fig. 171), and built by the North British Locomotive Co. Ltd. By skilful design a locomotive having a nominal tractive effort at 85 per cent. boiler pressure, of 48,090 lb. was produced, with a maximum axle load of 18 1/2 tons. The general proportions were very large, as follows:

Cylinder diameter, in.	24
Cylinder stroke, in.	28
Coupled wheels dia., ft. in.	4-9
Heating surfaces, sq. ft.	
Tubes	2554
Firebox	198
Arch tubes	23
Superheater	690
Combined total	3465
Grate area, sq. ft.	48.3
Boiler pressure, lb./sq. in.	200
Total weight of engine and tender (working order), tons	173.45

Large and powerful though these locomotives were, the year 1935 saw a further advance in capacity with the favourite 4-8-2 type, in the form of the first batch of the celebrated "15E" class (Fig. 172), 20 of which were built by Robert Stephenson and Co. These engines differed from the previous "15CA" class in having R.C. poppet valve gear; but although the nominal tractive effort was practically the same the capacity of the new engines was considerably greater. The cylinders had the same dimensions, but the coupled diameter was 5 ft. 0 in. instead of 4 ft. 9 in., and the boiler pressure was raised from 200 to 210 lb. per sq. in. The tractive effort at 85 per cent. boiler pressure was 47,980 lb., against 48,090 lb. The boiler and firebox in the "15E" was very much larger, providing an evaporative heating surface of 3414.5 sq. ft., against 2775 sq. ft. on the "15CA"; a combined total of 4075.5 sq. ft. against 3465, and a grate area of no less than 62.5 sq. ft. Although the nominal tractive effort of the locomotive was not increased, by use of large cylinders, it was considered that the better steam distribution afforded by R.C. poppet valve would result in a higher output from cylinders of the same size.

Although of such very large size these locomotives were hand fired. A steam operated firedoor was provided, also steam operated shaking gear for the large grate. A notable feature of the design was that the increases in boiler capacity over the "15CA" class, namely 17.6 per cent. in heating surface and 29.4 per cent. in grate area were obtained with practically no increase in the maximum axle load nor in the total engine weight. The latter came out at 107.7 tons, against 106.15 tons in the "15CA". As

Fig. 171.—*South African Railways: Class* "15CA" 4-8-2 *and* 2-8-2 *locomotive* (1930) *built by North British Locomotive Co. Ltd.*

Fig. 172.—*South African Railways: Class* " 15E " 4-8-2 *and* 2-8-2 *locomotive, with R.C. poppet valve gear. Built* 1935, *by R. Stephenson & Co. Ltd.*

will be seen from the accompanying illustration (Fig. 172) the large boiler was pitched almost to the limit of the loading gauge, and its centre line was 9 ft. 2 1/2 in. above rail level. At that time the South African Railways were engaged upon an important programme of boiler standardisation, and the new 4-8-2 engines had boilers, and many other parts, in common with some large new 4-6-2 locomotives built in Germany at the same time. The large increases in heating surface and grate area over previous boilers, with little or no increase in weight, were achieved mainly by the use of 0.2 to 0.3 per cent. nickel steel for the plates. A further development of this excellent design was the " 15F " (Fig. 173), in which a reversion was made to Walschaerts valve gear, but which included the important addition of roller bearings on the leading bogie and the trailing truck, and also on the tender bogies. A large order for 44 of these locomotives was completed by the North British Locomotive Co. Ltd. in 1939. A line diagram, as well as a photograph of the " 15F " class is reproduced herewith (Fig. 174).

In 1937 the North British Locomotive Co. built a remarkable engine for the South African Railways, which at the time was described as experimental. A very powerful unit was needed for working over track laid with 60 lb. rail, and the huge non-articulated 2-10-4 was built (Fig. 175), doubtless for comparison in operating costs, and otherwise, with the many types of articulated locomotive then working on the system. The production of an engine, which weighed 171 tons without its tender must have been something of a record for the 60 lb. rail track. Furthermore it was required to traverse curves of 275 ft. radius, at which points the gauge was widened by 3/4 in. A single engine, with two cylinders 24 in. diameter by 26 in. stroke, fed by very short and direct passages is clearly a great deal simpler from the viewpoint of design and construction than the double engine arrangement of articulated types with relatively long steam and exhaust connections, and the complication of flexible joints; and one can appreciate the desire of any railway administration to try the alternative.

The boiler and firebox was interchangeable with those of the " 15E " and " 15F " classes of 4-8-2, except that the 2-10-4 engine (Class 21) was equipped with a mechanical stoker (Fig. 178). The accompanying diagram (Fig. 176) shows the weight distribution

Fig. 173.—*South African Railways: Class* " 15F " 4-8-2, *built* 1939, *by North British Locomotive Co. Ltd.*

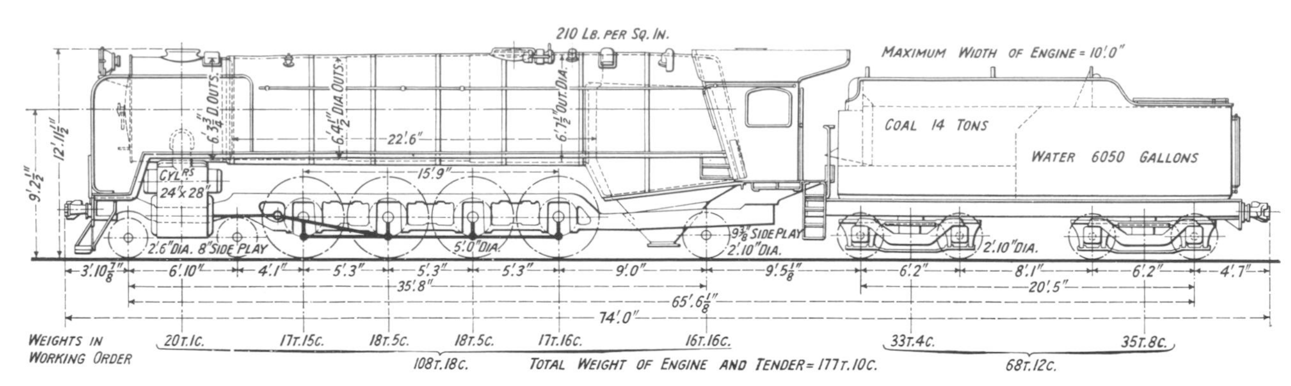

Fig. 174.—*South African Railways: Line diagram of Class* " 15F " 4-8-2.

Fig. 175.—*South African Railways:* 2-10-4 *locomotive Class* 21, *for working over lines laid with* 60 *lb. rails* (1937).

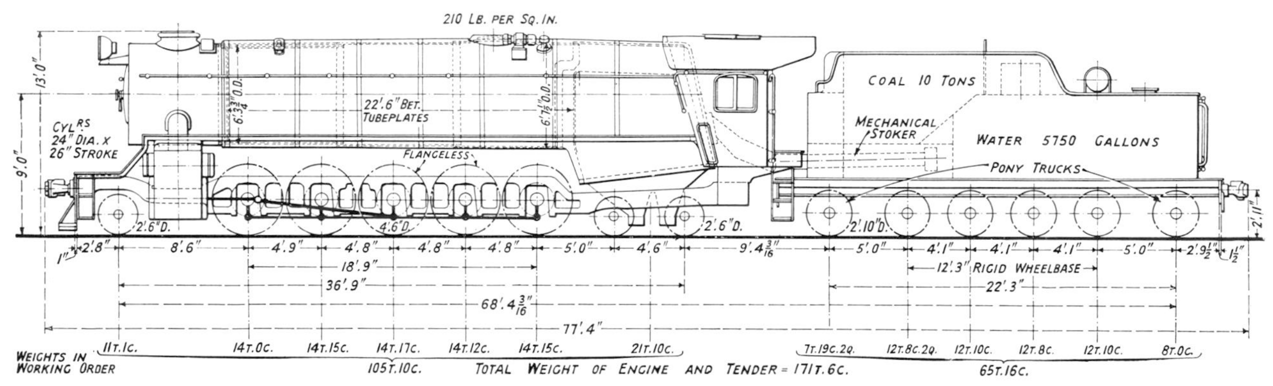

Fig. 176.—*South African Railways: Line diagram of Class* " 21 " 2-10-4.

Fig. 177.—*South African Railways:* 0-8-0 *shunting locomotive.*

Fig. 178.—*South African Railways: feed screw of mechanical stoker on* 2-10-4 *engine.*

of this fine engine, and it will be seen that to assist in negotiating curves the third and fourth pairs fo driving wheels were flangeless. The leading pair was provided with 7/8 in. side play on either side of the centre line, and the coupling rods between the first and second pair were hinged to accommodate this side play. The tender was large, having a capacity for 5,750 gallons of water and 10 tons of coal. It could be described as of the 2-8-2 type, because of the six axles the four centre ones formed a rigid wheelbase of 12 ft. 3 in. while the leading and trailing axles were carried in sliding trucks. The maximum engine axle load was 14.85 tons. The coupled wheels were 4 ft. 6 in. diameter, and in conjunction with cylinders 24 in. by 26 in. and a boiler pressure of 210 lb. per sq. in. the nominal tractive effort at 85 per cent. boiler pressure was 49,504 lb. In passing it may be noted that these large South African locomotives, both the British-built 4-8-2s and 2-10-4, and the German-built 4-6-2s, all had self-cleaning screens in the smokebox, very similar in layout to those subsequently adopted as standard on British Railways.

Another impressively massive locomotive built by the North British Locomotive Co. for the South African Railways was the 0-8-0 shunter illustrated herewith (Fig. 177).

On the Indian Railways, following the production of the standard range of locomotives described in Chapter Four, the period under review was a difficult one. On some of the Government-owned railways much trouble was being experienced with bad riding of the standard Pacifics, and on the Bengal Nagpur Railway which was then still privately owned there was a partial reversion to the 4-6-0 type, in 1937. That company previously used 4-cylinder compound Pacific engines for the heaviest express passenger and mail runs, but these were of non-standard design, so far as the rest of India was concerned. On the State-owned railways an interesting event was the construction in 1937, of two experimental Pacifics by the Vulcan Foundry for the Great Indian Peninsular Railway, with Caprotti valve gear and roller bearings throughout (Fig. 179). Following experience with the earlier standard " Pacifics " the main aim was to produce locomotives capable of working a monthly mileage of 10,000 and of running a total of 200,000 miles between major overhauls.

An engine of greater route-availability than the standard XC Pacific was desired, and so the " XB " boiler and firebox was taken as a basis though considerably modified as to be internal details, and designed for a higher working pressure. The respective dimensions were:

Class	" XB " Standard	" XP " Experimental
Heating surfaces, sq. ft.		
Tubes	1642	1543
Firebox	198	192
Thermic Syphons	—	50
Superheater	463	504
Combined total	2303	2289
Grate area	45	45
Boiler pressure, lb./sq. in.	180	210

With cylinders 21 1/2 in. diameter by 28 in. stroke, and coupled wheels 6 ft. 2 in. diameter, the nominal

Fig. 179.—*Great Indian Peninsular Railway: Experimental "XP"* 4-6-2 *locomotive, with Caprotti valve gear built by Vulcan Foundry Ltd.* 1937.

Fig. 180.—*North Western Railway of India:* 4–6–0 *locomotive Class* " XS2 ", *built by Vulcan Foundry Ltd.*

Fig. 181.—*South Indian Railway: Class* " XD3 " 2-8-2 *heavy freight locomotive.*

Fig. 182.—*Bengal Nagpur Railway:* " GSM " *class* 4-6-0 " *Mail* " *engine, built by R. Stephenson and Hawthorns Ltd.*

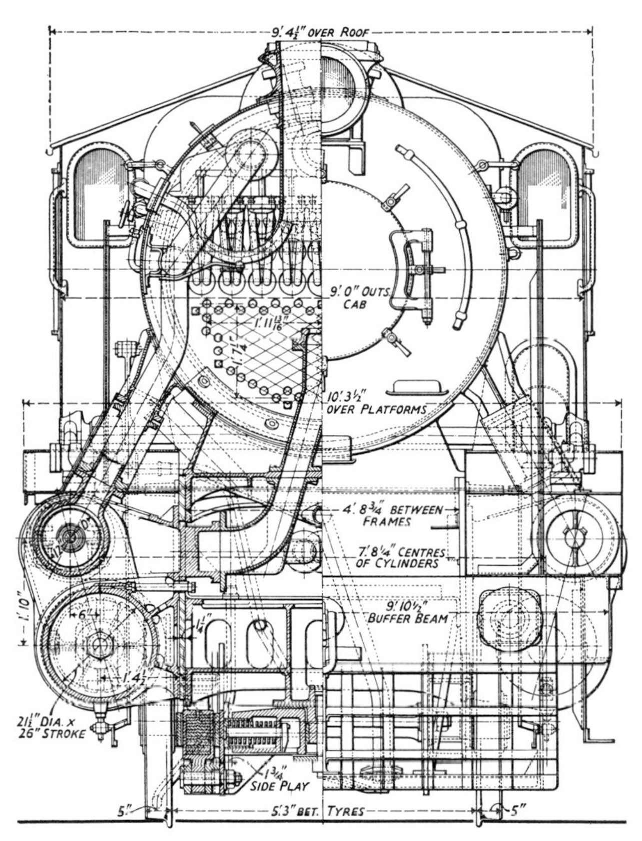

Fig. 183.—*Bengal Nagpur Railway: Cross-section of front-end "GSM" class 4-6-0.*

tractive effort at 85 per cent. boiler pressure was 31,220 lb., and as such greater than both the "XB" and the XC Pacifics, which were 26,760 and 30,625 lb. respectively.

Two engines of the "XP" class were built for the G.I.P.R., one equipped with Timken and the other with Skefko roller bearings. So far as riding was concerned the "XP" engines were better than the original standard types, but still not good enough to permit of their free-running capacity to be utilised to the full. Although these engines were no more than a partial success they did provide a stepping stone to the new standard "WP" class which was developed in India later.

In complete contrast to the general line of locomotive development in India during the period 1930–40, the Bengal Nagpur Railway ordered from Robert Stephenson and Hawthorns Ltd., two express passenger engines with tractive capacity slightly in excess of the standard "XB" class, with a maximum axle load of 17 tons, yet of the 4-6-0 rather than the 4-6-2 type. These two locomotives, of the "GSM" class were of advanced design, yet simple and straightforward in their detail. The basis was the very successful "GS" class 4-6-0 mail engines (Fig. 182), and particular attention was given to the boiler proportions to obtain maximum steaming capacity, without resorting to a wide firebox, and consequent use of the 4-6-2 type, which was not desired. Ratios obtained in these fine new engines were:

$$\frac{\text{Area of superheater elements}}{\text{main steam pipes}} = 1.078$$

$$\frac{\text{Evaporative heating surface}}{\text{grate area}} = 42.2$$

$$\frac{\text{Grate area}}{\text{Flue area}} = 6.78$$

$$\frac{\text{Firebox heating surface}}{\text{grate area}} = 5.3$$

$$\frac{\text{Tractive effort}}{\text{grate area}} = 731$$

$$\frac{\text{Firebox heating surface}}{\text{firebox volume}} = 0.994$$

In comparison with the standard Pacifics these engines had a very short boiler barrel, with only 13 ft. 6 in. between the tube plates, against 18 ft. 6 in. on the "XB" and "XC Pacifics"; but the tube diameter was itself large, and conducive to free steaming. The heating surfaces on the "GSM" engines were:

Heating surfaces, sq. ft.	
Small tubes (2 in. outside dia.)	937.2
Large tubes (5 $^1/_4$ in. outside dia.)	461.2
Firebox and arch tubes	201.6
Superheater	317.0
Combined total	1917.0
Grate area, sq. ft.	38.0

The boiler pressure was 200 lb. per sq. in., and with cylinders 21 $^1/_2$ in. diameter by 26 in. stroke and coupled wheels of 6 ft. 1 $^1/_2$ in. diameter, the nominal tractive effort at 85 per cent. boiler pressure was 27,787 lb. The total weight of engine in working order was only 74.9 tons; but the tender was very large, having a capacity of 4,750 gallons of water and 10 tons of coal, and weighed 65.65 tons in working order. A cross-sectional drawing of the front-end of these engines (Fig. 183) shows the extent to which internal streamlining has been applied to the steam circuit of these engines (see also drawing, Fig. 184).

A design of exceptional interest was prepared in 1934–35 for the Chinese National Railways, and 24 engines were duly built by the Vulcan Foundry Ltd. A powerful locomotive was required for general service and the 4-8-4 wheel arrangement was chosen (Fig. 185). They were required to work heavy trains over the steeply graded northern section of the Canton–Hankow line, negotiating gradients of 1 in 70 and curves of less than 500 ft. radius; they were also required for relatively fast running over the level stretches of the Shanghai–Nanking line. Above all was the necessity for burning low grade coal. Comments on this very notable design can start therefore, appropriately, at the firebox. This was very large, having a grate area of 67 sq. ft., and incorporates a combustion chamber of ample volume as shown in the accompanying

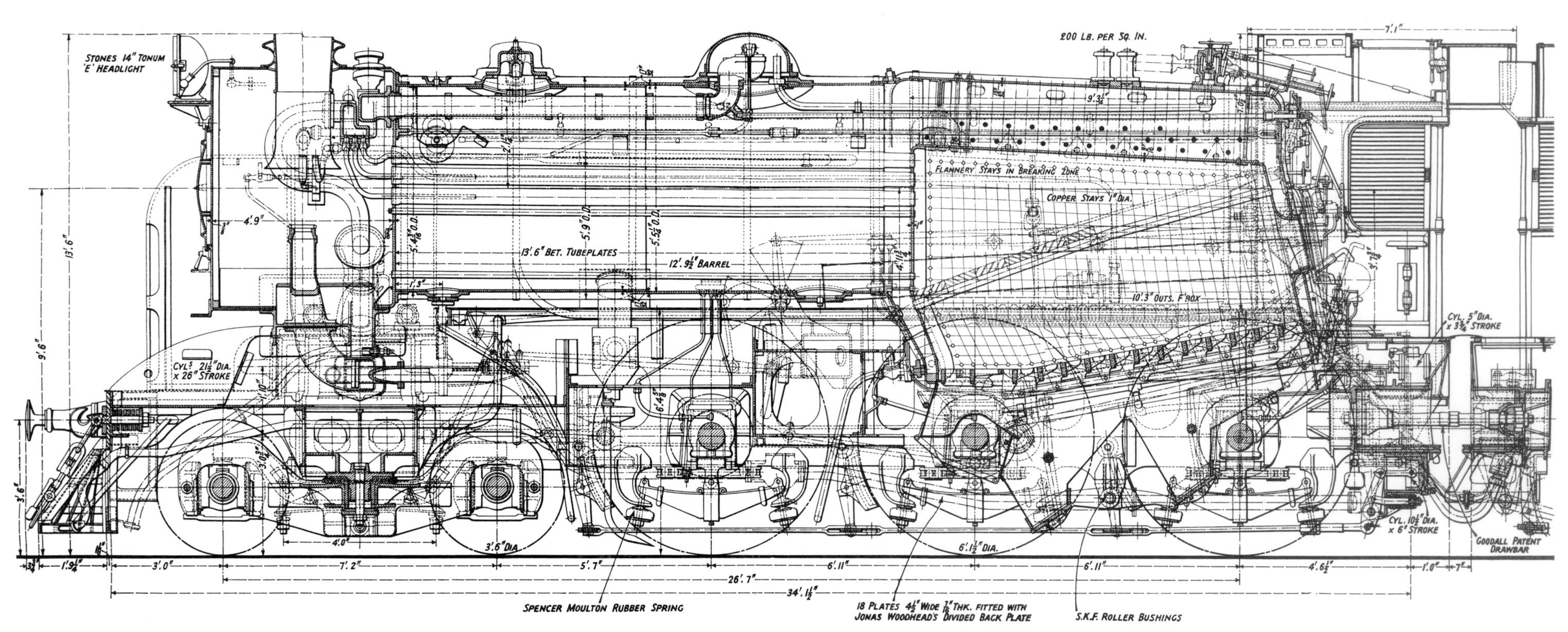

Fig. 184.—*Bengal Nagpur Railway: General arrangement of "GSM" class* 4-6-0.

Fig. 185.—*Chinese National Railways* 4-8-4 *locomotive for Canton Hankow line, built by Vulcan Foundry Ltd.* 1934-5.

drawing (Fig. 187). The inner firebox was of steel, so designed that thermic syphons could be added if required at a later date. As built, the engine was equipped with arch tubes, as shown in the drawing. A further drawing shows the arrangement of the mechanical stoker (Fig. 190). For such a large engine, having a combined total heating surface of over 4000 sq. ft., the boiler barrel is relatively short—an excellent feature, conducive to good steaming.

The longitudinal and transverse cross-sections of the smokebox and cylinders (Fig. 186) shows the very large superheater, which alone provided heating surface of 1076 sq. ft. There were 120 flue tubes each of 3 $^{1}/_{2}$ in. outside diameter, and the 33 elements of the superheater are each accommodated in four flues, except for six which occupy only two flues. The boiler proportions were as follows:

Boiler height of centre line from rail, ft. in.	10-6
,, length between tube plates, ft. in.	19-0 $^{3}/_{8}$
,, dia. at front end, ft. in.	5-10 $^{7}/_{8}$
,, dia. at firebox end, ft. in.	6-6 $^{3}/_{4}$
Heating surfaces, sq. ft.	
Small tubes (50)	560
Large tubes (120)	2088
Firebox	312
Arch tubes	28
Superheater	1076
Combined Total	4064
Grate area, sq. ft.	67.8
Boiler pressure, lb./sq. in.	220

The main frames were of the bar type, cut from rolled steel slabs, with a slab extension over the engine trailing bogie. A superstructure built up of welded plates and cross stretchers was carried by the bar frames for the support of the boiler barrel and smokebox. The cross-sectional drawing of the front-end

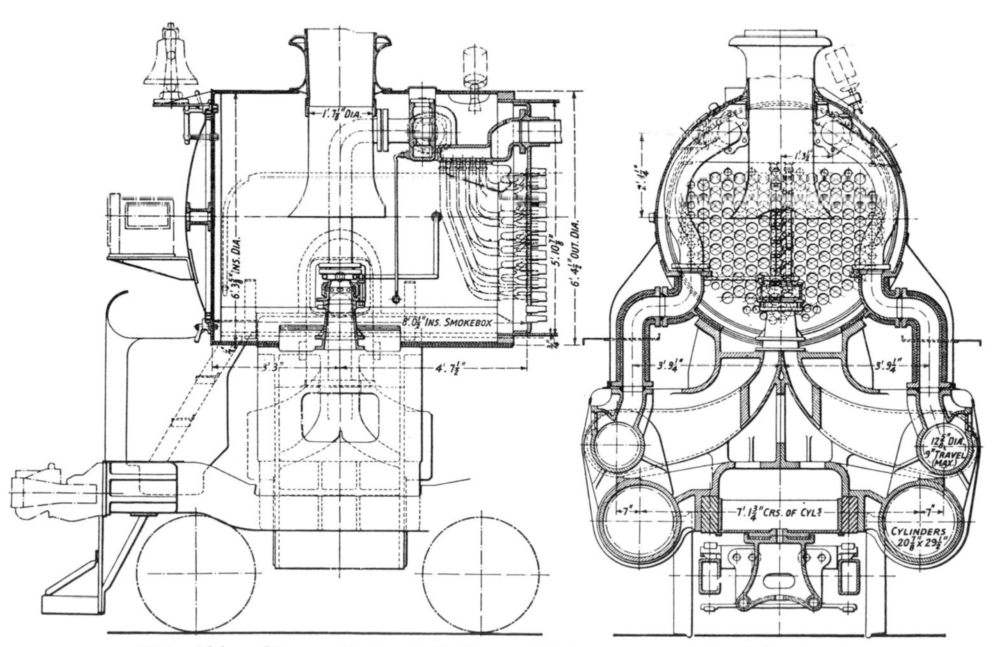

Fig. 186.—*Chinese National Railways* 4-8-4: *arrangement of front end.*

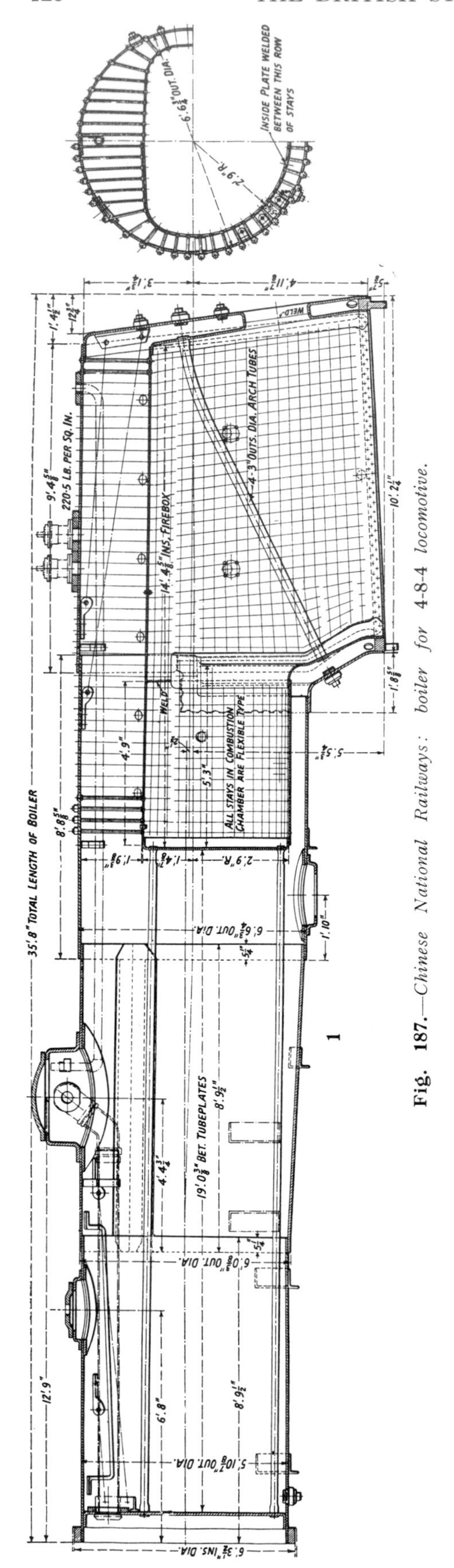

Fig. 187.—*Chinese National Railways: boiler for 4-8-4 locomotive.*

Fig. 188.—*Chinese National Railways* 4-8-4*: view of booster engine on tender.*

shows how admirably the bar-frame type of construction permitted of the utmost freedom in design of the exhaust passages from steam chests to the blastpipe. A three-point suspension system was provided for the engine as a whole by compensating the spring gear in two groups. The leading bogie formed one group and the coupled wheels and the trailing bogie the second group. The load from the main frames was conveyed to the trailing bogie at three points, one at the forward end and two at the rear. The load was applied at the rear end through rockers, the function of which was to stabilise the hind end and make the bogie self-centering without the use of spring centering devices.

The cylinders and valve chests formed a notably simple, and effective piece of casting design. Each cylinder and valve chest was cast in one with half the smokebox saddle, and the two bolted together on the centre line. Mention has already been made of the directness of the steam and exhaust passages. This careful attention to all features of the steam circuit was also seen in the exceptionally large diameter of the piston valves, namely 12 $^5/_8$ in., in relation to the cylinder dimensions of 20 $^7/_8$ in. diameter by 29 $^1/_2$ in. stroke. A further drawing (Fig. 189) shows the layout of the valve and drive gear, and the 2:1 lever extension of the valve rod used to provide a maximum travel, in full gear, of no less than 9 in. With 5 ft. 9 in. coupled wheels the nominal tractive

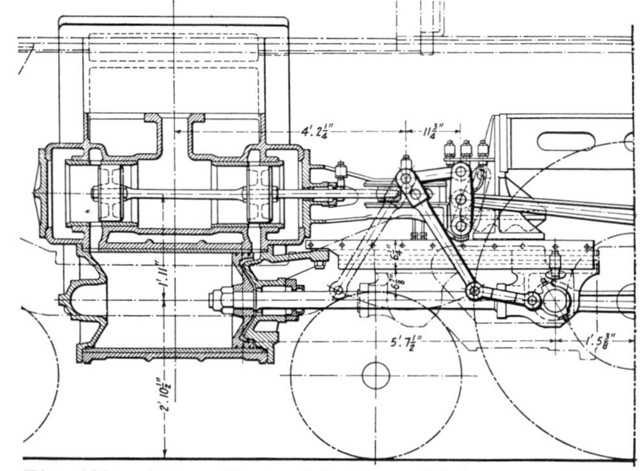

Fig. 189.—*Chinese National Railways* 4-8-4*: arrangement of valve and drive gear.*

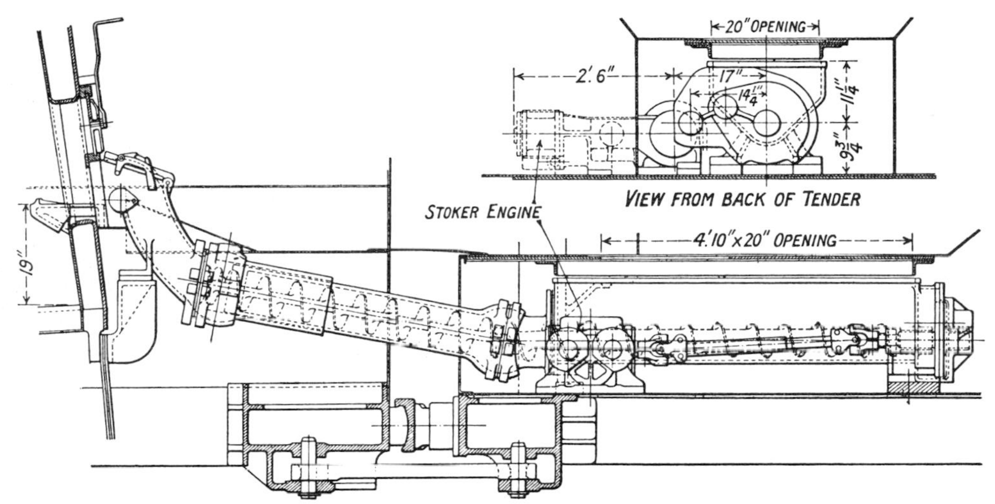

Fig. 190.—*Chinese National Railways* 4-8-4: *arrangement of mechanical stoker.*

effort at 85 per cent. boiler pressure was 32,920 lb. With these engines one feels that so many outstanding features had been packed into the design that the tractive effort gave no more than a modest impression of their capacity in traffic.

The tenders, providing space for 6,600 gallons of water, and 11 3/4 tons of coal, were carried on two 6-wheeled bogies. Six of the engines, those intended for the heavily graded northern section of the Canton–Hankow line, had booster-driven bogies on the tenders, and it is one of these engines that is illustrated in the photograph reproduced herewith (Fig. 188). The wheels were 2 ft. 9 7/8 in. diameter, and the booster of Messrs. Stone's Class " D " type (Fig. 191), had the following dimensions:

Cylinders, 7 in. dia. by 10 in. stroke
Cut-off, 75 per cent.
Gear ratio 3:1
Tractive force 7670 lb.
Factor of adhesion, 7.5

In one quite superficial detail, the outline of the chimney, these Chinese engines are reminiscent of the L.N.W.R. at Crewe. As such they undoubtedly reveal the affection for his *alma mater* of the engineer who was mainly responsible for the specification of these outstanding engines, Colonel K. Cantlie, a pupil of C. J. Bowen-Cooke. The Chinese 4-8-4s were still doing excellent work in 1966.

Fine examples of British locomotive construction to suit various road conditions were to be seen in " Pacifics " for the Gold Coast, and for the Malayan Railways and in the massive and powerful 2-8-0 for the Iranian State Railways. A further notable engine was the " streamlined " 4-8-2 for the New Zealand Government Railways. Taking the Iranian 2-8-0 first (Fig. 192), this design was prepared for working on a very heavily graded route, where there is no opportunity for anything approaching fast running. A high tractive effort at low speed was necessary, and with coupled wheels of no more than 4 ft. 0 in. diameter

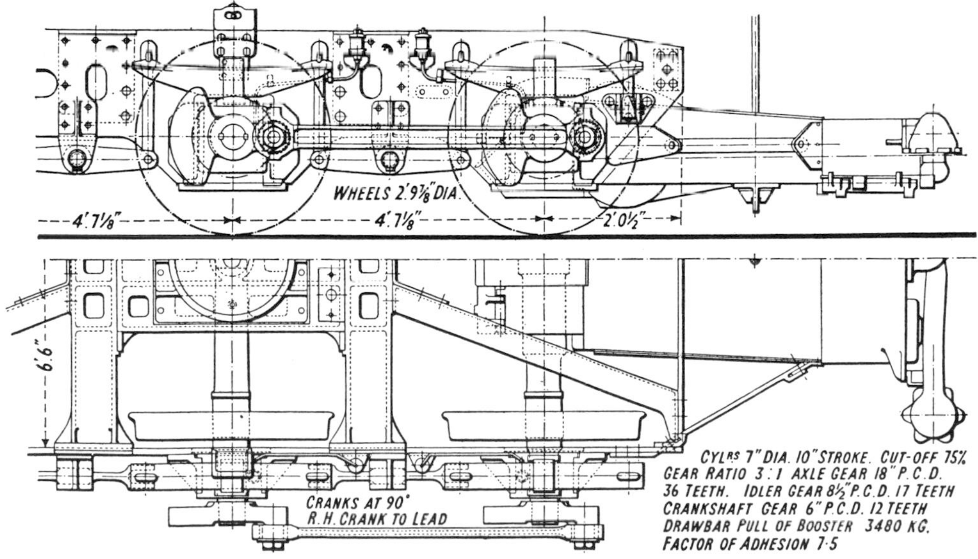

Fig. 191.—*Chinese National Railways* 4-8-4: *layout of booster engine.*

Fig. 192.—*Iranian State Railways:* 2-8-0 *locomotive for mountain gradients, built by Beyer-Peacock and Co. Ltd.* (1934).

Fig. 193.—*Malayan Railways:* 4-6-2 *locomotive with R.C. poppet valve gear* (1938).

these engines have a nominal tractive effort at 85 per cent. boiler pressure of 32,230 lb. As will be seen from the photograph the boiler was relatively short, though of large diameter, and the grate area, for oil firing, was 34 sq. ft. The general dimensions of these engines were:

Cylinders (2) dia., in.	20
Cylinder stroke, in.	26
Coupled wheel dia., ft. in.	4-0
Heating surfaces, sq. ft.	
Tubes	1658
Firebox	136
Superheater	322
Combined total	2116
Grate area, sq. ft.	34
Boiler pressure, lb. /sq. in.	175
Max. axle load, tons	14 $^3/_4$
Adhesion weight, tons	57.2
Total weight of engine and tender (working order), tons	116

These locomotives were built by Beyer-Peacock and Co. Ltd. in 1934, and run on the 4 ft. 8 $^1/_2$ in. gauge.

The Malayan Pacific illustrated (Fig. 193) built by the North British Locomotive Company Ltd. in 1940, which incorporates certain detail developments to a design introduced in 1938, was designed for the metre gauge, on a main route having a maximum permitted axle load of 12 $^3/_4$ tons. Very great care was therefore taken to reduce weight wherever possible. Bar frames were used, and the boiler shell was of nickel steel. The inner firebox was also of steel. An interesting feature of the design was the use of three cylinders, with the R.C. type of poppet valve gear. These engines were originally introduced for working the mail trains between Singapore, Kuala Lumpur and Prai, and despite the heavy gradients met on some parts of the line, particularly in the Taiping Pass, near Ipoh they did excellent work. Now that the principal passenger train workings on the Malayan Railways have been taken over by diesel electric locomotives these steam 4-6-2s are mostly engaged on goods and mixed trains. Their leading dimensions are:

Cylinders (3) dia., in.	$12\,^1/_2$
Cylinder stroke, in.	24
Coupled wheel dia., ft. in.	4-6
Heating surfaces, sq. ft.	
Tubes	960
Firebox and arch tubes	149
superheater	218
Combined total	1327
Grate area, sq. ft.	27
Boiler pressure, lb./sq. in.	250
Nom. tractive effort at 85% boiler pressure, lb.	22,130
Capacity of tender	
Coal, tons	10
Water, galls.	3500

The Gold Coast Pacific, built by Beyer-Peacock and Co. Ltd., in 1939, and illustrated herewith (Fig. 194) affords an interesting contrast in detail design to the Malayan engine just referred to, because the operating conditions were very similar. On a route laid with 60 lb. rail and having the 3 ft. 6 in. gauge. the maximum axle load permitted was in this latter case $12\,^1/_2$ tons. The engine was built on bar frames, but in view of the employment entirely of African enginemen the details were kept as simple as possible using two cylinders only, with Walschaerts valve gear outside and a moderate boiler pressure of 180 lb. per sq. in., against the Malayan use of 250 lb. The handsome, essentially British appearance of the engine rather belies its moderate size, for it was actually a small-power unit, having the following leading dimensions:

Cylinders (2) dia., in.	18
Cylinder stroke, in.	26
Coupled wheel dia., ft. in.	5-0
Heating surfaces, sq. ft.	
Tubes	1222
Firebox and arch tubes	136
Superheater	332
Combined total	1690
Grate area, sq. ft.	25
Boiler pressure, lb./sq. in.	180
Nom. tractive effort at 85% boiler pressure, lb.	21,480
Capacity of tender	
Coal, tons	$7\,^1/_2$
Water, galls.	4000

In 1939 the New Zealand Government Railways placed an order for 40 new mixed traffic locomotives of the 4-8-2 type with the North British Locomotive Co. Ltd. which proved to be another very interesting example of a medium powered machine for operating on light permanent way, in this case on rails of only 50 lb. per yard (Fig. 195). In general capacity the

Fig. 194.—*Gold Coast Railway.* 4-6-2 *locomotive built by Beyer Peacock & Co. Ltd.* 1939.

Fig. 195.—*New Zealand Government Railways: Lightweight* 4-8-2 *locomotive, for track laid with* 50 *lb. rail. Built* 1939, *by North British Locomotive Co. Ltd.*

Fig. 196.—*Buenos Aires Great Southern Railway: Vulcan* 4-6-2 *locomotive (oil fired) for heavy express passenger service* (1938).

Fig. 197.—*Buenos Aires Great Southern Railway: Vulcan* 4-8-0 *locomotive for fast freight and heavy long distance excursion trains* (1938).

engine bears a considerable resemblance to the Malayan and Gold Coast designs just discussed; but in working out the design, to produce an engine of 26,520 lb. tractive effort the approach was different. In order to obtain the necessary adhesion weight eight coupled wheels were necessary, and this had the effect of producing a much longer engine. Furthermore a very large firebox was provided, to burn soft, low grade coal. The grate was of the plain firebar type except that a drop grate was fitted at the rear end, immediately beneath the firedoor. There was also a large combustion chamber ahead of the firebox. The boiler barrel was long in relation to its diameter: 17 ft. 6 in. between the tubeplates against 4 ft. 9 in. diameter. The design was prepared at a time when locomotive streamlining was in fashion, and although not intended for high speed work a degree of external fairing was included, which gave the locomotive a very distinctive appearance. Other features included bar frames, Baker valve gear, and the Vanderbilt type of water tank on the tender. The leading dimensions were:

Cylinders (2) dia., in.	18
Cylinder stroke, in.	26
Coupled wheel dia., ft. in.	4-6
Heating surfaces, sq. ft.	
Tubes	1319.5
Firebox	149.5
Superheater	283
Combined total	1752
Grate area, sq. ft.	39
Boiler pressure, lb./sq. in.	200
Nom. tractive effort at 85% boiler pressure, lb.	26,520
Capacity of tender	
Coal, tons	6
Water, galls.	4000

In 1938 the Vulcan Foundry Ltd. built some fine locomotives of both 4-6-2 and 4-8-0 type for the Buenos Aires Great Southern Railway. The two types were very similar in appearance, and had inter-changeable boilers, motion, coupled axles, and tenders. The 4-6-2 (Fig. 196) had 6 ft. 0 in. diameter coupled wheels and the 4-8-0 (Fig. 197), 5 ft. 8 in. The latter engines had cylinders of 19 $^1/_2$ in. diameter against 19 in. on the 4-6-2. The respective nominal tractive efforts were 26,850 lb. and 29,940 lb. The main line over which both types were designed to operate is relatively level and very heavy trains were worked at moderately high speed. The seasonal fruit traffic, for which the 4-8-0s were particularly intended involved train loads of 1,000 tons, at overall booked speeds of 40 m.h.p., while the summer sleeping car expresses between Buenos Aires and Mar del Plata also loaded to much the same tonnage. As usual on the B.A.G.S. Rly. the locomotives were oil-fired. Their economical performance permitted of the making of through runs over the 740 miles between Buenos Aires and the Rio Negro valley, with one intermediate replenishment of the fuel supply. The design included many modern features that provided aids to high availability, and in cases where intense utilisation of stock was necessary locomotives of the 4-8-0 class had run through from the Rio Negro fruit loading stations to Buenos Aires, and then after brief inspection returned at once with the empty stock, making a virtually continuous run of nearly 1,500 miles. The 4-8-0 class proved particularly successful, and many more were built by the Vulcan Foundry Ltd. after the initial running of the first eight engines ordered in 1938.

The dimensions of the two classes were:

Type	4-6-2	4-8-0
Cylinders (2) dia., in.	19	19 1/2
Cylinder, stroke, in.	28	28
Coupled wheel dia., ft. in.	6-0	5-8
Nom. tractive effort at 85% boiler pressure, lb.	26,850	29,940
Max. axle load, tons	18	16.5

Dimensions common to both:

Heating surfaces, sq. ft.	
Tubes	1550
Firebox	190
Superheater	428
Combined total	2168
Grate area, sq. ft.	32.6
Boiler pressure, lb./sq. in.	225

Two interesting locomotives for service on 4 ft. 8 1/2 in. gauge lines in the Middle East are those illustrated for the Palestine and Egyptian State Railways. The former, of the 4-6-0 type (Fig. 198), was built in 1935, for main line passenger working between Haifa and El Kantara. For their overall weight (engine only) of 68.65 tons these locomotives had the high nominal tractive effort of 28,400 lb. derived from cylinders 20 1/2 in. diameter by 28 in. stroke; coupled wheels 5 ft. 6 3/4 in. diameter, and a boiler pressure of 190 lb. per sq. in. Furthermore the maximum axle-load was only 17 tons. The combined total heating surface was 1949 sq. ft. and the grate area 29 sq. ft.

The Egyptian 2-6-0 illustrated (Fig. 199) was one of a batch of 50 built by the North British Locomotive Co. Ltd. in 1935 and 1936. For purposes of extended trials 30 of these engines were fitted with Walschaerts valve gear and 20 with Caprotti poppet valve gear. In addition there were variations in the boiler feed arrangements as follows:

Davis and Metcalf Exhaust Steam Injectors	40
A.C.F.I. feed water heaters	5
Heinl feed water heaters	5

On five of the locomotives also a trial was made of the Kylchap type of double blastpipe and chimney. The accompanying illustration shows one of the Caprotti

Fig. 198.—*Palestine Railways:* 4-6-0 *of* 1935, *built by North British Locomotive Co. Ltd.*

Fig. 199.—*Egyptian State Railways:* 2-6-0 *locomotive with Caprotti valve gear and A.C.F.I. feed water heater, built* 1935-6 *by North British Locomotive Co. Ltd.*

Fig. 200.—*Leopoldina Railway: metre gauge* 4-6-2 *tank engine, built by Beyer Peacock & Co. Ltd. in* 1931.

engines, equipped with the A.C.F.I. feed water heaters. The basic dimensions of all 50 engines were:

Cylinders (2) dia., in.	17 $^3/_4$
Cylinders, stroke, in.	28
Coupled wheel dia., ft. in.	5-6 $^3/_4$
Heating surfaces, sq. ft.	
Tubes	1257
Firebox	149
Superheater	250
Combined total	1656
Grate area, sq. ft.	25
Boiler pressure, lb./sq. in.	225
Nom. tractive effort at 85% boiler pressure, lb.	25,270

In concluding these notes on some of the more important locomotives built in Great Britain for overseas service, during the period 1930–40, some reference to tank engine designs is necessary. In addition to the engines illustrated particular mention must be made of three interesting examples. The metre-gauge 4-6-2 engine for the Leopoldina Railway (Fig. 200), built by Beyer Peacock and Co. Ltd., in 1931, was designed for suburban passenger traffic requiring rapid acceleration from station stops. The maximum axle load permitted was 10 $^3/_4$ tons, but by the use of coupled wheels of no more than 4 ft. 4 in. diameter a tractive effort of 16,110 lb. at 85 per cent. boiler pressure was made possible, with such relatively small cylinders as 16 in. by 22 in. and a boiler pressure of 175 lb. per sq. in. The proportions were generally small, with a total heating surface of

Fig. 201.—*Sao Paulo Railway: Heavy express tank engine* 4-6-4 *type, built* 1936 *by North British Locomotive Co. Ltd.*

Fig. 202.—*Malayan Railways:* 4-6-4 *tank engine with R.C. poppet valve gear. Built* 1939 *by North British Locomotive Co. Ltd.*

Fig. 203.—*Sao Paulo Railway:* 2-8-4 *tank engine of* 1938.

Fig. 204.—*Buenos Aires Pacific Railway:* 4-6-4 *passenger tank engine of* 1930, *built by R. Stephenson & Co. Ltd.*

only 796 sq. ft. and a grate area of 15.8 sq. ft. The total weight of the engine in working order was $52\frac{1}{4}$ tons.

Contrasting to the foregoing, in their general massive proportions were some large 4-6-4 passenger tank engines for the Sao Paulo Railway (Fig. 201), working on the 5 ft. 3 in. gauge. These fine engines were built in 1936 by the North British Locomotive Co. Ltd. With the exception of the refinement of the A.C.F.I. feed water heater, prominently shown in the accompanying illustration the design was a very simple one. A noteworthy feature was the very large diameter of piston valves used, namely 11 in., in conjunction with cylinders $21\frac{1}{2}$ in. diameter by 26 in. stroke. The coupled wheels were 5 ft. 6 in. diameter; the boiler pressure 200 lb. per sq. in.; and the nominal tractive effort, at 85 per cent. boiler pressure, was 26,000 lb. The boiler and firebox, with a short barrel of large diameter provided a combined total heating surface of 2,030 sq. ft. and the grate area was 28.5 sq. ft. These locomotives were coal fired.

The Malayan 4-6-4 tanks (Fig. 202), built by the North British Locomotive Co. Ltd. in 1939, were similar in general design to the 4-6-2 tender engines previously referred to in this chapter. They were designed for local passenger working, but although of moderate size and built to have a maximum axle load of only $12\frac{1}{2}$ tons, their details of equipment were representative of first class modern practice. Like the main line 4-6-2s they had bar frames, and the R.C. poppet valve gear was used. The leading dimensions were:

Cylinders (2) dia., in.	$14\frac{1}{2}$
Cylinder stroke, in.	22
Coupled wheel dia., ft. in.	4-6
Heating surfaces, sq. ft.	
Tubes	717
Firebox and arch tubes	136.5
Superheater	180
Combined total	1033.5
Grate area, sq. ft.	28
Boiler pressure, lb./sq. in.	250
Nom. tractive effort at 85% working pressure, lb.	18,200

Chapter Eleven The Streamline Age

Until the year 1935 although there had been notable increases in the standards of performance on many routes in Great Britain there were still very few centres of population between which one could travel at an overall speed of 60 m.p.h. and the general standard of speed in the country remained largely that which the Great Western Railway had set in the great days of Churchward, and which the northern lines had at last emulated in 1932 and thereafter. There is no doubt that in the realm of speed Great Britain—once quite pre-eminent—was being left behind by both France and the U.S.A. Many factors in operation could be advanced in defence of this position, particularly in respect of dense line occupation; but in what might be termed the " shop window display " of British locomotive engineering practice the situation had shown little change for upwards of thirty years.

Individual locomotive performance on all four major railways of Britain had shown that speeds in excess of 80 m.p.h. could be sustained indefinitely on level track with trains of around 250 tons, and in contrast to the traditional practice on the principal expresses from London to the north, of running trains of very heavy formation, with through carriages for many different destinations, the trend that was developing elsewhere in the world of running high-speed trains of limited formation began to attract the attention of British railway managements. It could of course be argued that the Great Western Railway had already entered this particular field with the " Cheltenham Flyer " and its fast schedule of 71.4 min. from Swindon to Paddington; this was however no more than one intermediate booking of a train that did not give a particularly fast overall time from Cheltenham or Gloucester to London. Apart from this, the fastest runs on the four main line railways in the summer of 1934 were:

Railway	Route	Dist. miles	Time min.	Speed m.p.h.
L.M.S.R.	Crewe–Willesden	152.6	142	64.4
G.W.R.	Oxford–Paddington	63.5	60	63.5
L.N.E.R.	Grantham–King's Cross	105.5	100	63.3
Southern	Waterloo–Salisbury	83.8	87	57.7

Overall times to the largest provincial centres though mostly showing a marginal improvement over those of 1914 were still below, rather than above 60 m.p.h. as follows:

Fastest Times to Leading Provincial Cities

City	Dist. miles	Time min.	Average speed m.p.h.
Aberdeen	523.2	700	44.9
Birmingham	110.6	120	55.3
Bristol	118.3	120	59.2
Cardiff	145.1	161	54.1
Edinburgh	392.7	450	52.4
Glasgow	401.4	460	52.4
Leeds	158.8	193	57.7
Liverpool	193.7	200	58.1
Manchester	188.5	195	58.0
Newcastle	268.3	306	52.7
Norwich	115.0	144	47.9
Plymouth	225.5	240	56.4
Sheffield	161.2	180	53.7
Southampton	79.2	88	54.0

It was the introduction of the diesel-powered two-car high speed railcar service in Germany running the 178.1 miles between Berlin and Hamburg in 138 min. that sowed the seeds of future high speed development in Great Britain. This " limited " train provided a 77.4 m.p.h. service between these two major cities, and at the suggestion of Mr. H. N. Gresley serious consideration was given to the purchase of a similar train for experimental purposes on the L.N.E.R., and the makers of the German train were approached with a view to providing a high speed service between King's Cross and Newcastle. It was then that the hindrances to fast running inherent in many British main lines became more than ever apparent. Taking account of gradients, and all the regular speed restrictions the German manufacturers could not promise with a railcar service a higher average speed than 63 m.p.h. between King's Cross and Newcastle against the 77.4 m.p.h. of the " Flying Hamburger ". Moreover this was to be made with no more than a two-coach train with seating accommodation rather more restricted than that to which regular travellers were accustomed on the East Coast route.

In the autumn of 1934 preliminary trials were made on the L.N.E.R. towards a steam-hauled high speed service with a standard low-pressure "Pacific" engine No. 4472 *Flying Scotsman*, with results as follows, in each case running non-stop over the 185.8 miles between King's Cross and Leeds Central.

Direction	Load tons	Time min.	Average Speed m.p.h.	Max. speed m.p.h.
Northbound ..	147	152	73.3	95
Southbound ..	207	157 1/4	70.8	100

Gresley Pacific Locomotives

Fig. 205.—*L.N.E.R. One of the first four streamlined* 4-6-2s *Class* " A4 ", *for the* " *Silver Jubilee* " *service: No.* 2512 Silver Fox, *built Doncaster* 1935.

Fig. 206.—*L.N.E.R. Class* " A4 " *Pacific for the* " *Coronation* " *service, No.* 4489 Dominion of Canada, *with presentation bell.*

Fig. 207.—*L.N.E.R. Class* " A4 " *Pacific named after the designer, No.* 4498 Sir Nigel Gresley. *The 100th Pacific built at Doncaster.*

The locomotive was nearly twelve years old at the time of this round trip, and being one of the 180 lb. series did not represent the latest development in L.N.E.R. express passenger engine practice. Against the best promises made by the German manufacturers of the "Flying Hamburger" diesel train the results were most encouraging, for it showed that better time could be made by using standard locomotives and standard stock if the train load was strictly limited. A still more convincing test was made in March 1935 with engine No. 2750 *Papyrus*, one of the high pressure "Pacifics" of Class "A3" on a round trip from King's Cross to Newcastle and back, made in a single day's work, with a load of six coaches in each direction, 217 tons.

Direction	Actual time	Net time* min.	Actual Av. sp. m.p.h.	Max. speed m.p.h
Northbound ..	237	230	68.0	88 1/2
Southbound ..	231 3/4	228	69.5	108

* Allowing for severe out-of-course delay due to a derailment on adjoining line

Quite apart from demonstrating the feasibility of a regular 4-hour service between London and Newcastle—a clear hour faster than the best then in operation—the engine *Papyrus* made four new world records for steam traction, as follows:

(1) A distance of 12.3 miles run at 100.6 m.p.h.
(2) An aggregate of 500 miles in one day run at an average of 72.7 m.p.h., with a 217-ton train.
(3) An aggregate of 300 miles in one day's round trip at an average speed of 80 m.p.h.
(4) A maximum speed of 108 m.p.h.

As a result of this performance the decision was taken to inaugurate, in September 1935, a regular high speed service between London, Darlington and Newcastle, and an improved design of "Pacific" engine was prepared (Fig. 205), with the following modifications from the existing "A3" class:

L.N.E.R. Pacifics

Class	"A3"	"A4"
Cylinders (3) dia., in.	19	18 1/2
Cylinder stroke, in.	26	26
Dia. of piston valves, in.	8	9
Boiler pressure lb./sq in.	220	250
Distance between tube plates, ft.-in.	18-11 3/4	17-11 3/4
Total evaporative heating surface sq. ft.	2692	2576
Superheater, sq. ft.	703	748.9
Nom. tractive effort at 85% boiler pressure, lb.	32,909	35,455

The significant changes were the increase in boiler pressure, and reduction in cylinder diameter combined with the increased size of piston valve. These features incorporated into a design that already held the world speed record for steam made for an exceptionally free running engine. The decision to streamline the engine externally was a master stroke of publicity, but wind tunnel tests with scale models had shown that an appreciable saving in horsepower could be effected, at high speed by this additional feature. The following figures were obtained for the "A2" and "A4" classes:

Horsepower Saved by Streamlining

Speed, m.p.h.	Engine class	60	70	80	90
Horsepower to overcome head-on air resistance	"A3"	97.21	154.26	230.51	328.49
	"A4"	56.39	89.41	133.61	190.40
Horsepower saved by streamlining		40.82	64.85	96.90	138.09

The "A4" class fulfilled every expectation (Figs. 205, 206). On September 27, 1935 a trial run was made with the new Silver Jubilee train, weighing 230 tons, and although the first engine of the class, No. 2509 *Silver Link* was only three weeks out of the Doncaster works a maximum speed of 112 1/2 m.p.h. was attained twice, and an average speed of 100 m.p.h. maintained for 43 miles on end. Three days later the engine went into regular working, and for the first fortnight of the London–Newcastle 4-hour service this one engine worked the entire duty, 536 1/2 miles daily. The service actually demanded faster running than the trial trip of March 1935, because an intermediate stop was included at Darlington. This required the 232.3 miles between that station and King's Cross to be covered in 198 min. in each direction—a start to stop average speed of 70.5 m.p.h.

The success of the "A4" locomotives on the Silver Jubilee train and the vivid impression created by their appearance, paved the way for some further notable accelerations of service; to the application of the streamlined front to other L.N.E.R. locomotive (Figs. 208, 209) and to some still finer engine performance. In the meantime the locomotive world watched with intense interest to see how the L.M.S.R. would reply to this new-found high speed prowess on the East Coast Route. At first the indications were that the haulage of heavy trains would continue to be the principal tasks set to the Stanier "Pacific" engines of the "Princess Royal" class. On June 27, 1935 the dynamometer car was attached to the fast evening express from Liverpool to Euston, scheduled to average 64.5 m.p.h. start to stop from Crewe to Willesden Junction, and with a 15-coach train and a gross trailing load of 475 tons the schedule was cut by 12 1/2 min. to give an overall time, over 152.6 miles, of 129 1/2 min. and an average speed of 70.7 m.p.h. The engine was No. 6200 *The Princess Royal*, fitted with modified boiler and 32-element superheater. With such potentialities in weight haulage at high speed, some notable acceleration of the heavy day Anglo-Scottish expresses was made in the summer of 1936, and the schedule of the 2 p.m. from Euston to Glasgow, with one Pacific engine working throughout the 401.4 mile journey, represented about the hardest task that had then been set to a British locomotive. The first day's running however, served to show the margin Mr. Stanier's engines had in

Fig. 208.—*L.N.E.R. The pioneer* 2-8-2 *express locomotive No.* 2001 Cock o' the North *rebuilt with piston valves, Walschaerts valve gear, and streamlined front-end.*

Fig. 209.—*L.N.E.R. The characteristic Gresley streamlining applied to a "Sandringham" class* 4-6-0 *for the East Anglian service: No.* 2870 City of London.

Fig 210.—*L.N.E.R. Holder of the world's speed record with steam: engine No.* 4468 Mallard, *fitted with Kylchap blast arrangement.*

Fig. 211.—*L.M.S.R. Stanier* 4-6-2 *engine No.* 6201 Princess Elizabeth *as running at the time of the record London-Glasgow test of* 1936.

reserve, as will be seen from the accompanying summary of the journey.

L.M.S.R. 2 p.m. Euston to Glasgow

Section	Dist. miles	Sch. min.	Actual m. s.	Net time min.	Load, gross trailing tons
Euston–Crewe	158	163	160 55	153 $^1/_2$	465
Crewe–Lancaster	71.9	79	77 19	76	520
Lancaster–Penrith	51.2	59	59 26	59 $^1/_2$	520
Penrith–Carlisle	17.9	19	18 26	18 $^1/_2$	520
Carlisle–Glasgow	102.3	116	114 29	106 $^1/_2$	295

The aggregate net time on this journey showed an economy of 22 min. on schedule; but the most difficult part of the journey was that from Lancaster to Penrith, over Shap Summit, where an allowance of only 37 min. was tabled for the 31.4 miles from Carnforth to Shap Summit in which distance there is a vertical rise of 885 ft. The average gradient over this distance is thus 1 in 187, and the average speed demanded is 51 m.p.h. When this schedule was first introduced the minimum trailing load was one of 470 tons, but it was usual for extra coaches to be added at weekends and loads of 520 to 530 tons were common. So that the work entailed can be better appreciated details of an excellent run made by engine No. 6208 *Princess Helena Victoria* are shown in graphical form, alongside in relation to the gradients of the line. It will be seen in particular that on the long ascent of the Grayrigg bank it was only in the last two miles, where the gradient is 1 in 106, that speed fell below 45 m.p.h. On this journey there was a gain of 2 $^3/_4$ min. on schedule from Lancaster to Penrith.

Another very fine performance was put up by engine No. 6206 *Princess Marie Louise* on the corresponding southbound train, with a gross trailing load of 570 tons. From a standing start at Carlisle the 31.4 miles to Shap Summit were climbed in 42 $^1/_2$ min., and a speed of 41 m.p.h. was sustained unvaryingly on the 1 in 125 gradient from Clifton to Shap station. The equivalent d.h.p. involved was 1800. As usual the engine worked through from Glasgow to Euston, with this very heavy load unchanged south of Carlisle, and the final 158 miles from Crewe to Euston were covered in 163 $^1/_4$ min. But more significant than any feats of individual performance was the general behaviour of these large " Pacific " engines in sustaining their quality of performance, over the entire span of their service from one general overhaul at main works to the next. The following results were obtained on a dynamometer car test with engine No. 6210, when the mileage since last major overhaul was 98,977.

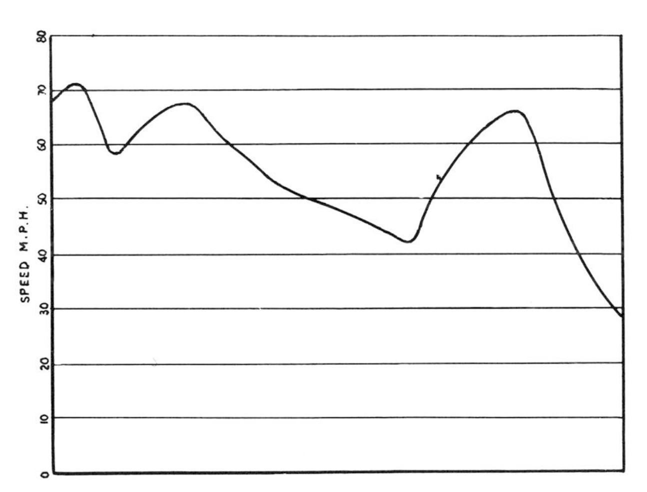

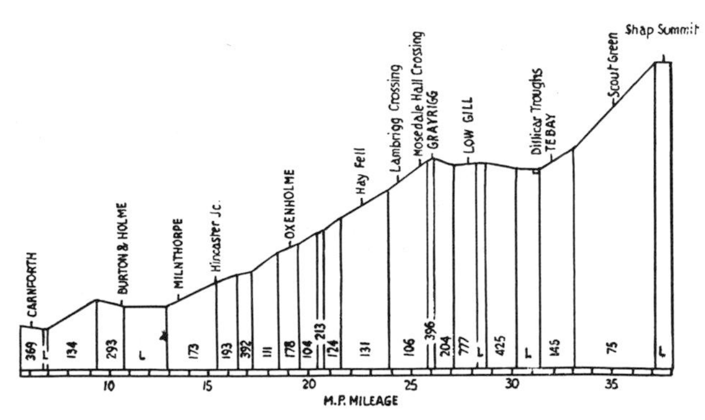

Euston to Glasgow Test Run

Mileage on test	402
Load, tons tare	522
Average speed, m.p.h.	52.0
Coal consumption	
lb. per train mile	45.0
lb. per ton mile (including engine)	0.068
lb. per d.h.p. hour	2.98
lb./sq. ft. of grate area per hr.	62.2
Water consumption	
Gallons per mile	37.2
lb. per d.h.p. hour	24.7
Evaporation	
lb. of water per lb. of coal	8.30
lb. per hour (average)	19,400

The above figures are particularly interesting as showing that the failing of former locomotives on this route, in deteriorating to high coal consumption with increasing mileage, had been virtually eliminated. The coal consumption of 2.98 lb. per d.h.p. hour would have been excellent on any reckoning, but from a locomotive that had run nearly 100,000 miles since last general overhaul it was indeed remarkable.

As the year 1936 drew towards its close both East Coast and West Coast routes were engaged in active preparation for further accelerations between London and the Scottish cities. The L.N.E.R. had fully consolidated the working of the 4-hour service between Newcastle and King's Cross, and with the same standard of locomotive performance north of Newcastle a 6-hour service between King's Cross and Edinburgh was an easy possibility. A dynamometer car test run on a Saturday in September 1936 with the Silver Jubilee train and one of the streamlined "A4 Pacifics" amply confirmed the feasibility of regular running over the 124 1/2 miles from Newcastle to Edinburgh in 115 min., if need be. On the West Coast route equally careful consideration was given to the introduction of a 6-hour service between Euston and Glasgow. This was likely to involve much more arduous working. The distance was nine miles greater than that from King's Cross to Edinburgh, and on the West Coast route the altitudes of Shap and Beattock summits had to be crossed.

In November 1936 some trial runs were made with the dynamometer car and engine No. 6201 (Fig. 211), and on two successive days the journey between London and Glasgow was made non-stop with the following remarkable results:

	Down Nov. 16	Up Nov. 17
Load (tons tare) behind tender	225	255
Actual running time (min.)	353.7	344.1
Average speed, m.p.h.	68.2	70.15
Coal		
lb. per mile	46.8	44.8
lb. per d.h.p. hr.	3.68	3.48
lb./sq. ft. of grate area per hr.	70.8	69.9
lb. per hour	3,190	3,130
Water galls. per mile	34.5	30.2
Evaporation		
lb. of water per lb. of coal	7.36	6.70
lb. per hr.	23,500	21,100

L.M.S.R. Euston–Glasgow Trial Runs Engine 6201

	Down Run		Up Run	
Incline	Tebay to Shap Summit	Beattock Bank	Lamington to Beattock Summit	Clifton to Shap Summit
Length (miles)	5.5	10.0	13.5	11.3
Av. d.h.p. (actual)	1187	1241	1117	1180
Max. d.h.p.	1251	1350	1260	1260
Max. calculated 1 h.p.	2413	2428	2448	2343
Av. speed, m.p.h.	64.5	62.5	74.8	60.2
Speed at summit, m.p.h.	57.0	56.0	70.0	70.0
Boiler pressure, lb./sq. in.	245	240	240–245	220–240
Cut-off range, %	25–32	30–37.5	20–28	30–35

Details of the performance on the Shap and Beattock banks are given a separate table, and indicate clearly the very strenuous nature of the work done. In view

Fig. 212.—*One of the "Duchess" class* 4-6-2s, *in British Railways livery No.* 46245 City of London.

of the length of run made non-stop it should be added that the engine was manned by a crew of three, driver, fireman, and locomotive inspector who also took a turn with the firing. The average speeds of 68.2 m.p.h. northbound and 70.15 m.p.h. south-bound on the following day represented a remarkable performance, and overall times 6 1/4 and 15 3/4 min. inside the 6-hour schedule. The runs were nevertheless the product of a locomotive in superb condition, expertly handled, and were hardly to be considered possible of repeating in ordinary service as a day-to-day proposition—let alone non-stop from end to end. But the runs had nevertheless demonstrated the quality of performance that a high speed service would demand, and as on the L.N.E.R. in 1935 following the London–Newcastle trials with the "A3" engine *Papyrus*, so on the L.M.S.R. the design of a special high speed Pacific was put in hand.

Fig. 213.—*L.M.S.R. Stanier streamlined* 4-6-2 *No.* 6222 Queen Mary *for the "Coronation Scot" service* (1937).

The resulting locomotive, the *Coronation* of 1937 (Fig. 213), was one of the best express passenger types to run in this country. It included many features of design and metallurgy that were outstanding; but before coming to points of detail the general proportions of the new locomotives as compared with those of the "Princess Royal" class must be set down. In addition it is important to mention that whereas the original Stanier Pacifics had four sets of valve gear the later class had Walschaerts gear outside, and the valves of the inside cylinders actuated by rocking shafts from the outside gear. There were originally two varieties of the "Duchess" class: the "Coronation" series ,streamlined (Fig. 219), and the "Duchess" series of 1938, non-streamlined (Fig. 220). The weight quoted in the accompanying table is for the non-streamlined variety to make a true comparison with the "Princess Royal" class.

Comparative Dimensions: L.M.S.R. Pacifics

Class ..	*"Princess Royal"*	*"Duchess"*
Cylinders:		
Dia., in.	16 1/4	16 1/2
Stroke, in.	28	28
Coupled wheel dia., ft. in.	6-6	6-9
Dia. of piston valves, in.	8	9
Max. travel of valves, in.	7 1/4	7 1/32
Steam lap, in.	1 3/4	1 3/4
Exhaust clearance, in.	Nil	1/16
Lead, in.	1/4	1/4
Cut-off in full gear, %	73 1/2	75
Boiler:		
Tubes, small	123	129
Outside dia., in.	2 3/8	2/3/8
Flues	32	40
Outside dia., in.	5 1/8	5 1/8
Distance between tube plates, ft. in.	19-3	19-3
Heating surface, sq. ft.		
Small tubes	1272	1545
Flues	825	1032
Firebox	217	230
Superheater elements	653	830
Total	2967	3637
Grate area, sq. ft.	45	50
Boiler pressure, lb./sq. in.	250	250
Nom. tractive effort at 85% boiler pressure, lb.	40,300	40,000
Wt. of engine in working order, tons	104 1/2	105 1/4

The use of 9 in. instead of 8 in. diameter piston valves, and careful design of ports and passages, and every detail of the steam circuit facilitated the free flow of steam, and contributed to the exceptional speed-worthiness of these engines, but perhaps the feature that immediately commands attention in the table is the very large increase in heating surface, while securing this with a very small increase in overall weight. This was achieved largely by the use of nickel steel having the following chemical and physical properties:

Analysis	*Per Cent*
Carbon	0.2 to 0.25
Silicon	0.1 to 0.15
Manganese	0.5 to 0.7
Sulphur	0.04 max.
Phosphorous	0.04 max.
Nickel	1.75 to 2.0
Physical properties	
Maximum stress	34-38 tons/sq. in.
Yield point	17–19 tons/sq. in.
Elongation	22–24 % on 8 in. gauge length
Reduction in area	50% min.

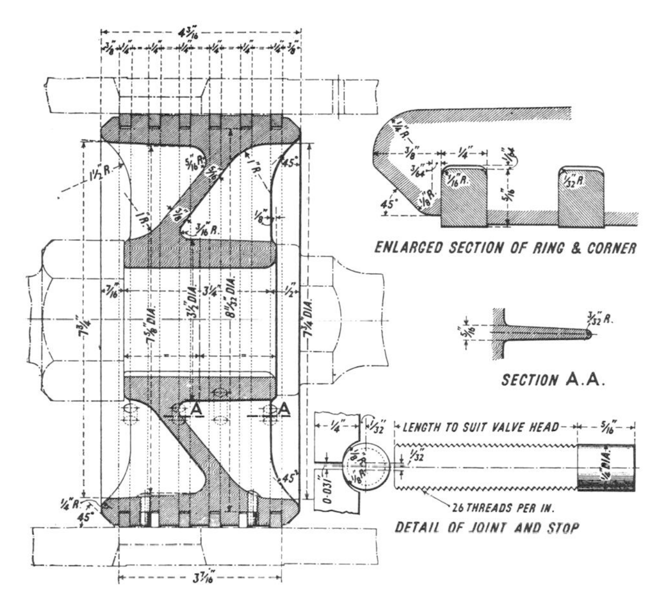

Fig. 214.—*L.M.S.R.* " *Coronation* " *class* 4-6-2, *detail of piston valve.*

The increased tensile strength and superior all round qualities of this material over 26–30 ton carbon steel made possible the following reductions in plate thickness:

	Nickel steel boiler as built in.	Boiler if made of mild steel in.
First barrel	5/8	13/16
Inside cover strap ..	9/16	3/4
Outside cover strap ..	9/16	3/4
Liner for top feed seating	9/16	11/16
Second barrel	11/16	7/8
Inside cover strap ..	9/16	3/4
Outside cover strap ..	9/16	3/4
Liner for dome seating ..	13/16	1
Throat plate, top half ..	3/4	7/8
Throat plate, bottom half	3/4	7/8
Firebox backplate ..	9/16	5/8
Steel wrapper	1/2	5/8

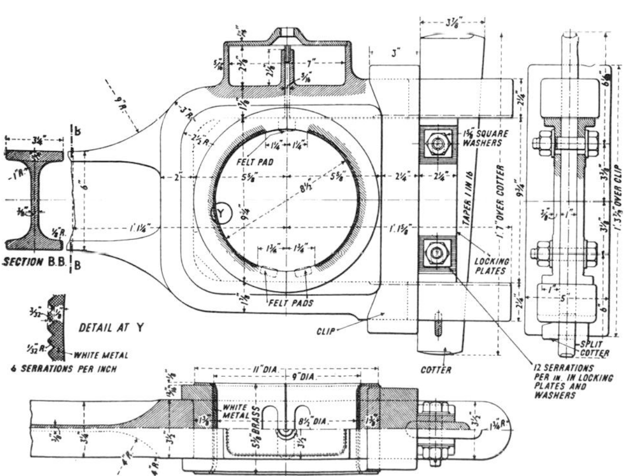

Fig. 215.—*L.M.S.R.* " *Coronation* " *class* 4-6-2, *detail of inside connecting rod big end.*

By this use of 2 per cent. nickel steel over two tons was saved in the construction of the boiler. Detail drawings are also reproduced showing the construction of the piston valve head (Fig. 214) connecting rods (Fig. 215) built up crank axle (Fig. 218), and the driving axle boxes (Fig. 216). The connecting rods were of Vibrac steel with ultimate tensile strength of 50–60 tons per sq. in. In view of the trouble-free records of these locomotives, from bearing and big end failures the design of the axle boxes will be studied with interest. The material used for bearings in both driving boxes, small ends, and big ends had the following chemical analysis:

	Per cent.
Tin	85
Copper	5
Antimony	10

A point of difficulty sometimes experienced with the " Princess Royal " class engines, on long mileage duties was the getting of coal forward from the back of the tenders. The " Coronation " class engines

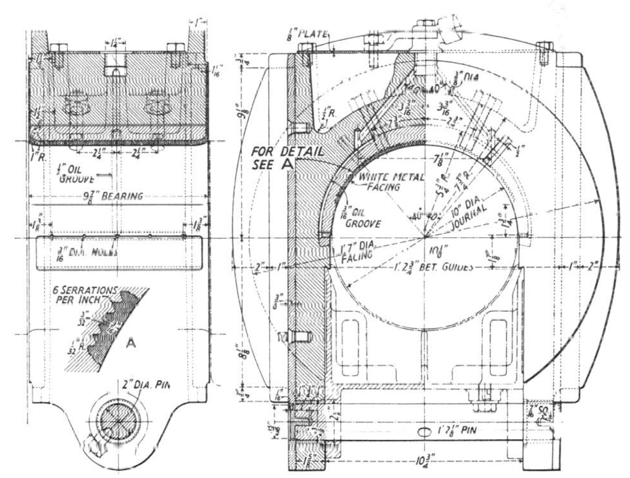

Fig. 216. *L.M.S.R.* " *Coronation* " *class* 4-6-2, *detail of driving axle box.*

were accordingly fitted with tenders having steam operated coal pushers, as shown in the accompanying drawing (Fig. 217) and these proved most effective in service.

When the " Coronation Scot " service was introduced in 1937 there was some disappointment that the schedule was fixed at 6 1/2 hours from Euston to Glasgow. Having regard to the striking performances made by the engine *Princess Elizabeth* in November 1936, and the advances in design incorporated in the new engines it was felt that 6 1/4 hours at the most should have sufficed, inclusive of a stop at Carlisle. But high speed running of this kind was new to L.M.S.R. traditions and no doubt the policy was to advance gradually towards a faster service. Nevertheless, the first engine of the class, the streamlined *Coronation* had not been long out of Crewe works

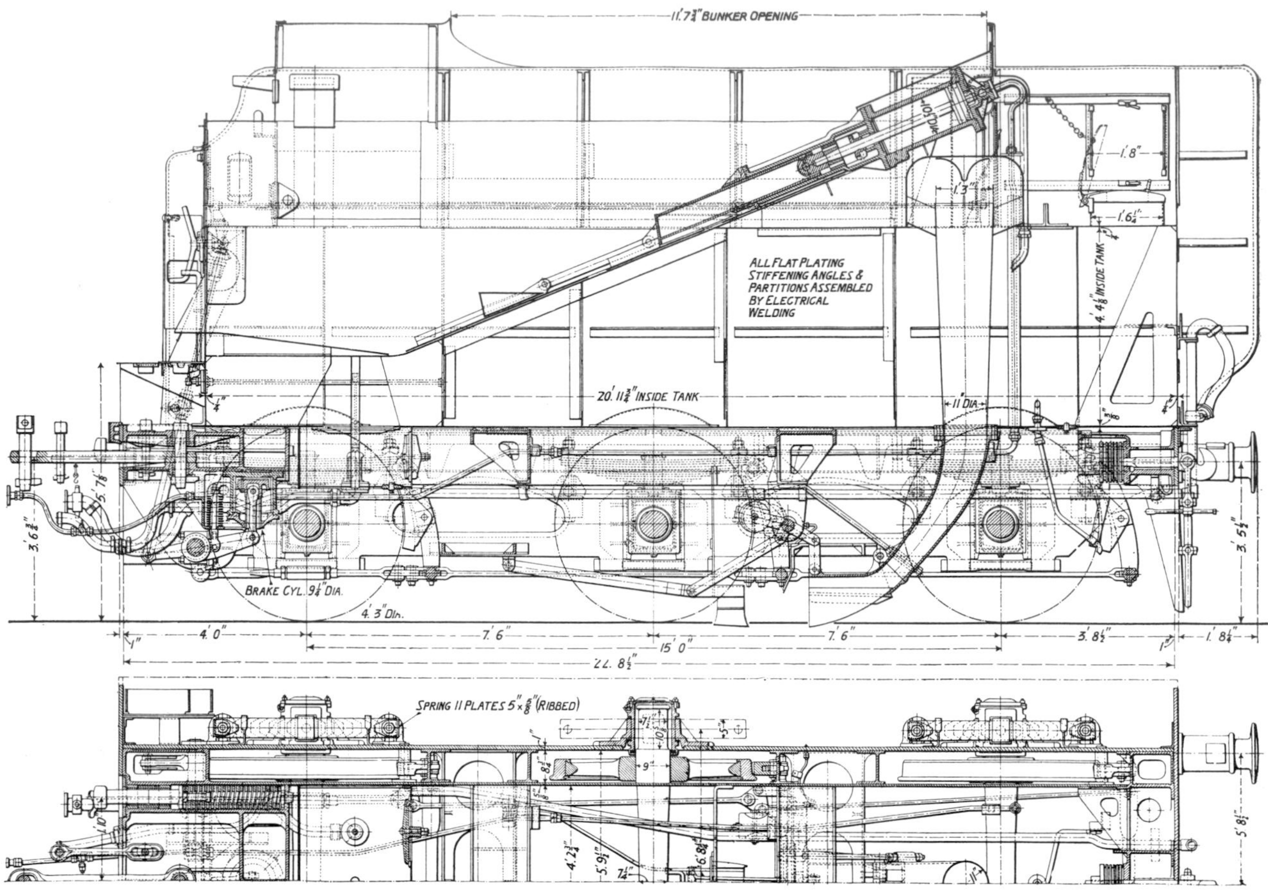

Fig. 217.—*L.M.S.R. " Coronation " class* 4-6-2. *Drawing of tender, showing layout of coal pusher.*

before some notable records had been made in the course of a return trip from Euston to Crewe and back with a party of invited guests. On the outward journey a new British speed record of 114 m.p.h. was set up in the descent of the Madeley bank, while on the return journey, one of the longest exhibitions of sustained high speed yet seen in this country resulted in a start-to-stop average speed of almost 80 m.p.h. over the 158 miles from Crewe to Euston.

Engine No. 6220 *Coronation*
Load: 263 tons tare, 270 tons full

Dist. Miles		Actual m.	s.	Av. speed m.p.h.
0.0	Crewe	0	00	—
10.4	Whitmore	10	41	58.4
19.1	Norton Bridge	16	48	85.5
24.4	Stafford	21	05	74.2
33.7	Rugeley	29	13	68.6
48.0	Tamworth	38	50	89.4
60.9	Nuneaton	48	23	82.5
75.5	Rugby	59	20	79.7
88.3	Weedon	69	32	75.3
111.3	Bletchley	85	00	89.5
126.3	Tring	95	12	88.3
140.5	Watford Junc.	104	39	90.0
152.6	Willesden Junc.	112	39	90.8
158.0	Euston	118	57	

In view of the present day electric running over this route an abbreviated log of the record run of June 29, 1937 is of interest. The maximum speed was 100 m.p.h.

In normal service with the " Coronation Scot " express, a test with the dynamometer car gave the following results in running from London to Glasgow and back 401 $^1/_2$ miles each way.

" Coronation Scot " test runs

Weight of train, tons tare	331
Average running speed, m.p.h.	60
Coal consumption	
lb. per train mile	39.2
lb. per d.h.p. hour	3.03
lb./sq. ft. grate area per hr.	47.3
Water consumption	
Gall. per mile	32.3
lb. per d.h.p. hour	25.0
Evaporation	
lb. of water per lb. of coal	8.24
Average d.h.p.	825

Prior to the decelerations of service compelled by the outbreak of war in 1939 the most severe test put to one of these locomotives was on February 26, 1939, when a special test train of 20 coaches, 604 tons tare, was hauled from Crewe to Glasgow and back without any assistance on the Shap and Beattock inclines in either direction. The engine concerned was one of the

L.M.S.R. Tests Runs, February 1939 *: Engine* 6234

	Crewe to Glasgow						Glasgow to Crewe					
	Carnforth—Shap			Gretna—Beattock Summit			Motherwell—Beattock Summit			Carlisle—Shap		
	Carn-forth Oxen-holme	Oxen-holme Tebay	Tebay Sum-mit	Gretna Locker-bie	Locker-bie Beat-tock	Beat-tock Sum-mit	Mother-well Law Junc.	Law Junc. Car-stairs	Syming-ton Sum-mit	Car-lisle Plump-ton	Plump-ton Penrith	Pen-rith Sum-mit
Length of ascent miles	12.96	13.08	5.69	17.27	13.96	10.13	5.42	10.53	17.28	13.03	4.77	13.68
Average drawbar horsepower	1870	1668	1830	1598	1609	1724	1923	—	1860	1822	2000	1560
Maximum drawbar horsepower	2120	1934	2065	1733	1823	2081	1998	1978	2282	2511	2394	2331
Maximum I.H.P. (calculated)	3209	2806	2963	2236	2556	2761	2583	2567	3333	3248	3241	3021
Average rate of ascent (m.p.h.)	68.0	53.0	47.9	59.3	72.5	36.8	46.7	49.4	63.4	43.9	71.4	44.4
Cut off range per cent.	20 25	25	25 35	20 25	20 25	30 40	20 30	30 35	30 35	30 35	20 30	30 40
Boiler pressure p.s.i.	250	245	240	250	245	245	250	250	245	245	230	245

non-streamlined variety, No. 6234 *Duchess of Abercorn*, equipped with a double blastpipe and chimney. With this very heavy train the net running average over the complete round trip of 487.2 miles was 55.2 m.p.h. throughout. The coal consumption was 68.7 lb. per train mile, and 3.12 lb. per d.h.p. hour. The very fine performance of the locomotive on the principal gradients is shown in the accompanying table.

Rivalry between the East Coast and West Coast routes to Scotland in 1937 reached an intensity not seen since the Race to the North in 1895, and while the L.M.S.R. had produced a very fine locomotive in the " Duchess " class Pacific the speed honours undoubtedly lay with the L.N.E.R. In the summer of 1937 the long expected 6-hour service from King's Cross to Edinburgh became an actuality; but the particular point of interest was that although the trailing load of the " Silver Jubilee " express of 1935 was only 230 tons, the " Coronation " of 1937 carried a tare trailing load of 312 tons. On similar timings this naturally involved very much harder work than with the earlier streamlined train. With the " Silver Jubilee " and its load of only 230 tons the uphill work was usually so fast as to make any exceptional speed downhill unnecessary, but with the " Coronation " speeds of 100 m.p.h. and more became quite frequent, if not necessarily common-place (Fig. 206). An example of the superb running capacity of the Gresley " A4 Pacifics " occurred in the author's own travelling experience in the summer of 1937, on the descent from Stoke summit—5.4 miles south of Grantham—towards Peterborough, and this is of particular interest in view of the deliberate, and brilliantly successful attempt on the world record speed for steam made roughly a year later over this same stretch of line.

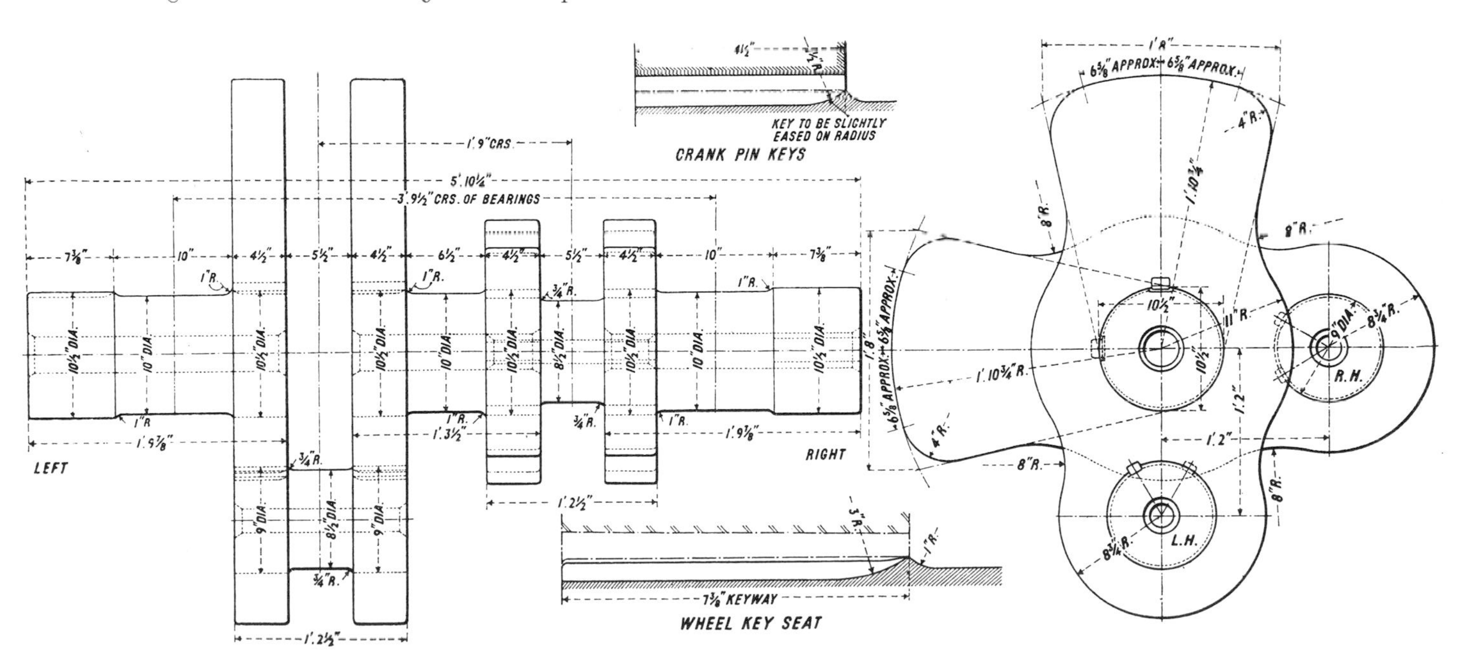

Fig. 218.—*L.M.S.R. " Coronation " class* 4-6-2, *detail of built-up crank axle.*

Fig. 219.—*L.M.S.R. One of the later " Princess Coronation " class* 4-6-2s, *painted standard " red " with gold stripes.*

Fig. 220.—*L.M.S.R. The first of the non-streamlined " Princess Coronation " or " Duchess " class, No.* 6230 Duchess of Buccleugh; *built Crewe* 1938.

Fig. 221 (right).—*L.M.S.R. Access doors in streamlining, to reach smokebox.*

Fig. 222.—*L.M.S.R. A striking comparison: a streamlined " Coronation "* 4-6-2 *alongside a full-sized replica of the* Rocket.

Fig. 223.—*L.N.E.R.* Mallard *with dynamometer car and test train just prior to the run on which the record speed of* 126 *m.p.h. was attained.*

The logs of the two runs are shown alongside in the accompanying table, and in studying the figures it must be borne in mind that engine No. 4491 was working a service train through from Edinburgh to King's Cross, and from the one intermediate stop, at Newcastle, had covered 162.8 miles in 151 $^3/_4$ min. when the tabulated part of the run commenced, on passing Grantham. Engine No. 4468 (Fig. 210) was specially prepared to make an attempt on the speed record, and had worked down from King's Cross to Grantham on a brake test special earlier in the same day (Fig. 223).

L.N.E.R. Grantham–Peterborough

	4491		4468	
Engine No.	4491		4468	
Engine Name	*Commonwealth of Australia*		*Mallard*	
Load tons E/F	312/325		236/240	
Dist. Miles	Actual m. s.	Speeds m.p.h.	Actual m. s.	Speeds m.p.h.
0.0 Grantham	0 00*	69	0 00*	24
5.4 *Stoke Box*	4 55	64 $^1/_2$	5 57	74 $^1/_2$
8.4 Corby	7 21	85	7 58	104
13.3 Little Bytham	10 32	98	10 33 $^1/_2$	122 $^1/_4$
16.9 Essendine	12 37	106/104	12 21	125†
20.7 Tallington	14 47	106	15 15	70
23.6 *Helpston Box*	16 27	104/106	17 37	75
26.0 *Werrington Jc.*	17 54	—	eased	
29.1 Peterborough	20 52			
Average speeds	Little Bytham to Helpston 10.3 miles 5m. 55s. = 104.5 m.p.h.		Mileposts 94 to 89 2m. 29 $^1/_2$s. = 120.4 m.p.h.	

* Speeds on passing through Grantham station
† Max. speed near Milepost 90

The difference between the two circumstances lay in the fact that on the service train the locomotive was being worked in 15 per cent. cut-off, with the regulator full open from the point of passing Grantham, and made her own speed on the varying gradients that ensued, falling away to a minimum of 64 $^1/_2$ m.p.h. on the 5 $^1/_2$ miles of 1 in 200 gradient to Stoke box, and then running very swiftly down to Peterborough maintaining a very high rate to the point where it was necessary to slow down for the water troughs at Werrington Junction. From a point about half way between the stations of Essendine and Tallington the gradients are indeed little easier than level, and here the speed continued unbrokenly at 104 to 106 m.p.h. In 1938 the celebrated engine No. 4468 *Mallard* was pressed to a very rapid acceleration from Grantham, and was worked continuously at a very high steam rate—full regulator and 40 per cent. cut-off—until the maximum speed record was attained near the 90th mile-post. The dynamometer car record showed a peak of 126 m.p.h. and this provided a record that has not been surpassed with steam traction.

Although the record of *Mallard* was a great triumph for Sir Nigel Gresley, and the London and North Eastern Railway, and for British locomotive engineering in general, so far as practical railway operating was concerned the very free running of engine No. 4491 in normal steaming conditions has probably a greater significance. Furthermore, at the time this particular run was made the engine had the original single-orifice blastpipe, whereas *Mallard* had the Kylchap double blastpipe and chimney. There is every reason to suppose that even faster running could have been made, in 15 per cent. cut-off with a locomotive having a double blastpipe than that actually achieved by No. 4491.

Although this chapter has been concerned almost entirely with the prowess of the northern lines in high speed running an individual performance by a " Castle " class 4-6-0 of the G.W.R. in June 1937 is important as indicating that the record run of the Cheltenham Flyer on June 6, 1932, and described in Chapter 6, did not by any means represent the ultimate in the high speed capacity of these engines. The run of June 30, 1937 was the product of a keen driver and a locomotive in first-class condition, and

Fig. 224.—*L.N.E.R. An "* A4 *" Pacific as modified during World War II, valances removed, and painted plain black.*

its leading features compare with the "World Record" of 1932 as shown on accompanying table:

Engine No. 5039 was being eased even before Southall was passed, because the train was by then well ahead of time. There would have been no difficulty in repeating the finishing time into Paddington and the record of 1932 could thus have been equalled, or marginally beaten with a load of 235 tons, against 195—a significant increase of 20 per cent.

The engine in question, No. 5039 *Rhuddlan Castle* belonged to a batch of engines that had the reputation of being the best and freest running of the entire "Castle" stud; and this reputation is certainly borne out by the relative ease with which this exceedingly fine run was made.

Date	6/6/32	30/6/37
Engine No.	5006	5039
Load gross trailing tons	195	235
Dist.	Actual	Actual
Miles	m. s.	m. s.
0.0 Swindon	0 00	0 00
10.8 Uffington	9 51	10 10
41.3 Reading	30 11	30 27
68.2 Southall	48 51	48 46
76.0 Westbourne Park	54 40	57 40
77.3 Paddington	56 47	61 07
Average speed, m.p.h. Uffington–Southall	88.3	89.3
Max. speed, m.p.h.	$92^1/_2$	95

Fig. 225.—*A Stanier "Duchess" class* 4-6-2 *on test, showing wind gauges mounted ahead of the smokebox.*

Chapter Twelve Testing of Locomotives
The development of new techniques

The period from 1930 onwards had marked a further advance in steam locomotive design techniques and although not yielding quite so spectacular an increase in thermal efficiency as that resulting from the introduction of superheating, or from the general adoption of long-lap, long-travel valves in the 1920's it was nevertheless a time of notable achievement. The advances from the simple non-superheated locomotives of the early 1900s could be broadly stated thus:

Period	Change	Result
1907–1914	Introduction of superheating	Reduction in coal consumption: Increased power per unit of engine weight
1921–1927	General introduction of long-lap long-travel valves	Reduction of 10 to 20 per cent. in coal consumption
1933–1938	Improved front-end designs: better draughting, internal streamlining	Greater power per unit of engine weight: higher speeds

The work of the 1930s, particularly on the L.M.S.R. and L.N.E.R. included the nearest approach to what could be strictly termed "research" in locomotive design, and Sir Nigel Gresley in particular frequently expressed his disatisfaction with the facilities for truly scientific testing of locomotives in Great Britain. It was not without significance that he arranged for the *Cock o' the North* locomotive (Fig. 226) to be tested on the Vitry plant of the French National Railways, because no suitable plant was available at home. It is true that three out of the four main line companies owned well-equipped dynamometer cars; but the results of road tests at variable speed, whether carried out on special or service trains frequently displayed variations in overall performance, as measured by the coal consumption per d.h.p. hour, that did not admit of any explanation from the running conditions existing at the time the tests were made. Sir Nigel Gresley was only one of many British engineers who felt that the basic features of locomotive performance could be ascertained with exactitude only in the closely controlled conditions of a stationary testing plant, as already existed in France.

In the meantime some important work that was little known at the time was being done on the Great Western Railway, by the testing section of the Swindon drawing office. The early 1930s were a time of much scheme-making towards new and faster schedules, while the world-wide economic depression made it necessary to reduce expenditure in every way possible. Any proposed acceleration of service had thus to be considered against the probable increase in track maintenance charges, and while the average speeds involved in existing schedules had been accredited the civil engineer was interested to know

Fig. 226.—*L.N.E.R. The* Cock o' the North *as originally built when on trial on the G.N. line, at King's Cross.*

Fig. 227.—*G.W.R. The Swindon stationary testing plant with engine No.* 1007 County of Brecknock.

the actual speeds run regularly in ordinary service. It had been accepted for many years by the locomotive department that variations from strict point-to-point time-keeping on certain sections of line were the rule rather than the exception, and that while the schedule times were traditional in many cases the actual running times made in daily service approximated to a constant rate of evaporation. This was the logical result of experienced drivers running their trains in a manner likely to result in minimum coal consumption. End to end times were observed, but there was divergence from scheduled times intermediately.

This analysis led to a revision of certain time-honoured point-to-point bookings, but it led to a much more critical and scientific analysis of locomotive performance in general. To those concerned it seemed that the performance and efficiency of a steam locomotive could be summarised in three basic relations:

(a) the coal rate, in relation to the steam rate.
(b) the steam rate in relation to power.
(c) the power in relation to the coal rate.

These relations were capable of establishment on a stationary testing plant, at constant speed; but in this respect the plant in existence at Swindon was totally inadequate for the testing of modern locomotives because it was not capable of absorbing a greater output than 500 h.p. This limitation was well known among British locomotive engineers, and at one time it was thought that the Great Western Railway would join with the L.M.S.R. and the L.N.E.R. in the construction of a new stationary testing plant. Unbeknown to the northern companies however the Great Western undertook the modernisation of the Churchward plant, in order to develop the testing practices that were then felt desirable at Swindon, and when representatives of the L.M.S.R. and the L.N.E.R. visited the works in connection with the proposals for a national testing plant the Great Western authorities were able to demonstrate the workings of a modern express locomotive at maximum output (Fig. 227). The alterations had been relatively cheap and simple to carry out; but although effective in its new form the northern companies felt that the facilities available were inadequate for the large-scale testing programmes they had in mind, and the decision was taken jointly by the two companies to build a new testing station at Rugby, that would be readily accessible from the main lines of both the L.M.S.R. and the L.N.E.R.

The building of the fine modern locomotive laboratory at Rugby, and the development of testing practice on the Great Western Railway were much delayed by the onset of the Second World War, in 1939; but rather than take events in true chronological order it will be convenient to review the development of testing practice as a whole, at this stage, although the actual results in the form of published data did not begin to appear until after the nationalisation of the railways in 1948. Then the practice developed on the Great Western Railway was adopted as a national standard, and the plants at both Swindon and Rugby

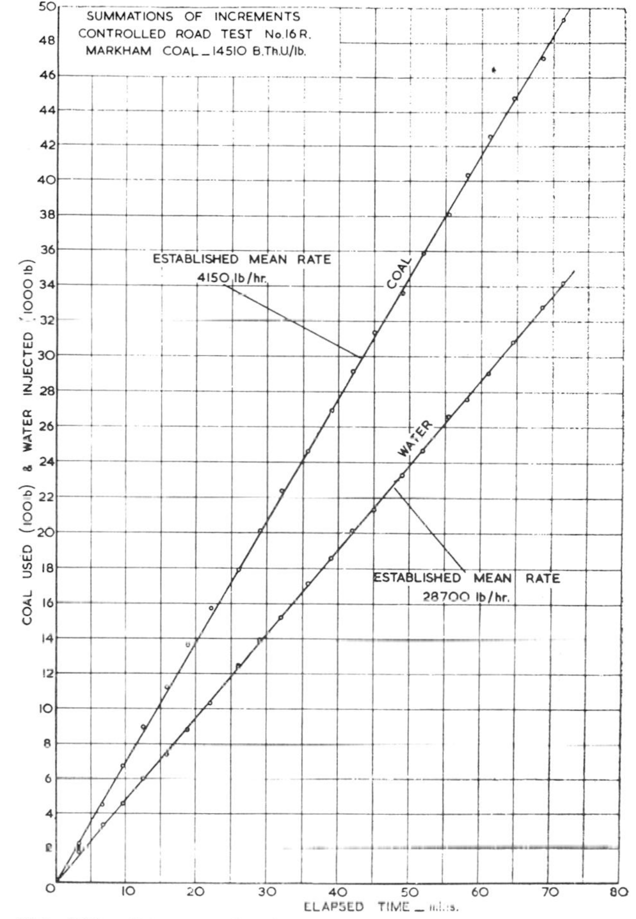

Fig. 228. *Diagram showing summation of increments in coal fired during controlled road test.*

worked in cooperation in the testing of many different types of locomotive, the results of which will be discussed in a later chapter.

Turning now to the three basic relations concerned in locomotive performance attention was first given to means for ensuring a perfectly constant rate of evaporation in the boiler, on a constant speed test on the stationary plant, and of metering the fuel supply accordingly. This required, first, the building of an optimum firebed, appropriate to the rate of evaporation proposed for the test, and varying according to the type of fuel. Then as the test proceeded the coal would be supplied to the fireman in increments as he needed it, each increment being recorded and plotted on a graph (see Fig. 228). The increment actually used took the forms of bags of 1 cwt., and unless the plotting of the increments revealed a straight line relation between coal consumption and time, for a constant rate of evaporation the test was discarded as unreliable. At Swindon a tolerance of no more than $^1/_8$ in. was allowed in the boiler water level. If this were maintained it was considered that conditions of constant evaporation were being obtained. A series of tests at varying rates of evaporation would then establish, with precision never previously attained in British locomotive practice,

Fig. 229.—*Southern Railway: The* Lord Nelson, *fitted with indicator shelters, for Waterloo–Exeter trial runs.*

Fig. 230.—*L.M.S.R. The Stanier* 4-6-2 *No.* 6201 Princess Elizabeth *as originally built, fitted with indicator shelters, for "full-dress" trials.*

the principal boiler relations, and ultimately the limit of performance of the locomotive in question.

The analysis of smokebox gases and determining of the relations between the coal fired and the coal fully burnt partook of a reliability never previously known, and this has enabled some important work in the improvement of front end draughting to be carried out. In the past many locomotive types have gone down in history as indifferent in steaming. The rough and ready methods of improvements, such as "darts" in the blastpipe and so on, have been adopted, with damaging effects upon the economy; but a closer analysis had not been possible because of an absence of precise and consistent data to work upon.

The resistance characteristics of conventional boilers of good general proportions followed a pattern which was well defined under the analytical testing conditions already mentioned. By the use of certain blastpipe and chimney proportions in single orifice designs, it was found that over a wide range of boilers, draught and discharge limits could be made to coincide approximately, and this enabled a limiting relation to be found between boiler size and orifice dimensions that was valuable as a first approximation in design. Expressing boiler size by evaporative heating surface, the unit rate of discharge at the discharge limit was investigated for a wide range of boilers in which the temperature of admission steam which affects steam exhaust temperatures rose somewhat with boiler size as was usual.

Equating discharge and draught limits and bringing in the applicable ejector proportions which could be expressed in terms of the orifice dimensions resulted in the relationships given in the accompanying diagram (Fig. 231). The orifice diameter should be regarded as a minimum. On checking by experiment, one inch of smokebox vacuum should be obtained for each inch of back pressure measured just below the orifice and

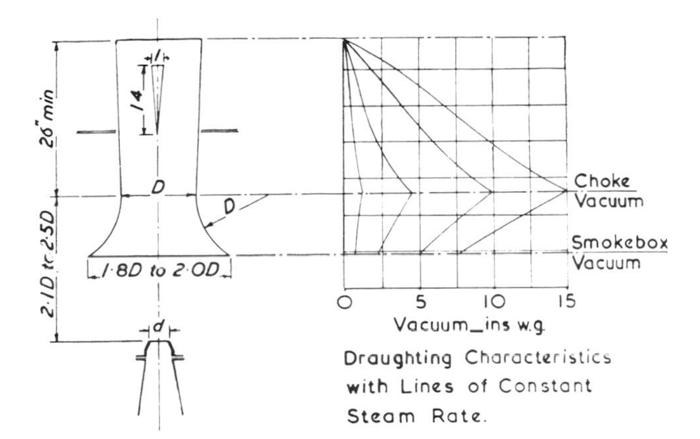

Limiting Diameter of Blast Orifice.

$$d = 1{\cdot}128\sqrt{9{\cdot}5 + {\cdot}0059S} \text{ ins.}$$

Where S = Evap. Heating Surface in Sq. Ft.

Diameter of Choke.

$$D = 2{\cdot}85d \text{ to } 2{\cdot}95d \text{ ins.}$$

Fig. 231.—*Principles of draughting.*

the vacuum at the surface of the chimney choke should not be less than twice the smokebox vacuum at all rates of discharge. If the engine steamed too freely, this indicated the discharge limit was below the draught limit. It was then necessary to make these two values coincident by increasing the orifice diameter; on the other hand, if steaming was unsatisfactory it was unlikely that improvement would result from decreasing the orifice diameter and it was necessary to examine the resistance of the boiler. In narrow firebox boilers, draught and discharge limits may approach and sometimes equal the grate limit. A good front end was a vital necessity; many an otherwise good boiler has been condemned by its inferior front end. In many cases, improvements could be carried out at little cost, even for a single orifice design, and could place the boiler limit of continuous evaporation over 30 per cent. higher than its normal working rate. In so doing the combustion efficiency and the ability to burn inferior coal could be improved.

One of the great difficulties associated with engine testing in the past had been the reconciling of the designed features of performance with what was actually obtained in working service trains. Furthermore, a locomotive under good conditions might give a most gratifyingly efficient performance, whereas the overall results from the class might quickly prove to be disappointing. Many attempts at explanation—some no more than spontaneous conjecture—have been made to account for the behaviour of locomotives that have not fulfilled their early promise, and simultaneously with their work towards the precise determination of the boiler and front-end characteristics, on the stationary plant, the test section of the Swindon drawing office set out to devise a way in which the results could be corroborated on the road. It was a case of scientifically linking the thermodynamics of the locomotive with the mechanics of the train. Thermodynamic analysis had, in the past, always been confined to testing at constant speed, and where stationary plants have not been available ingenious outfits have been devised to run locomotives at constant speed over routes of undulating gradient. But in actual service a train running for any appreciable time at a constant speed is so rare as to constitute an unrealistic aspect of railway working, and on the Great Western Railway means were sought of obtaining scientifically controlled road tests *at variable speed*.

This part of the work centred upon the development of an instrument that would measure the rate at which steam flowed to or from the cylinders. The blast-pipe tip was used as an orifice meter, which indicated the rate of flow by the pressure difference across it. With such a means of indication, suitably relayed to the engine cab, the driver could be instructed to run continuously at a certain pre-arranged steam rate, and by adjusting his cut-off to keep this rate constant whether the train were running uphill, on the level, or downhill. In applying this principle to the practical work of road testing there were two considerations to be borne in mind: first, the problem of conveying the indication from the smokebox to the cab, and in so doing preventing the setting up of false heads by condensation; and secondly the damping of pulsations, due to the exhaust beats of the locomotive. These problems were surmounted in the following way.

Air at very low velocity was allowed to filter into a pipe which was closed at one end and open to the exhaust steam just below the orifice at the other; the rate of air infiltration was controlled and indicated with sufficient accuracy by passing the air through a needle valve and bubble jar. Air pressure in the pipe line attained the pressure of the steam below the orifice and excess air filtered away in the exhaust steam. Pulsations were cushioned; the air line was always dry and any number of indicators—usually mercury manometers—could be connected into the pipe line. In practice on the stationary plant, it was found that speed could be altered at the control table, and the driver in turn could alter the cut-off so that the meniscus of the mercury could be kept at a predetermined level. The apparatus was used only on constant evaporation tests and, though a connection could be readily established between steam rate and pressure, this was used only to estimate conditions that would lead to a desired end, never to establish them (Fig. 232).

In general it has been found that boiler efficiency is not affected by changes in road speed and variations in cut-off, when the steam rate is constant, though in road testing it was sometimes found that the effect of vibration and so on have been such as to disturb the firebed to such a degree as to cause a break-down in the coal-steam relation. In such cases of course the test had to be abandoned. Apart from such exceptional cases however, the establishment, on the stationary plant, of the fact that boiler efficiency did not vary with speed, or cut-off, paved the way for the introduction of the Controlled Road Testing System,

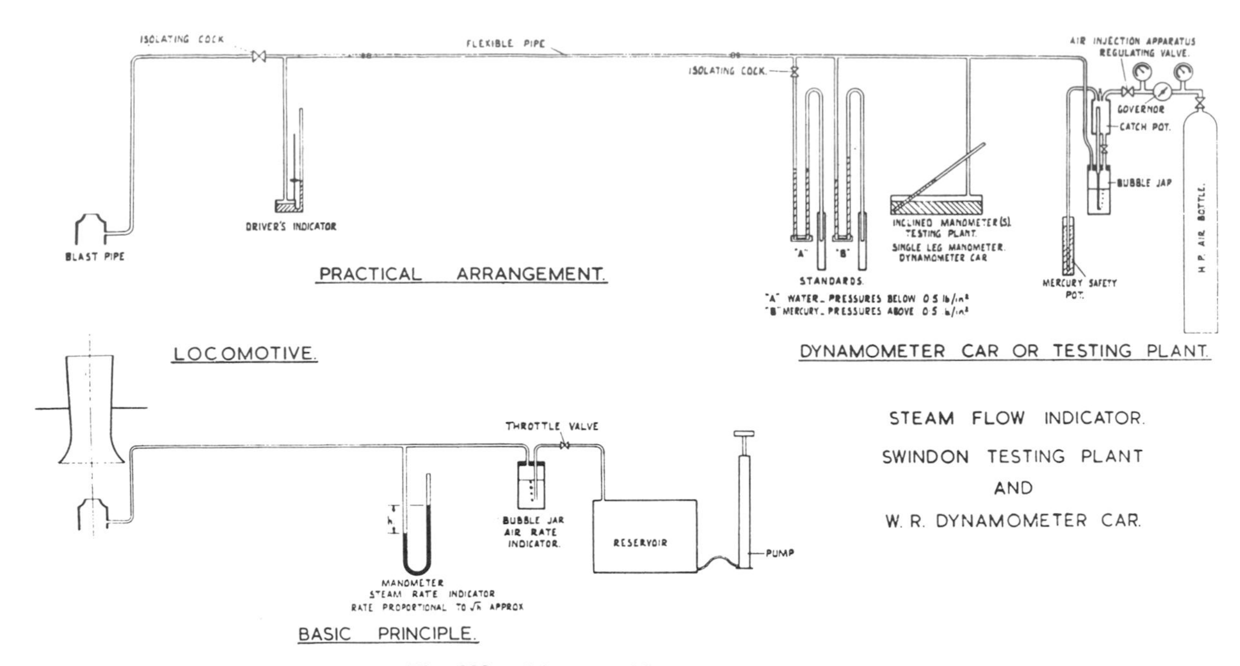

Fig. 232.—*Diagram of Steam Flow Indicator.*

in which the control of the rate of evaporation and firing by methods just described, is incorporated in a variable speed run on the line. On such a test the load consisted of ordinary coaching stock, with a dynamometer car as the leading vehicle, and this formed a natural " live " load that would provide resistance to traction according to the rise and fall of the gradients along the line (Fig. 234). The air injection apparatus for the flow meter was installed in the dynamometer car, and two manometer indicators

Fig. 233.—*Controlled road test in progress, with Gresley " V2 " class 2-6-2 engine, on Western Region.*

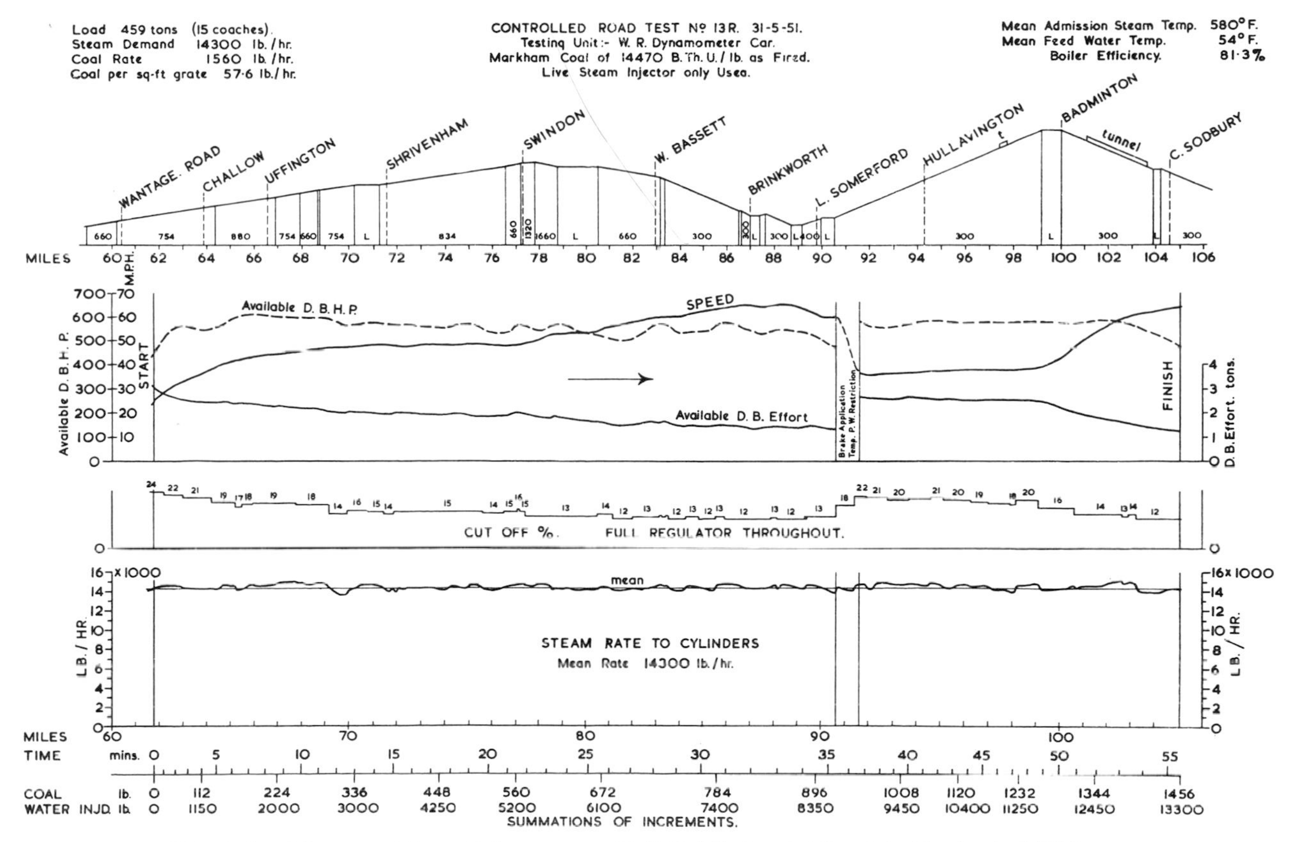

Fig. 234.—*Diagram of Controlled road test, with "Hall" class* 4-6-0 *and trailing load of* 459 *tons.*

were normally used, one in the engine cab, and one for general information and supervisory purposes in the dynamometer car itself.

In the practical operation of a test run the pointer of the driver's manometer was set at a pressure which was known to correspond with a certain steam rate, and once under way the driver worked the locomotive so as to keep the meniscus of the mercury against the set pointer. All the incidentals of practical railway working were accepted and included in the carrying out of a controlled road test. If the route should include a speed restriction, which would normally have been observed by shutting off steam and application of the brakes, a constant rate of steaming was maintained by steaming the engine against the brakes. A particular instance will make this technique clear. A test train was running at 75 m.p.h. on a falling gradient of 1 in 754 and approaching a point where a temporary speed restriction to 15 m.p.h. was enforced. The engine was working with regulator full open, and cut-off 23 per cent. The brakes were applied two miles before the point of restriction, but the regulator was kept full open, and cut-off increased, against the brakes till it was 65 per cent., keeping the steam rate absolutely constant throughout. Once the restricted length was passed the brakes were released, and again with a constant steam rate linking up commenced, and the train was restored to normal running speed. A favourite length for testing was over the Great Western main line between Reading and Stoke Gifford (Fig. 234) which included no more than a moderate variation in gradient; but other tests were made from Swindon to Westbury, via Reading, and this involved slow running for about two miles between Reading West Junction and Southcote Junction.

The full theory behind the development of modern locomotive testing methods in Great Britain was described in a paper read before the Institution of Locomotive Engineers in November 1953, by S. O. Ell,

Fig. 235.—*Swindon test plant: ex-G.W.R. "Hall" class* 4-6-0 *running at speed.*

Fig. 236.—*Controlled road test with ex-G.W.R. "County" class* 4-6-0.

and in addition to the actual testing methods employed emphasis was laid in this paper upon the need for careful verification of the results. If this were done, and the validity of the data obtained proved beyond question then it was also made clear that no difference existed between the basic test results obtained on the stationary plant in Swindon works (Fig. 235) and the results of a test at the same rate of evaporation at variable speed with a live load out on the line. By carefully controlling the steam rate and the firing rate results of test house precision and consistency could be obtained. Furthermore, in relation to the practical task of operating the traffic of a railway another important point began to emerge. Comparison of numerous records of train running in regular service showed that in a majority of instances where a high output was being obtained from the locomotive for any appreciable time the driver's method of working corresponded very closely to what could be expected with controlled testing at constant evaporation. This had a two-fold significance: first, that working at constant evaporation was not merely a convenient yardstick of the test engineer, but approximated also to the logical technique of an experienced engineman running his train in the most economical manner; and secondly that by revising the intermediate timings of express trains to correspond with constant evaporation working one was likely to secure more consistent running. Of course, in replanning schedules the observance of speed restrictions was expected to be made in the normal manner, with steam shut-off.

The construction of the joint L.M.S.R. and L.N.E.R. testing station at Rugby was so delayed by the war, and by the many restrictions of the period of austerity that immediately followed that it was not until the year 1950 that it commenced work. By that time the engineers of British Railways had examined the Great Western system of testing and decided to use it as a national standard. Thereafter some locomotives were tested at Swindon (Fig. 236) and some at Rugby, but whereas the Western Region dynamometer car, and test crews were available on the spot, in the case of locomotives tested on the plant at Rugby the controlled road tests made in verification of the "stationary" trials were mostly conducted on that favourite test route of early L.M.S.R. days, between Carlisle and Skipton. Although this route had the advantage of providing ready-made "paths" for the special test trains, and in its gradients a severe test of locomotive tractive capacity, the entire route between Skipton and Carlisle could not be run under conditions of constant evaporation. The steep gradients between Aisgill summit and Carlisle going north, and between Blea Moor and Settle Junction going south would have resulted in excessive speed downhill, and therefore the tests were limited to the sections Skipton–Aisgill, northbound, and Carlisle–Blea Moor southbound.

Scientific and successful though this method of testing proved to be it came very nearly too late in the history of the British steam railway locomotive to have any lasting benefits. The concluding chapters of this book include some notable examples of locomotive modernisation, and of greatly improved performance in consequence. But the historian, having in mind various cases where relatively minor changes in detail design, very cheaply made, revolutionised engine performance, is inclined to look back wistfully to many locomotives the moderate performance and heavy coal consumption of which led to their early withdrawal from the duties for which they were designed. No less one can recall comparative tests on famous locomotives which at different times have given widely varying results over the same section of line with trains running to the same booked times. Quite clearly extraneous factors that seriously affected the overall results were being allowed to intrude, to such an extent as to make the tests valueless. Perhaps the most extraordinary case of all was that of a locomotive that in competitive trials yielded a coal consumption of more than 5 lb. per d.h.p. hour, and yet, on a later occasion, another member of the same class gave a result of 3.7 lb. Unfortunately the earlier trials were, for a time, taken as representative of the large class as a whole.

The precise results obtained from the Swindon method of testing enabled many useful deductions to be drawn concerning the economical use of locomotives, and particularly with regard to varying values of coal consumed per d.h.p. hour. With the Great Western "Hall" class mixed traffic locomotives, for example, at a speed of 65 m.p.h. on level track the coal per d.h.p. hour is at a maximum when hauling a load of between 400 and 500 tons—slightly under 3 lb. with first-grade Welsh coal. The value rises to 3.4 lb. with a train of 250 tons, and to 4.4 lb. with a load of 150 tons. These results show how necessary it is for the duties to be strictly comparable before any comparisons are made of coal consumption between locomotives of different railways, and also the grade of coal in use. The "Hall" class engine provides a very interesting example of this, as shown in the following table.

G.W.R. 4-6-0 " Hall " class
Coal per d.h.p. hour at 65 m.p.h. on Level Track

		1A	2B
Coal	Grade	1A	2B
	Colliery	Markham	Blidworth
	B.Th.U. per lb.	14330	12500
Load	tons	lb./d.h.p.hr.	lb./d.h.p.hr.
	200	3.55	4.5
	300	3.2	3.85
	400	3.0	3.8
	450	2.95	3.9
	500	2.98	4.15

Another interesting feature brought out is the power developed at any given speed by the use of differing values of the cut-off. In the case of the " Hall " class engine the characteristics are perhaps not generally representative of British locomotive practice, because the Stephenson link motion is used rather than the Walschaerts gear. But the "Hall" was one of the first types of locomotive to be subjected to this form of analysis.

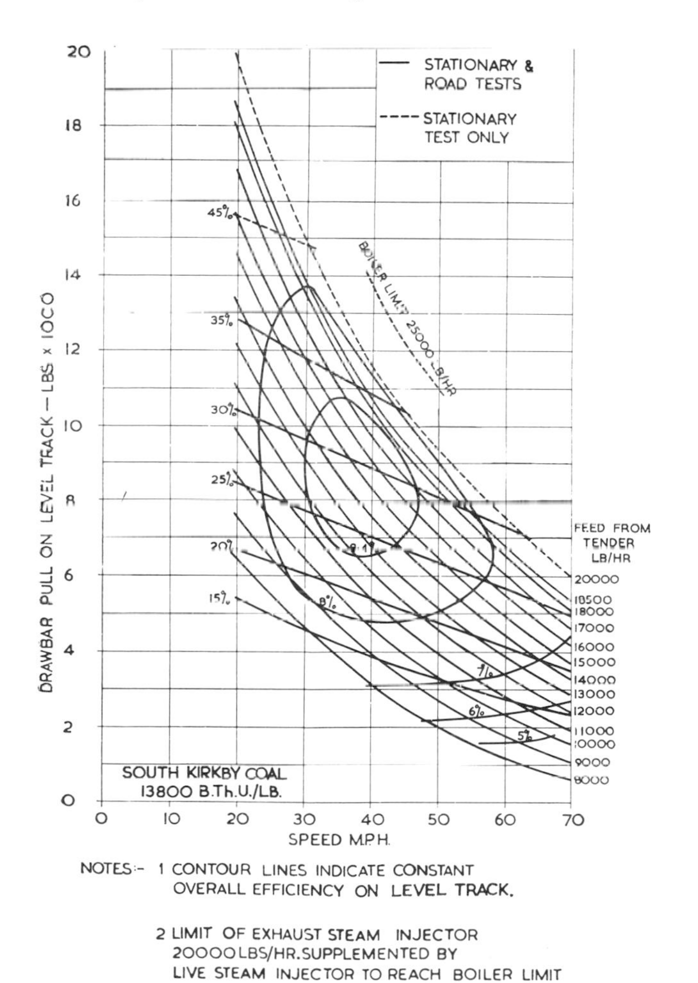

Fig. 237.—*Scientific testing methods: diagram of drawbar pull characteristic, " Hall " class 4-6-0.*

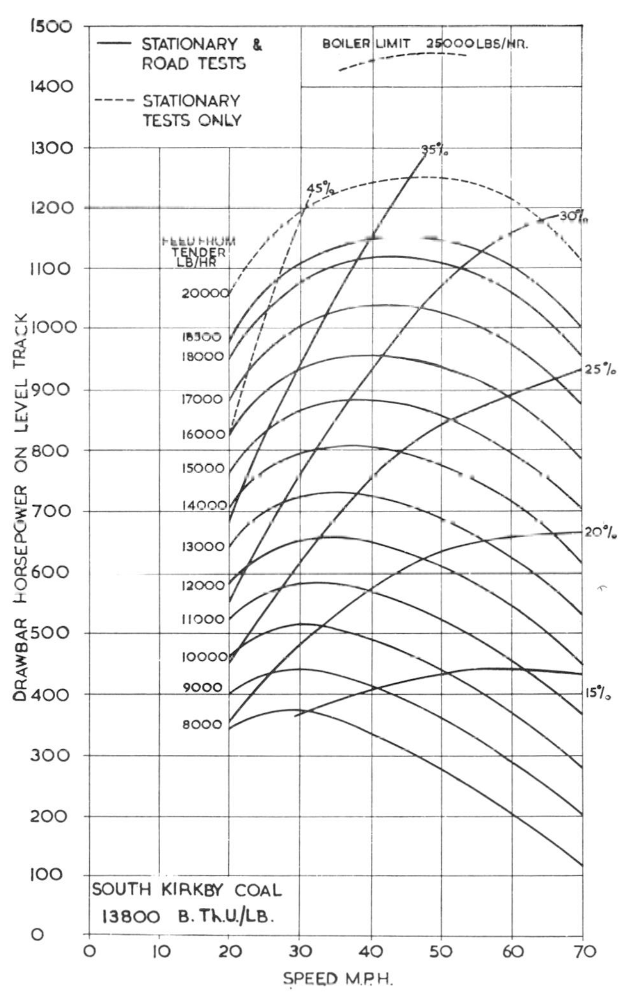

Fig. 238.—*Diagram of drawbar horsepower characteristics, " Hall " class 4-6-0.*

Fig. 239.—*Heavy load trials on Western Region: the author (left) with* S. O. Ell, *Assistant Engineer Locomotive Testing, at Swindon, before a run with engine No.* 6001 King Edward VII, *trailing load* 795 *tons.*

Power at 50 *m.p.h. on Level Track*

Cut-off %	d.h.p.	Steam rate lb./hr.	Coal rate lb./hr.
15	655	15,000	1530
20	890	18,000	1968
25	1090	21,500	2580
30	1280	24,000	3275

It was not usual to link the Great Western 2-cylinder engines up to much less than 22 per cent. cut-off, and these test results show how rapidly the power output falls off below 20 per cent. cut-off.

An example of one of the tests on a "Hall" class engine at a steam rate such as to give the most economical coal consumption is shown in the accompanying diagram (Fig. 234). This includes the gradients of the line, the speed, the drawbar pull and shows the cut-offs used, and the recorded increments of coal and water used. The following summary details apply to this particular test:

Load, trailing, tons	459
Steam demand, lb. per hr.	14,300
Coal rate, lb. per hr.	1560
Coal, per sq. ft. of grate, lb. per hr.	57.6
Boiler efficiency, per cent	81.3

Although it is taking the subject rather out of the period of this chapter, when controlled road tests came to be carried out over the Settle and Carlisle line it was interesting to compare the variation of speed under controlled conditions with that which was recorded many years earlier when the running of test trains was left entirely to individual drivers; perhaps with the occasional guidance of a running inspector who would in any case be riding on the footplate. The accompanying table compares the speeds run on three dynamometer car test runs, and one run over the same route in ordinary service, when the speeds were noted by stopwatch by an observer travelling as a passenger in the train.

Run No.	1	2	3	4
Conditions	Dyna. car passenger train	Dyna. car passenger train	Ordinary passenger train	Controlled Road Test
Load, tons	370	370	335	343
Engine	M.R.	L.N.W.R.	L.N.W.R.	L.N.E.R.
Class	Compound	"Prince of Wales"	"Claughton"	"B1"
Location	Sp. m.p.h.	Sp. m.p.h.	Sp. m.p.h.	Sp. m.p.h.
Appleby	52 1/2	47	48	40
Ormside	64 1/2	62 1/2	63 1/2	60
Griseburn Box	37	36 3/4	35 1/4	38
Crosby Garrett ..	47	47	46 1/4	47
Kirkby Stephen ..	35	32 1/2	34 1/2	38
Birkett Tunnel, South end	27	29 1/4	31 1/4	32
Mallerstang Box ..	37 1/2	40	39	41
Aisgill Box ..	27 1/4	32 1/4	34	33
Hawes Junc. ..	60 1/2	60	61 1/2	68

The foregoing details show very clearly that on this hard section of line all drivers on the service passenger trains were working their engines in such a way as to produce speeds remarkably close to those run on the controlled road test. These three service runs were all made very many years before the controlled road system of testing was inaugurated; runs 1 and 2 were made in 1923, and run 3 in 1931, and they amply confirm the contention of the Great Western Railway testing engineers than in continuous heavy duty, no matter how fluctuating the intermediate gradients may be a good engineman will work at an approximately constant steam rate. It is only when the demands upon the boiler are light that one can note very considerable divergence in driving techniques.

Chapter Thirteen — 1938–45: Locomotives for the War Effort

In locomotive development for the British home railways the year 1937 represented the high water mark in the period that ended with the outbreak of the Second World War. Locomotive performance on the northern lines, reaching new levels of achievement in 1937 has been described in Chapter Eleven; but the remarkable thing is that despite the successes attained with the " A4 Pacifics " on the L.N.E.R. and with the " Duchess " class on the L.M.S.R. both Sir Nigel Gresley and Mr. W. A. Stanier, as he then was, were planning still larger express passenger locomotives: a 4-8-2 on the L.N.E.R. and a 4-6-4 on the L.M.S.R. A marked change in the Southern Railway practice was foreshadowed in 1938 after R. E. L. Maunsell had been succeeded as Chief Mechanical Engineer by O. V. S. Bulleid—formerly assistant to Sir Nigel Gresley on the L.N.E.R.; but on the Great Western Railway, so far as new developments in design were concerned, things were virtually at a standstill. Improvements in shop practice were constantly being made at Swindon; but these concerned the precision of repair and constructional work rather than features of design.

Several new locomotive designs not yet covered in this book were produced in the last years before the war, and reference to these is necessary before coming to the war locomotives proper. One of the most important of these new designs, particularly in view of the part it came to play in dealing with wartime traffic, was Sir Nigel Gresley's large 2-6-2 mixed traffic engine of Class " V2 " (Fig. 240), of which four were built for trial purposes in 1936. Having regard to their intended use on the sharply timed fitted freight services of the L.N.E.R., some of which loaded up to well over 500 tons, the first engine of the class, No. 4771, was named *Green Arrow* after the special express goods service being developed by the company. This design could be correctly described as a synthesis of standard parts, in that the boiler was a shortened version of that carried by the latest of the " A3 Pacific " engines, and having the same large firebox; the coupled wheels, 6 ft. 2 in. diameter were the same as those of the " Cock o' the North " class 2-8-2s, while the leading pony truck was that used on the " K3 " class 2-6-0. But while the broad features of the design followed well established Gresley practice there were a number of interesting design features.

The three cylinders, steam chests, smokebox saddle and all steam and exhaust passages were cast as one unit. The outside steam pipes were also integral with the main " monobloc " casting. The piston heads and rods were combined in a single steel forging. The whole of the revolving and 40 per cent. of the reciprocating masses were balanced. The revolving mass at the centre crank was balanced by means of extensions to the crank webs, while the remainder was balanced by crescents cast in the wheel centres. The balancing was effective in producing an exceptionally smooth and steady riding locomotive. As usual on the L.N.E.R. at that period the connecting and coupling rods were of nickel chrome steel, thus permitting of a considerable reduction in weight of these parts. The boiler had a steam collector, instead of a dome, as on the latest Pacific express locomotives. The entrance to the steam collector was through a series of circumferential slots cut in the top of the barrel plate, and to avoid wire-drawing through the slots the total area of these slots was made twice the cross-sectional area through the regulator.

An interesting point was to be noted in the wedge-shaped front of the cab. This form was originally

Fig. 240.—*L.N.E.R. The Gresley* " V2 " 2-6-2 *for general passenger and goods service, first built* 1936.

Fig. 241.—*L.N.E.R. 3-cylinder 2-6-0 Class " K4 " for service on West Highland line: built Darlington* 1937.

used on the streamlined engines, as an integral part of the aero-dynamic screening adopted to reduce air resistance at high speed. It proved beneficial in another way, in that having the glasses inclined, instead of at right angles to the centre-line of the locomotive, it eliminated all reflection from the cab interior when running at night. The difference could be noted particularly when changing from one class of locomotive to another with a wedge-shaped cab front. Whereas the older large engines usually carried reflections of the fire-glow and objects in the cab, which could prove a distraction from the view of the line ahead at night, the " A4 " and " V2 " engines had a completely clear lookout.

One of the reasons for the very free running of the streamlined " A4 " engines was undoubtedly the large size of piston valve in relation to cylinder diameter. The previous standard 8 in. diameter piston valves had been replaced by 9 in. diameter valves in the " A4 " engines, while the cylinder diameter was reduced, as compared with the "A3" class, from 19 in. to 18 1/2 in. In the " V2 " engines the piston valves were made still larger, namely 9 1/2 in., and in conjunction with 18 1/2 in. diameter cylinders, this produced a very fast engine, even though the coupled wheels were reduced from 6 ft. 8 in. to 6 ft. 2 in. diameter. The nominal tractive effort of the " V2 " engines was 33,730 lb. at 85 per cent. boiler pressure. The basic dimensions can be summarised thus:

Cylinders (3) dia. in.	18 1/2
Cylinder stroke, in.	26
Dia. of piston valves, in.	9 1/2
Boiler pressure, lb./sq. in.	220
Coupled wheel dia., ft. in.	6 2
Heating surfaces, sq. ft.:	
Small tubes	1211.5
Superheater flues	1004.6
Firebox	215
Superheater elements	722
Combined total	3153.1
Grate area, sq. ft.	41.25

Although designed for mixed traffic duties these engines, even before the war, were regularly used in fast express passenger service, such as the Yorkshire Pullman, between Doncaster and King's Cross. This train, with a load of about 400 tons, was booked to average 60 m.p.h. over this distance of 156 miles, and speeds of more than 90 m.p.h. were frequently attained. On a well maintained track the riding of the locomotives at these high speeds was very good. This point needs emphasis, because in later years when neither track conditions nor engine maintenance was so good the class fell somewhat into disrepute, at any rate for fast passenger workings. After successful trials of the first four engines a total of 184 was built, and they proved invaluable during the war.

Another interesting Gresley 3-cylinder engine first introduced in 1937 was the " K4 " 2-6-0 specially designed for the West Highland line (Fig. 241). The combination of exceptionally severe gradients, constant curvature, and a limitation in axle loadings had hitherto precluded the use of any standard L.N.E.R. locomotive larger than the " K2 " Mogul of the former Great Northern Railway. These engines had a nominal tractive effort of 22,070 lb. at 85 per cent. boiler pressure, and they were limited to a maximum unpiloted tonnage of 220. This entailed a great deal of double heading, and to reduce this the " K4 " class was designed. The cylinders were made the same as the main line " K3 " engines, but considerations of weight prevented the use of the very successful 6 ft. diameter boiler that originated on the G.N.R. engine No. 1000. But by the use of coupled wheels only 5 ft. 2 in. diameter a high tractive effort was obtained. The dimensions of these new engines, in comparison with the " K3 " are shown in the accompanying table. Although the nominal tractive effort was 36,600 lb. the size of the boiler and firegrate did not permit of a proportionate increase in rated load over Class " K2 ". Nevertheless the " K4 " engines, in being capable of taking 300 tons without assistance, did materially reduce the amount of double heading on the West Highland line, and the design formed the basis of a new standard L.N.E.R. class in later years.

Fig. 242.—*Southern Railway:* " *Schools* " *class* 4-4-0 *rebuilt with multiple jet blastpipe.*

Fig. 243.—*Southern Railway:* " *Lord Nelson* " *class* 4-6-0 *rebuilt by O.V.S. Bulleid with new cylinders and multiple jet blastpipe.*

Fig. 244.—*Southern Railway: Urie class* " N15 " 4-6-0 *rebuilt with multiple jet blastpipe.*

L.N.E.R. 3-cylinder 2-6-0 locomotives

Class	K3	K4
Cylinders (3) dia., in.	18 1/2	18 1/2
Cylinder stroke, in.	26	26
Coupled wheel dia., ft. in.	5-8	5-2
Heating surfaces, sq. ft.:		
Tubes	1719	1253.6
Firebox	182	168
Superheater	407	310
Combined total	2308	1731.6
Grate area, sq. ft.	28	27.5
Boiler pressure, lb./sq. in.	180	200
Nom. tractive effort at 85% b.p., lb.	30,031	36,600
Max. axle load, tons	20.35	19.85

On the L.N.E.R., except for the streamlined " A4 Pacifics ", " P2 " the " Cock o' the North " class and one experimental application of the Kylchap double-blastpipe to an " A3 Pacific ", Gresley was content with single-orifice blastpipes on all his standard locomotives. Generally the steaming was reliable, and the Kylchap arrangement was used only when exceptional power was needed. On the Southern Railway however, O. V. S. Bulleid on taking over from Maunsell was not satisfied with the performance of the " Lord Nelson " class 4-cylinder 4-6-0s, the steaming of which was not always reliable. After a number of experiments a multiple-jet blastpipe with five nozzles was fitted to all engines of the class. The arrangement is shown in the accompanying drawing (Fig. 245) and the altered appearance of the engines can be appreciated from the photographic illustrations (Figs. 242, 243, 244).

The five 2 5/8 in. diameter nozzles discharge into a chimney having a diameter of 2 ft. 1 in. at the choke. As compared with the original single blastpipe lips with a 5 1/2 in. diameter orifice and a chimney 1 ft. 3 in. diameter at the choke the five small nozzles had a cross-sectional area 13 per cent. greater than that of the 5 1/2 in. diameter orifice while the new chimney had a cross-sectional area 2 3/4 times greater than the old one. The larger blastpipe area decreased the back pressure in the cylinders, and the increased diameter of the chimney reduced the velocity of the mingled smokebox gas and exhaust steam discharged

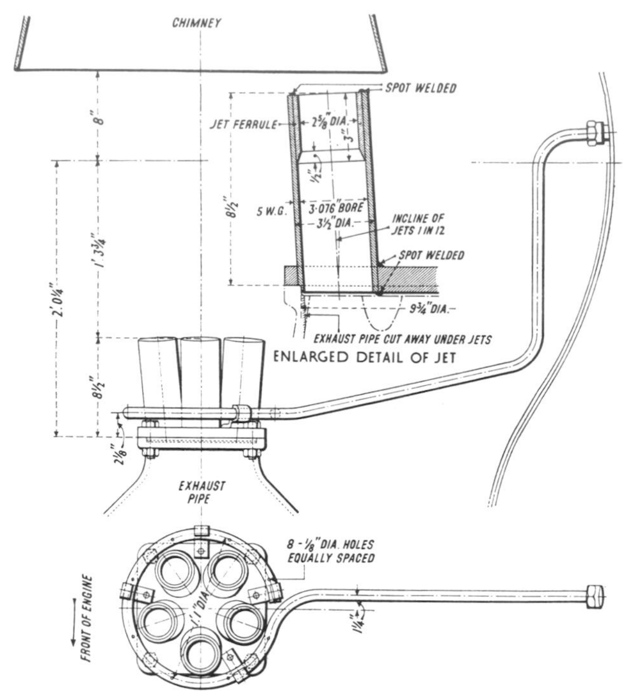

Fig. 245.—*Southern Railway: 5-nozzle multiple jet blastpipe applied to " Lord Nelson " class 4-6-0s in 1939.*

to atmosphere. Notwithstanding this lower velocity, a higher smokebox vacuum resulted, on account of the large surface area presented by the multiplicity of jets of exhaust steam and their greater entraining action.

The chimney and bell were built up in steel plate, electrically welded in one piece. The hole on top of the smokebox was made large enough to admit the bell, and this was covered by the wide baseplate attached to the chimney. Although by conventional standards the resulting chimney looked disproportionately large it was handsomely finished off with an ornamental rim at the top, formed from rings of steel plate welded together.

Accompanying the change in draughting arrangements came also a redesign of the cylinders, with

Fig. 246.—*Stanier 2-8-0 as adapted for war service overseas*

an increase in piston valve diameter from 8 in. to 10 in., and an increase in steam port area from 24.8 sq. in. to 35.5 sq. in. This change made a notable improvement in the performance of the "Lord Nelson" class locomotives, but it was more important in foreshadowing a new standard feature for future Southern Railway locomotives.

On the outbreak of war in September 1939 and the subsequent despatch of the British Expeditionary Force to France, preparations were immediately put in hand for the provision of rolling stock to support the army in what the military strategists felt would be a long campaign following somewhat in the pattern of the First World War on its western front. As previously, a British design of 2-8-0 was chosen as a Ministry of Supply standard for large scale production. Among modern designs the Stanier "8F" 2-8-0 of the L.M.S.R. was the obvious choice (Fig. 246). At that time in history a locomotive with inside valve gear would not have been generally acceptable, on grounds of accessibility, and this ruled out the otherwise excellent G.W.R. "28XX" class. Similarly the L.N.E.R. 3-cylinder "O2", with the Gresley conjugated gear for the inside cylinder was considered too complicated for service in the "rough and tumble" conditions likely to exist on military railways behind a battle front. The Stanier 2-8-0, apart from needing a few small modifications to suit French conditions and rolling stock, required only the substitution of Westinghouse for vacuum braking.

By the time the first locomotive of an order for 230 was completed however, the course of the war had changed swiftly from the expected pattern. The Western Front had collapsed completely, and these "W.D." 2-8-0s eventually took up service in the Middle East. A batch of 90 was sent to the Iranian State Railways, and equipped for oil-firing, while others were used in Egypt, Palestine and in the Western Desert campaigns. In these duties they rendered good service and were readily acceptable locomotives on account of their low maximum axle load of 16 tons. At a later stage in the war by direction of the Ministry of War Transport, all the British main line companies were required to build engines of the standard L.M.S. "8F" design for service at home (Fig. 247). In addition to construction at Crewe and Horwich, further batches were built at Ashford, Brighton, Darlington, Doncaster, Eastleigh and Swindon. Eventually, when most of the "W.D." engines had been returned from service overseas at the end of the war, the class totalled more than 700 locomotives. Their basic dimensions were: Cylinders, 18 1/2 in. diameter by 28 in. stroke; coupled wheels 4 ft. 8 1/2 in. diameter; combined total heating surface 1895 sq. ft.; grate area 28.65 sq. ft.; boiler pressure 225 lb. per sq. in.; nominal tractive effort, at 85 per cent. boiler pressure, 32,438 lb.

During the years 1941 and 1942 a few new locomotive designs of a distinctive character made their appearance on the home railways, notably the first batch of Mr. Bulleid's revolutionary "air-smoothed" "Pacifics" on the Southern Railway. These, of course, were primarily intended for heavy express passenger service to pre-war standards of speed; but at the time they were officially designated "mixed-traffic" and some of their earliest work was performed in heavy goods service. There was thus an opportunity to try out, in arduous conditions, the many novel features incorporated in the design of these locomotives, while fulfilling a useful wartime function, so that when hostilities ended the Southern Railway would have its new express passenger design ready for quantity production, if the necessary authorisation was forthcoming. Although the broad traffic specification for these new engines stemmed from pre-war conditions, and although the detailed design was worked out in wartime, these engines really belong to the post-war era, and they are dealt with in a subsequent chapter. Here, locomotives built particularly for war service in many parts of the world are described.

Outstanding among these were the Ministry of Supply "Austerity" 2-8-0s (Fig. 248) of which the first was formally handed over by the North British Locomotive Co. on January 16, 1943. The need to economise in every way in constructional practice, to achieve a rugged simplicity in design, and a high degree of reliability in wartime service made it necessary to work out an entirely new design for the job, rather than take an existing standard from British

Fig. 247.—*L.M.S.R. Stanier* "8F" 2-8-0 *for service on home railways.*

Fig. 248.—*War Department: R. A. Riddles' " Austerity "* 2-8-0 *for war service overseas, first built by North British Locomotive Co. Ltd.* (1943).

locomotive practice. Basically the new engines were of proportions generally similar to those of the Stanier " 8F " 2-8-0 of the L.M.S.R. with a maximum axle load of 15.6 tons, and a nominal tractive effort of 34,215 lb. at 85 per cent. boiler pressure. As such it was capable of meeting main line operating requirements, and would handle loads of 1,000 tons at speeds of about 41 m.p.h. on level track. The basic dimensions were:

Cylinders (2) dia., in.	19
Cylinder stroke, in.	28
Piston valve dia., in.	10
Coupled wheel dia., ft. in.	4-8 1/2
Heating surfaces, sq. ft.:	
Tubes	1512
Firebox	168
Superheater	310
Combined total	1990
Grate area, sq. ft.	28.6
Boiler pressure, lb./sq. in.	225
Nom. tractive effort at 85% boiler pressure, lb.	34,215
Max. axle load, tons	15.6
Total engine weight, tons	70 1/4
Total tender weight, tons	55 1/2

Every opportunity was taken to make the design an economical one from the manufacturing point of view (see drawing, Fig. 249). To minimise the requirement for labour all parts were kept as simple as possible. Here and there some slight elaboration was deemed desirable to make the engine accessible for cleaning and overhaul, but generally only the most straightforward principles of construction were employed. Fabrication was the method used in the construction of many components that would normally be made from steel castings and forgings. This process was not necessarily economical in labour; the reason for its adoption was to limit the demand for steel castings and forgings, manufacturing facilities for which were already fully engaged in meeting other wartime requirements. Certain steels and non-ferrous materials were in short supply; their use in the new locomotives was restricted to the minimum.

As an illustration of simplicity in design leading to economies in labour the boiler makes a good example; the round-topped firebox and the parallel barrel made it a particularly straightforward type for quantity production. Its clothing, moreover, was of the simplest, for except on the backplate it consisted merely of steel plates carried on crinolines; sufficiently good insulation against heat loss was provided by the air space between these plates and the boiler proper.

Interesting instances of the elimination of steel castings and forgings by the use of fabricated parts were to be found all over the engine, but four examples must suffice. The axlebox guides, which would normally have been cast were made from flanged plate reinforced by triangular ribs. Brake-hanger and spring-link brackets were made from strip material; a bent piece and a straight piece welded together to form a single bracket of the type that was usually forged. The reversing rod was tubular with the ends welded on and the reverse shaft was likewise a tube with forged ends and levers welded in place.

Cast iron was used in the normal way for the cylinders, for the blastpipe which was in one piece, and for the smokebox saddle which had the exhaust passages formed in it. The front end cylinder covers were of cast iron and so was the chimney. Cast iron was used to replace steel in the manufacture of certain of the wheels. The centres of those on the first two-coupled axles and on the trailing coupled axle were of high duty cast iron; the main driving wheel centres were of steel castings; the leading truck wheels and the tender wheels were rolled with tyres in one piece, but there was an ample rim section so that they could be re-turned after wear had taken place.

The steel and iron centres of the coupled wheels were made from similar patterns and were pressed on to their axles. Rather less pressure, about 6 instead of 10 tons per inch of axle diameter was used to press on the iron centres. Balance weights were cast integral with the centres and no allowance was made for reciprocating parts. The rear cover casting for the

Fig. 253.—*L.N.E.R. 3-cylinder* 2-6-2 *locomotive Class* " V4 " *built Doncaster* 1941.

general service at home, two for the L.N.E.R. and one for the Southern. In 1941 there appeared from Doncaster Works the last design for which Sir Nigel Gresley was responsible, a light 2-6-2 tender engine for mixed duties having a high route availability (Fig. 253). It must be said nevertheless that this design, of which two prototypes were built, was conceived in the pre-war tradition of the L.N.E.R. with three cylinders and the conjugated valve gear, and achieving a low axle loading by use of high quality materials, and a relatively small boiler using an unusually high steam pressure for a locomotive intended for such duties. The death of Sir Nigel Gresley in April 1941 led to a complete change in the locomotive policy of the L.N.E.R. and in 1942, his successor, Edward Thompson, had produced a general utility engine of comparable weight, capacity and route availability by very different means.

The Class " B1 " 4-6-0 (Fig. 254), of which a large number was subsequently built, was a synthesis of existing standard parts, and required a minimum of new machinery in order to produce it. Thus the boiler, and firebox were identical to those of the " B17 ", or " Sandringham " class of 3-cylinder express passenger 4-6-0; the coupled wheels, 6 ft. 2 in. diameter were the same as those of the " V2 " and " P2 " classes, while the two cylinders, with their steam chests were made from the patterns of the ex-G.N.R. " K2 " 2-6-0, which was then being maintained as a standard L.N.E.R. class. The tender was the normal pattern for medium-powered locomotives carrying 4,200 gallons of water and 7 $^1/_2$ tons of coal. Although departing from the Gresley tradition of 3-cylinder propulsion the 2-cylinder " B1 " was a neat and compact looking locomotive, wholly British in its outward appearance, and refreshingly welcome in that respect at a time when the austere conditions of wartime were being sometimes quoted as the reason for the production of other locomotives that were less pleasant to the eye. The basic dimensions of the " V4 " and " B1 " locomotives of the L.N.E.R. make an interesting comparison.

L.N.E.R. General Utility Locomotives

Class	" V4 "	" B1 "
Type	2-6-2	4-6-0
Cylinders, number	3	2
Cylinder dia., in.	15	20
Cylinder stroke, in.	26	26
Coupled wheel dia., ft. in.	5-8	6-2
Heating surfaces, sq. ft.:		
Tubes	1292.5	1508
Firebox	151.6	168
Superheater	355.8	344
Combined total	1810.4	2020
Grate area, sq. ft.	28.8	27.9
Boiler pressure, lb./sq. in.	250	225
Max. axle load, tons	17.0	17.75
Total engine weight, tons	70.4	71.1
Nom. tractive effort at 85% boiler pressure lb.	27,420	26,878

No more than the original two examples of the " V4 " class were built. After some initial running in East Anglia they were drafted to Scotland and for some time worked on the West Highland line between Glasgow and Fort William. It cannot be said that they were markedly successful, though with only two engines in the class it was no more than natural that they should have had little attention technically. The detail adjustments that are frequently necessary with a new locomotive class to bring it up to the standard envisaged by the designer were not forthcoming in the case of the " V4s ", and their full potentialities were never revealed. The " B1 " 4-6-0s on the other hand, in their extreme simplicity and use of well tried components at once became most useful traffic machines (Fig. 255). Some reference to their work in post-war years is made in Chapter Sixteen.

The third general utility locomotive class of this period was the Bulleid " Q1 " class 0-6-0 on the Southern (Fig. 256). New engines were required for general freight service, to have as high a route availability as possible. The need could have been met by building more of the Maunsell " Q " class, but Mr. Bulleid thought that larger boilers were desirable, and in planning the new engines to have the largest firebox

Fig. 254.—*L.N.E.R. One of a large post-war batch of* " B1 " *class* 4-6-0s *built by North British Locomotive Co. Ltd.*

Fig. 255.—*L.N.E.R. Edward Thompson's Class* " B1 " 4-6-0, *as originally built.*

Fig. 256.—*Southern Railway. O. V. S. Bulleid's* " *Austerity* " 0-6-0 *for general service, Class* " Q1 ", 1942.

Fig. 257.—*L.M.S.R. The Stanier streamlined* 4-6-2s *in wartime livery: No.* 6246 City of Manchester in *plain black.*

Fig. 258.—*L.M.S.R. One of the later wartime Stanier Pacifics, No.* 6252, *after streamlining had been abandoned for new locomotives.*

in cross-section that could be accommodated consistent with providing adequate look-outs from the restricted widths of cab imposed by the composite loading gauge of the Southern Railway, it was found that the flanging blocks of the " Lord Nelson " class could be used. In providing this large boiler and firebox a higher proportion than usual of the maximum permissible weight was used, and consequently the most drastic and unconventional economies had to be made elsewhere, such as discarding the running plates, and fabricating the cab from thin sheet, suitably reinforced. The result was a locomotive having a nominal tractive effort of 30,000 lb., weighing no more than 51.2 tons in working order, and able to run over 93 per cent. of the entire Southern Railway network. The other basic dimensions were: cylinders, 19 in. diameter by 26 in. stroke; coupled wheels 5 ft. 1 in. diameter; total heating surface 1690 sq. ft. grate area 27 sq. ft.; and boiler pressure 230 lb. per sq. in. The smokebox was fitted with the multiple-jet 5-nozzle blastpipe, and the piston valves had long laps and long travel and were actuated by Stephenson's link motion.

Turning now to locomotives built for war service overseas there are first of all two remarkable examples of the Beyer-Garratt type to be described. Numerous locomotives were required for various phases of the campaigns in the Far East, and among these there was a need for units of maximum power to run on metre gauge lines. It had been agreed early in the war that any new locomotives designed for service overseas should have the widest possible sphere of activity, and accordingly Messrs. Beyer Peacock and Co. Ltd. were entrusted with the design of a general purpose Beyer-Garratt locomotive of maximum power, to operate on a 50 lb. rail, and having a maximum axle load of 10 tons. The locomotives were also required to be capable of running on any metre gauge system likely to be involved, and consequently had to be built to the most restricted loading gauge,

Fig. 259.—*War Department. Beyer-Peacock's light war-service Garratt locomotive as used on the Kenya-Uganda Railway.*

Fig. 260.—*War Department. Heavy war-service Beyer-Garratt locomotive, as working on the French Congo-Ocean Railway.*

Fig. 261.—*War Department. Heavy Beyer-Garratt locomotive as allocated to the Rhodesia Railways.*

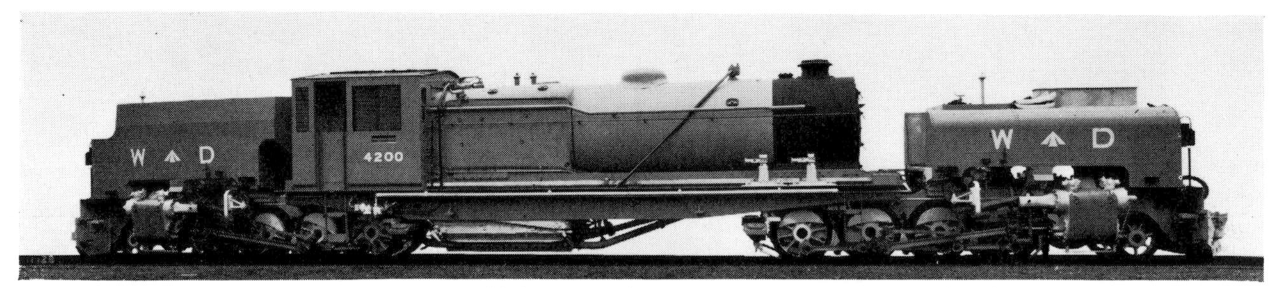

Fig 262.—*War Department. Light war-service Beyer-Garratt locomotive* 2-8-0 + 0-8-2 *types.*

namely the Indian standard, which permitted maximum height and width as 11 ft. 3 in. and 8 ft. 6 in. respectively. Despite these restrictions a splendid locomotive having a nominal tractive effort of 43,520 lb. (at 85 per cent. boiler pressure) was produced (Fig. 259), and examples of this class, known as the War Standard Light Type were eventually to be found in service on the Bengal Assam Railway, in Burma, and on the Kenya and Uganda Railways, now the East African Railways.

The War Standard Heavy Freight Garratt (Fig. 260) was designed for general use on railways with 3 ft. 6 in. gauge, 60 lb. rails, and a maximum axle load of 13 $^1/_4$ tons. They were intended mainly for African railways, with severe ruling gradients, and were allocated to the Gold Coast, the French Congo Ocean, and to Rhodesia. These locomotives were of the 2-8-2 + 2-8-2 wheel arrangement, and a nominal tractive effort at 85 per cent. boiler pressure, of 58,260 lb. In the Gold Coast territory they were used on very heavy trains of manganese and bauxite, which were of great importance to the war effort. On the Congo-Ocean Railway, with its mainline of 300 miles these locomotives took loads of 400 tons on gradients of 1 in 43, and 1,000 tons elsewhere. The Rhodesia Railways were already extensive users of Beyer-Garratt locomotives, but to meet the needs of the heavy increased traffic in the Wankie coalfield area nine of the War Standard Heavy Freight engines were allocated to Rhodesia. They were used principally between Wankie and Dett, 51 miles, where the maximum grade is 1 in 61 uncompensated for curvature. On this section the new Beyer Garratts took a maximum load of 1,100 tons.

The basic dimensions of the light and heavy WD Beyer-Garratts were as follows:

War Standard Beyer-Garratt Locomotives

Class	Light	Heavy
Rail gauge	Metre	3 ft. 6 in.
Type	4-8-2+2-8-4	2-8-2+2-8-2
Max. axle load, tons	10	13.125
Cylinders dia., in.	16	19
Cylinder stroke, in.	24	24
Coupled wheel dia., ft. in.	4-0	3 9$^1/_2$
Boiler pressure, lb./sq. in.	200	180
Heating surfaces, sq. ft.:		
Tubes	1813	2328
Firebox	183	212
Superheater	399	470
Combined total	2395	3010
Grate area, sq. ft.	48.75	51.3
Total wt. in working order, tons	136.8	151.8
Nom. tractive effort at 85% boiler pressure, lb.	43,520	58,260

Another very interesting locomotive built in connection with the war, or rather to meet the conditions anticipated in many European countries after the end of hostilities was the "Liberation" 2-8-0 (Fig. 263, 264), of which 110 were built by the Vulcan Foundry Ltd., to the order of "U.N.R.R.A.". This was surely a design having a history without parallel in the long story of British locomotive engineering. The requirements of seven countries had to be coordinated and fused into a design which, as nearly as possible, was to be able to "go anywhere and do anything" in European rail transport. The language difficulty alone at the meetings of the international Technical Advisory Committee on Inland Transport (T.A.C.I.T.) was formidable. After considerable initial work the Vulcan Foundry staff drafted a preliminary design, and it is certainly a resounding tribute to their appreciation of the problems involved that the trial effort needed little modification to satisfy the exacting requirements of T.A.C.I.T. The European countries involved originally were France, Belgium, Holland, Czechoslovakia, Poland, Yugoslavia and Greece. The basic requirements laid down were:

(1) Tractive effort to be in the range 40,000–50,000 lb.
(2) Boiler to have wide firebox and grate area of 40-50 sq. ft.
(3) Axle load not more than 18 tons.
(4) Wheelbase suitable for curves of 330 ft. radius.
(5) Profile to conform to the "Berne" international railway loading gauge.

Fig. 263.—*Cab view of Vulcan "Liberation" locomotive.*

Fig. 264.—*The Vulcan "Liberation" 2-8-0 locomotive, built for post war service in Europe.*

The resulting locomotive, of the 2-8-0 type, had a nominal tractive effort of 43,800 lb. The boiler was designed to steam freely on low grade fuel, and had a wide round-topped firebox with copper inner shell and three arch tubes. The continental engineers required rhomboidal steam ports and as a consequence the piston valves had the large diameter of 12 in., in conjunction with cylinders 21 $^3/_4$ in. diameter by 28 in. stroke. The design as a whole (Fig. 265) represented a judicious blend of British, Continental and American practice, including plate frames of traditional British type, while American influence was to be seen in the 4-wheeled cast steel bogies of the plankless type.

A total of 120 was built, all in 1946. 110 of these were distributed between Czechoslovakia, Poland and Yugoslavia, while 10 were delivered to Luxemburg. From all countries where they were used these locomotives made an excellent impression, and their delivery fitly concluded a very notable chapter in the history of British locomotive engineering. During the war more than 1,000 main line locomotives were shipped overseas and of these no fewer than 733 of the Austerity 2-8-0 type were subsequently purchased by the nationalised British Railways in 1948, together with 25 of the 2-10-0 Austerity type. These 2-8-0 and 2-10-0 Austerity locomotives were built mainly by the Vulcan Foundry and the North British Locomotive Co. The basic dimensions of the Vulcan "Liberation" locomotives were:

Cylinders dia., in.	21 $^5/_8$
Cylinder stroke, in.	28
Coupled wheel dia., ft. in.	4-9 $^1/_8$
Boiler pressure, lb./sq. in.	227
Heating surfaces, sq. ft.:	
Tubes	2098
Firebox	175
Superheater	660
Combined total	2933
Grate area, sq. ft.	44
Total wt. of engine in working order, tons	84.3
Nom. tractive effort at 85% boiler pressure, lb.	43,800

Before concluding this chapter reference must be made to the very good impression made in Great Britain by certain features of the American-built Austerity 2-8-0s which were in service on British railways for a short time towards the end of the war. The experience gained with certain items of their equipment, particularly rocking grates, self-cleaning smokeboxes, and vertical injectors undoubtedly influenced the adoption of these features in subsequent British designs.

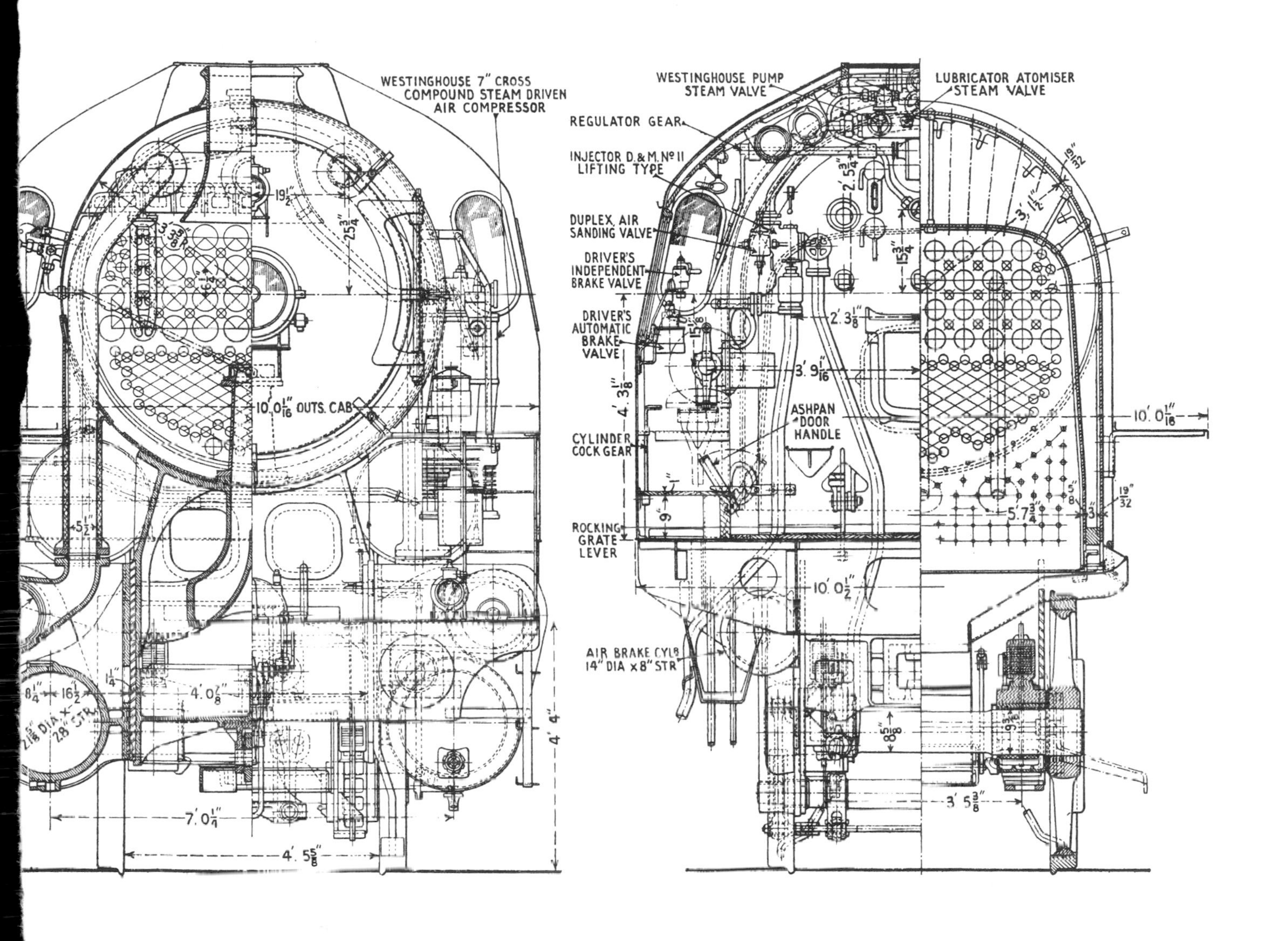

Fig. 265.

LOCOMOTIVE BUILT BY THE VULCAN FOUNDRY LIMITED FOR SERVICE IN EUROPE

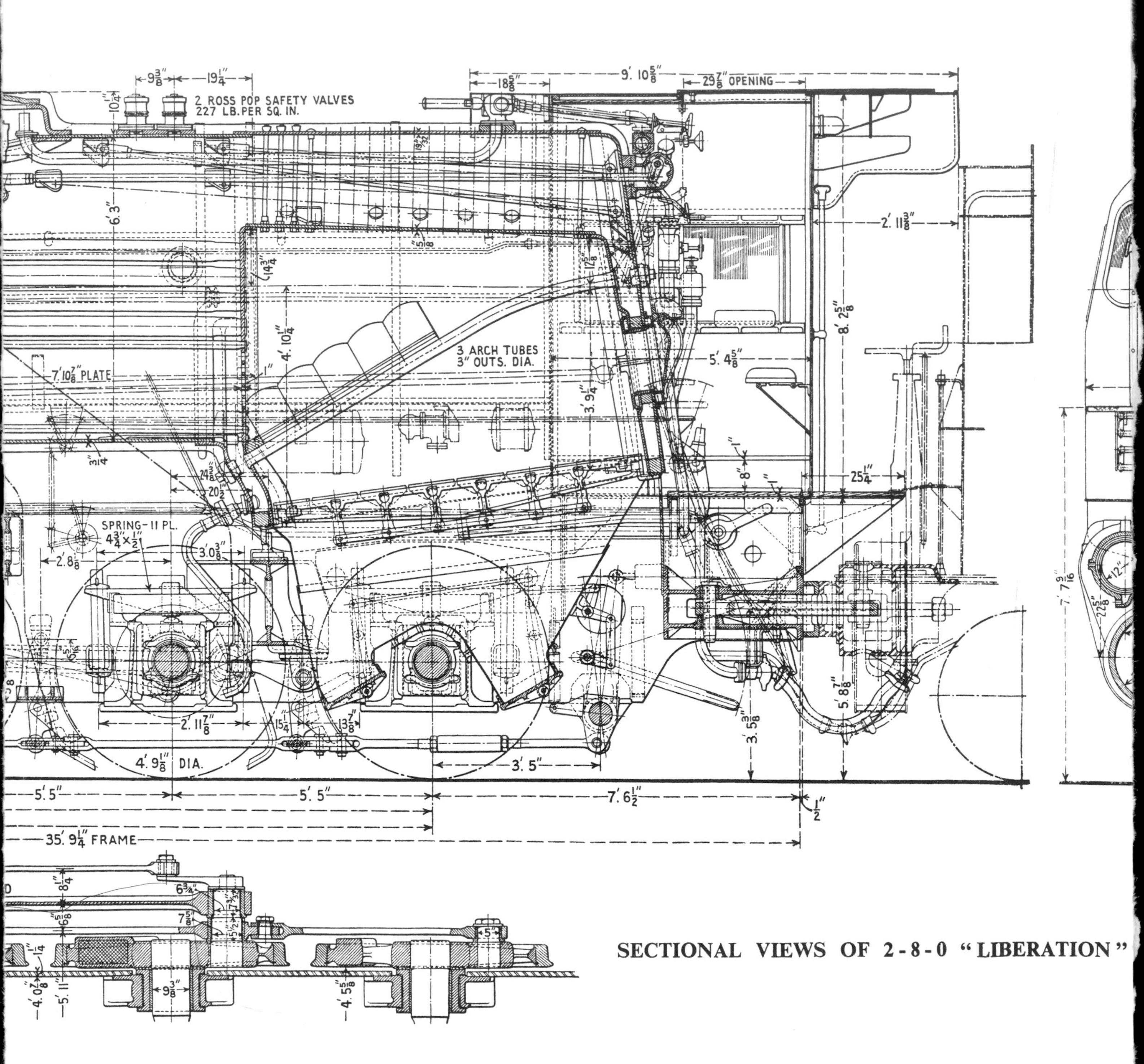

SECTIONAL VIEWS OF 2-8-0 "LIBERATION"

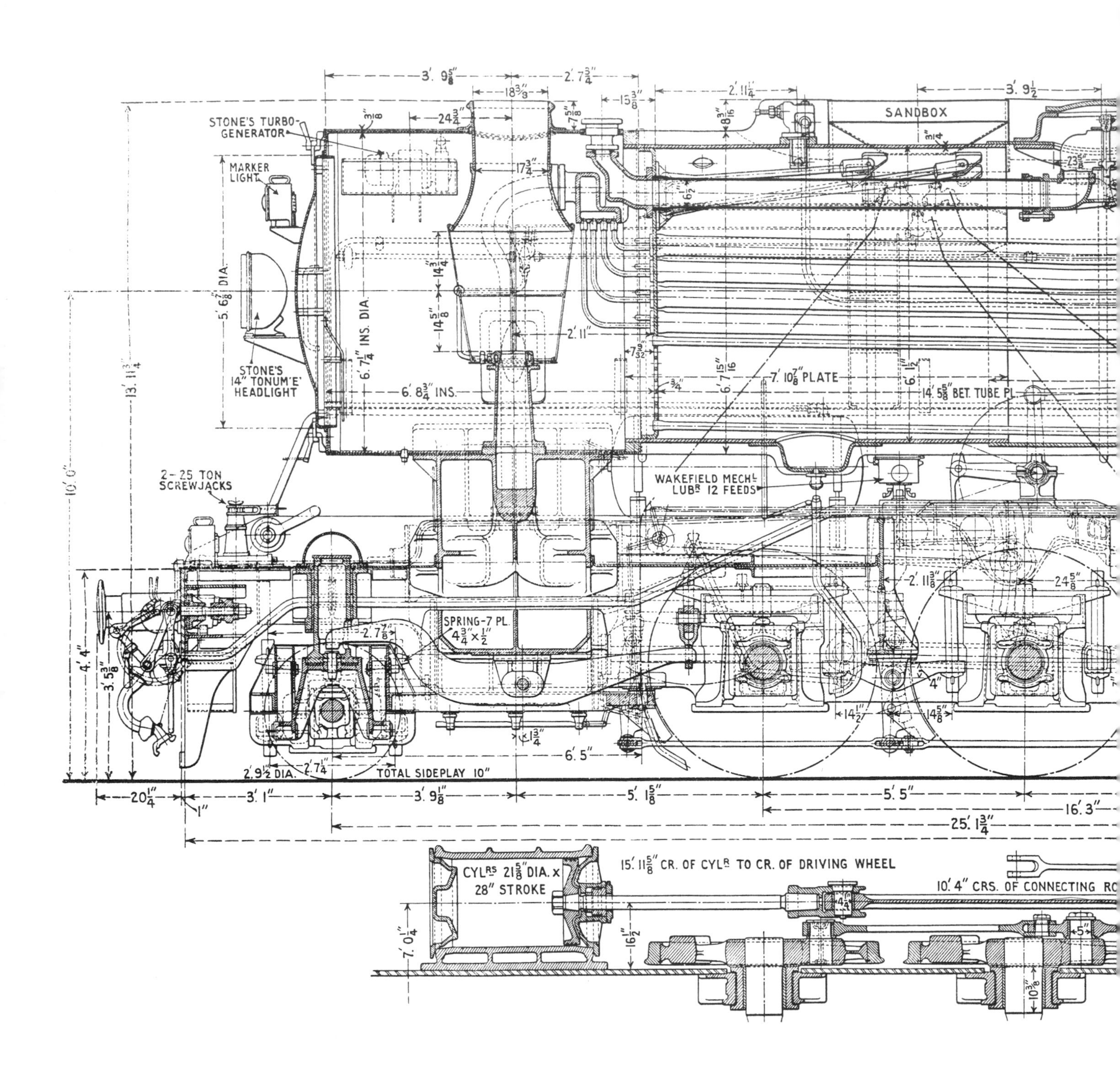

STONE'S TURBO-GENERATOR
MARKER LIGHT
STONE'S 14" TONUM'E' HEADLIGHT
2-25 TON SCREWJACKS
SANDBOX
WAKEFIELD MECHL LUBR 12 FEEDS
7' 10⅞" PLATE
14' 5⅝" BET. TUBE PL.
6' 8¾" INS.
6' 7¼" INS. DIA.
5' 6⅞" DIA.
SPRING-7 PL. 4¾"×½"
2' 9½" DIA.
TOTAL SIDEPLAY 10"
CYLRS 21⅝" DIA. x 28" STROKE
15' 11⅝" CR. OF CYLR TO CR. OF DRIVING WHEEL
10' 4" CRS. OF CONNECTING RO
25' 1¾"
16' 3"

Chapter Fourteen — Post-War Development
The Last Years of Private Ownership

During the last stages of the war it became evident that railway working conditions after the end of hostilities would be much changed, and that a return to the situation existing in 1937–39 would not be possible for many years. The coming state of affairs was indeed anticipated by locomotive designers, and a number of interesting developments were actually in progress before the end of the war. Outstanding in this forward-looking trend was the work of O.V.S. Bulleid on the Southern Railway, which produced the much discussed " Merchant Navy " class " Pacific " locomotives (Fig. 266) as early in the war as 1941. Considerations towards this revolutionary design had of course begun some little time before the war; but the fact that these powerful machines could be used on freight service enabled work on them to continue during the war, even though their intended use was, of course, on express passenger service both heavier than, and accelerated from pre-war standards. But the fact that the project was pushed forward during the war naturally involved certain handicaps and disadvantages particularly with regard to materials.

The design considerations and details of these locomotives were fully described in a paper presented before the Institution of Mechanical Engineers by Mr. Bulleid on December 14, 1945, to which reference may be made; but the major points are of such importance that extended reference is also made to them in this chapter. In designing these engines the specification given to R. E. L. Maunsell prior to the design of the " Lord Nelson " class 4-6-0s in 1925, namely the haulage of 500-ton express trains at start-to-stop average speeds of 55 m.p.h., was stepped up very considerably. The train loads envisaged in future were in the 550–600 ton range, and the start-to-stop average speeds likely to be demanded were 60 m.p.h. on the shorter runs, such as with the boat trains between London and Dover, and as much as 70 m.p.h. on longer non-stop runs, from Waterloo to the West. The haulage of 600-ton trains at start-to-stop average speeds of 70 m.p.h. was something unprecedented in British railway service. To meet such a demand Bulleid estimated he needed a 4-6-2 with a nominal tractive effort of 35,000–40,000 lb., an adhesive weight of at least 63 tons, and a grate area of about 50 sq. ft. There were restrictions in length, width and height, as well as in maximum weight and the design was therefore based on the following parameters:

Tractive effort, lb.	37,500
Coupled wheel dia., ft. in.	6-2
Boiler pressure, lb./sq. in.	280
Grate area, sq. ft.	48 $^1/_2$
Total weight of engine and tender, tons ..	145
Total weight of engine only, tons	94 $^3/_4$

In working out the detail design the first consideration was to provide the largest possible boiler and firebox, to facilitate free steaming on relatively low grade fuels; and then to include every possible feature that could be calculated to simplify maintenance, and actual handling by the engine crews. The great size of the proposed boiler made it necessary to reduce weight wherever possible elsewhere to keep within the overall limit of 94 $^3/_4$ tons. The boiler pressure of 280 lb. per

Fig. 266.—*Southern Railway: Bulleid's " Merchant Navy " class* 4-6-2 (1941).

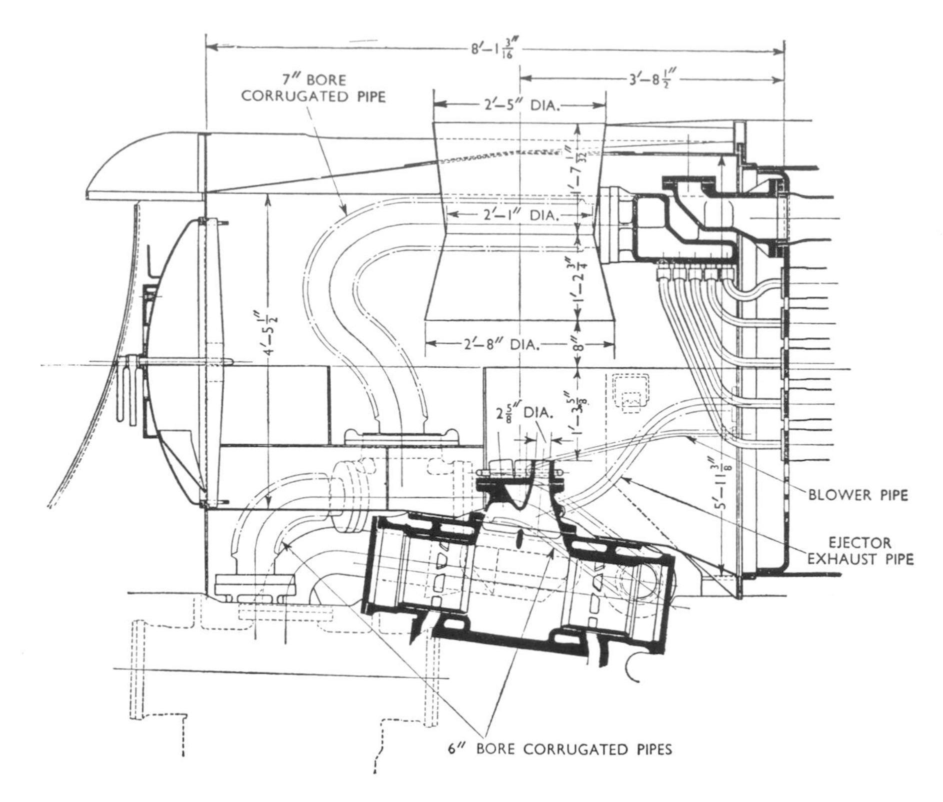

Fig. 267.—*" Merchant Navy " class; cross-section of smokebox.*

sq. in. was the highest used up to that time in orthodox British practice, though before the war Sir William Stanier had on the drawing board a design for a high-speed 4-6-4 locomotive using a pressure of 300 lb. per sq. in. Bulleid adopted 280 in order to keep down the cylinder dimensions, and these latter—18 in. diameter by 24 in. stroke—were unusually small for a locomotive of such size. Experience with copper fireboxes on the Southern Railway had not been satisfactory, by the standard Mr. Bulleid felt ought to be maintained by the time the " Merchant Navy " class engines were designed, and he therefore decided upon steel, completely welded.

The plate material used had the following chemical composition:

Carbon	0.15 per cent max.
Manganese	0.55 ,, ,,
Phosphorus and Sulphur	0.03 ,, ,,

Mechanical properties were:
Ultimate tensile strength: 24–28 tons per sq. in.
Elongation on 8 in.: 25 per cent. min.

The particular attraction of this material was that it was almost free from creep, and was readily weldable. With it a completely welded inner and outer firebox was a practical proposition, and a special feature of welded construction was that there would be no cases of double thickness of metal that are necessary in riveted construction, and thus, it was thought, would eliminate trouble through local overheating and burning away. The use of thin steel plates, welded, instead of a conventionally designed copper firebox, saved about 1 1/2 tons in weight. The boilers and fireboxes of the first ten locomotives of the new class were built by the North British Locomotive Co. Ltd., the experience of which firm in design and construction of steel fireboxes was invaluable to the Southern Railway. The firebox was equipped with two Nicholson thermic syphons to assist in rapid circulation of the water. As finalised the boiler had the following dimensions:

Barrel:	
Length, ft. in.	16-9 5/8
Outside dia. max., ft. in.	6-3 1/2
Inside dia. min., ft. in.	5-9 3/4
Length between tube plates, ft. in.	17-0
Large tubes: number	40
outside dia., in.	5 1/4
Small tubes: number	124
outside dia., in.	2 1/4
Heating surfaces, sq. ft.:	
Tubes (large and small)	2175.9
Firebox	275.0
Superheater	822.0
Combined total	3272.9
Grate area, sq. ft.	48.5

The relatively short distance between tube-plates, and the large diameter of the small tubes combined with the excellent design of firebox and the 5-nozzle multiple-jet blastpipe combined to produce a unit

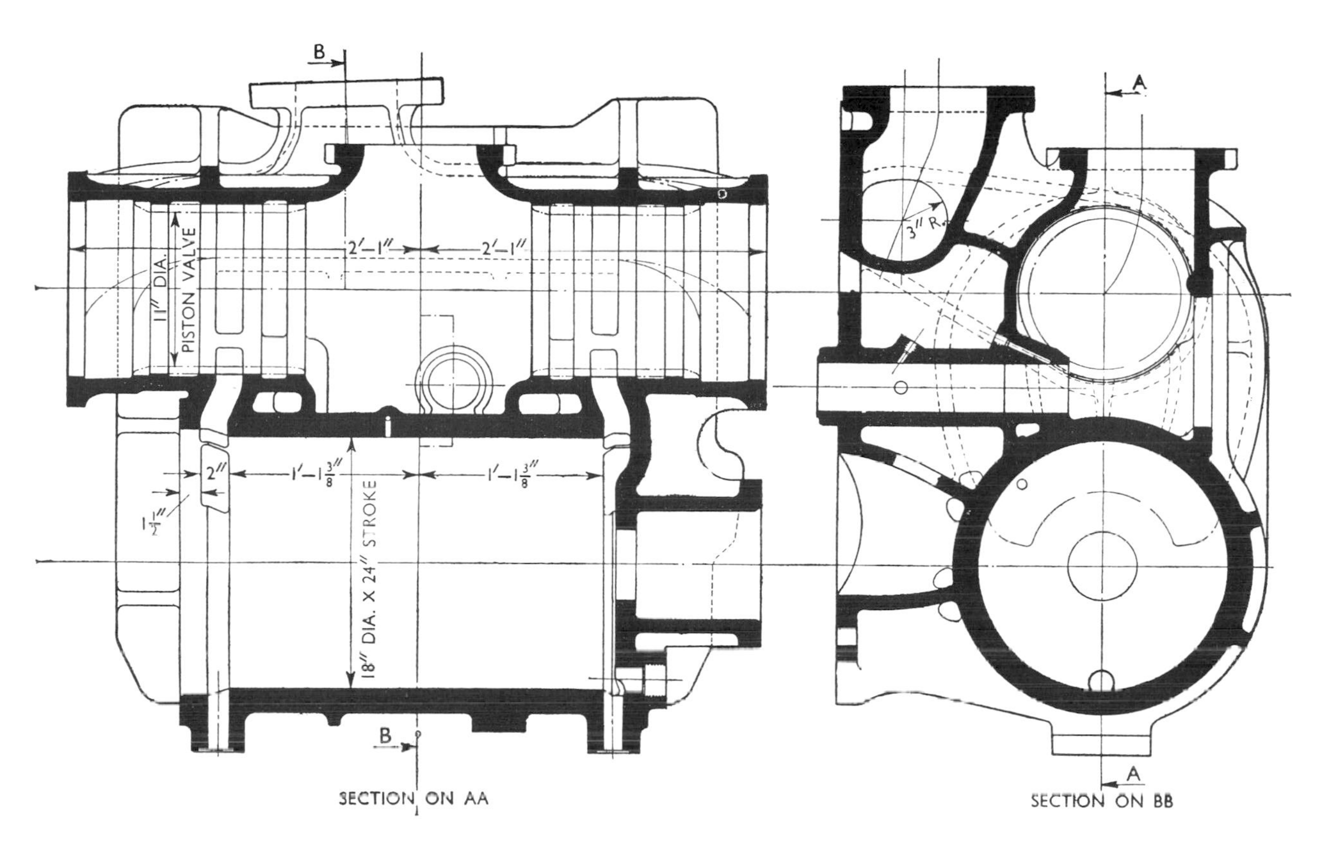

Fig. 268.—*" Merchant Navy " class; section of outside cylinders and valve chests.*

that was remarkably free in steaming, though in actual working on the road it was rare to see the full advantage taken of the high boiler pressure to secure a high rate of expansion. More usually drivers tended to run with little more than about 200 lb. per sq. in. steam chest pressure.

It will be appreciated that some apprehension existed in certain quarters as to the performance that could be obtained from the steel fireboxes of the " Merchant Navy " and of the later " West Country " class boilers, in view of the previous limited experience in this country with steel as a firebox material. In the event, both classes of locomotives gave excellent service so far as the boilers were concerned, and in the light of experience in heavy traffic conditions only two relatively minor modifications in design were found necessary, namely the substitution of monel metal for some of the steel stays in the breaking zones, and in the case of the first ten " Merchant Navy " boilers the insertion of syphon diaphragm plates in the throat plate. Nevertheless experience with the first ten " Merchant Navy " class locomotives indicated that corrosion of the water side of the inner firebox plates was more rapid than desirable, and in 1947 a comprehensive and fully controlled system of water treatment was introduced. This was based on the French T.I.A. system which provides for water treatment on the tender itself, adding compounds to the water to form a sludge instead of scale, and maintain in the water a sufficient alkalinity to avoid ordinary corrosion. This process virtually doubled the life of the steel fireboxes.

A cross-section of the smokebox is shown in the accompanying drawing (Fig. 267), while a second drawing shows a detail of the outside cylinders and valve chests (Fig. 268). It was in connection with the design of the valve gear that one of the greatest novelties of these locomotives was introduced. While the general trend of design practice in steam locomotives all over the world had been, in R. E. L. Maunsell's expressive phrase, " to make everything get-at-able " Bulleid decided to go in exactly the opposite direction, and completely enclose as much of the motion as possible, in order to provide continuous flood lubrication of the moving parts. In making this decision he set himself a number of major problems, in providing compensation for the relative movements of the frame in relation to the crank axle, and in safeguarding against the escape of oil. In the " Merchant Navy " class engines it was decided not only to enclose the three sets of valve motion, but also the middle connecting rod, crosshead, slide bar, and crank-pin. Thus the three sets of valve gear had all to be located between the frames, and the inside piston rod, and all three valve rods had to be taken through the forward end of the oil bath casing.

The arrangement devised for meeting these conditions is shown in the accompanying drawing

(Fig. 269). The special conditions were such that no existing valve gear could be accommodated in the very confined space available. Moreover it was desirable to keep down the unsprung weight of the driving axle. Again, if it should be necessary to remove the driving axle, as little as possible of the sump should be disturbed. A new valve gear was therefore designed specially to do the job, in which each piston valve was operated by an independent set of motion. The three sets of gear were operated by a three-throw secondary crankshaft. Each throw of this crankshaft oscillated its quadrant link by a vertical connecting rod pinned to an arm extended backwards from the link; at the same time it reciprocated the foot of the combination lever by a horizontal link pinned to the big end of the vertical connecting rod. The quadrant link operated the upper end of the combination lever in the usual way. The combined motion was conveyed, through a plunger working in a guide, by the valve rods to the valve-operating rocker shaft. Provision was made for the frame to rise 2 in. and fall $1\,^1/_2$ in. relative to the crank axle, and for that axle to move as much as $^1/_4$ in. from its mid-position, by using a chain drive. This drive was in two parts, through an intermediate lay-shaft 4 ft. ahead of the crankshaft centre. As the chain wheels are equal in diameter the three-throw crankshaft was driven truly in phase with the main crank axle.

Another novel feature concerned the method of driving the piston valves. With high steam pressures the conventional arrangement had been to use inside admission and outside exhaust, to reduce the pressures on the valve spindle packings. But in the "Merchant Navy" class locomotives a new method of operation was used whereby each pair of piston valves was driven by a rocker in the exhaust cavity. No valve spindles were used, and so the need for glands was eliminated, and with them the objections to outside admission. The rocker was arranged across the cylinder, and when uncoupled the arm in the exhaust cavity drops clear to allow the piston valve to be withdrawn. The piston valves, of 11 in. diameter were exceptionally large for cylinders of such relatively small size, and with large port areas contributed to the very free-running of the locomotives. The accompanying drawing (Fig. 270) shows the detail of the piston valves.

Externally the locomotives were rendered most distinctive by the air-smoothed casing. This was a functional design, cleverly adapted to catch the eye

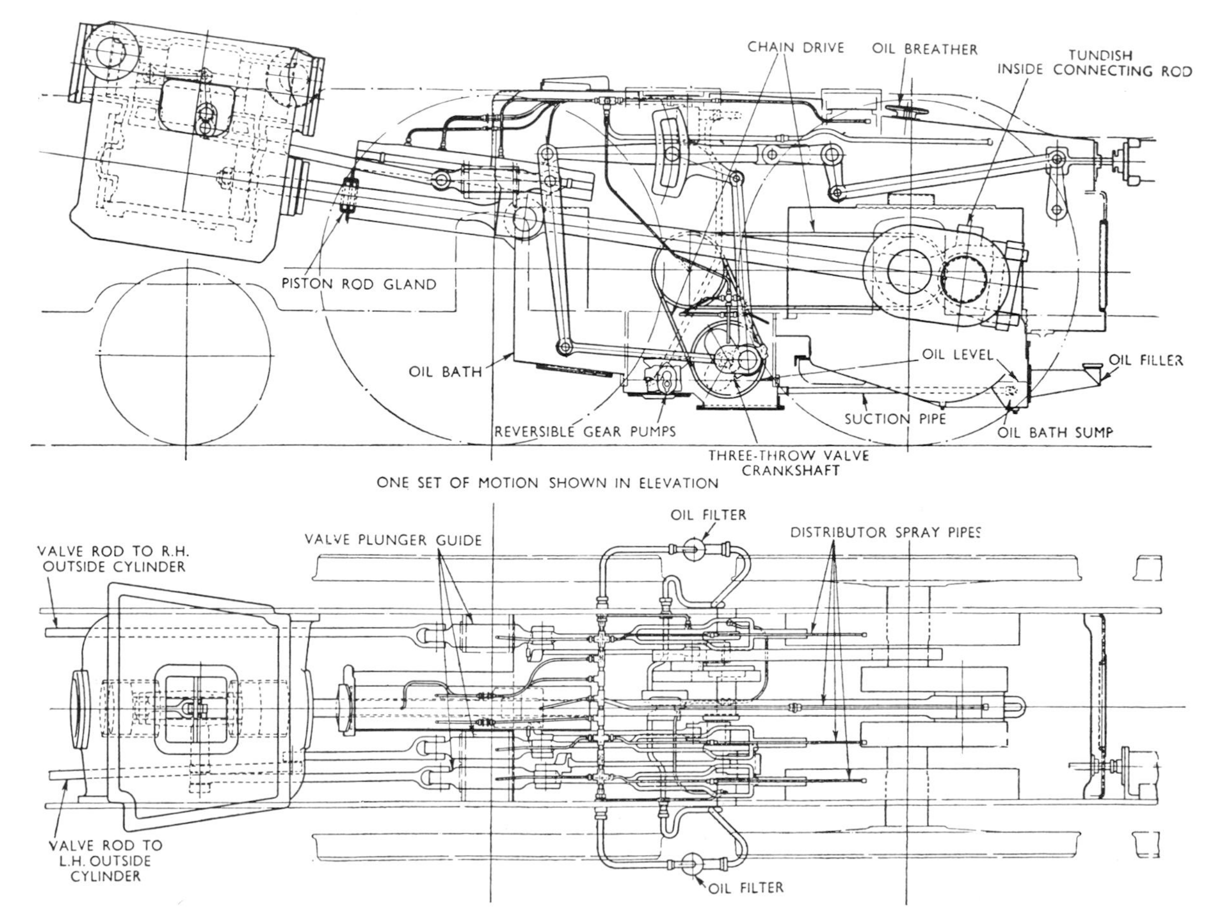

Fig. 269.—*"Merchant Navy" class; layout of valve gear oil bath and inside cylinder.*

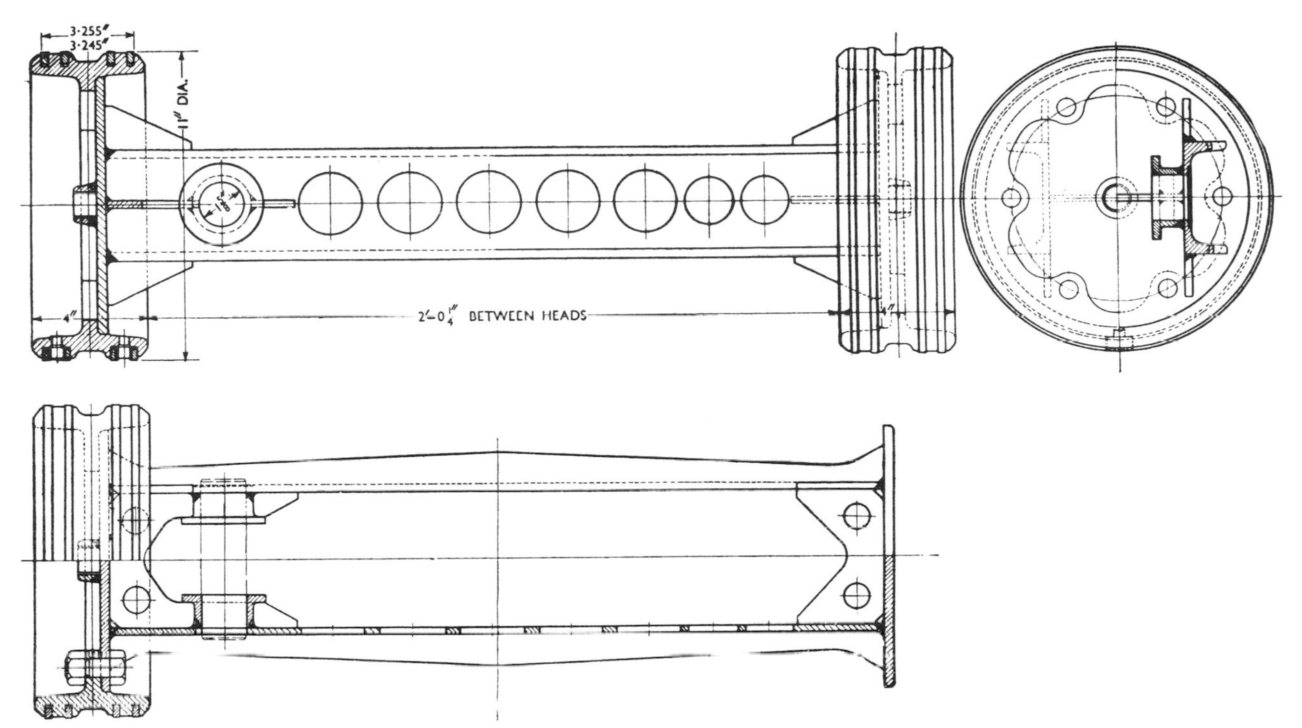

Fig. 270.—*" Merchant Navy " class; detail of piston valve.*

by its novelty. The lagging casing was carried on the main frames instead of on the boiler. A very light welded frame of cold-rolled sections was used to support this casing, which was itself built up from 20-gauge plate. This form of construction while providing a smooth, if not streamlined exterior, saved about 17 cwt., as compared with conventional British practice in the provision of boiler lagging. Taken all round, the " Merchant Navy " class 4-6-2s, and the smaller " West Country " class (Fig. 271) that followed in 1945 represented a remarkable departure from normal practice, and some outstanding feats of weight-haulage at high speed stand to the credit of both classes. The basic dimensions of the two series, as originally built were:

Southern Railway 4-6-2*s*

Class	" Merchant Navy "	" West Country "
Cylinders dia., in.	18	16 3/8
Cylinder stroke, in.	24	24
Coupled wheel dia., ft. in.	6-2	6-2
Combined total heating surface, sq. ft.	3272.9	2667
Grate area, sq. ft.	48.5	38.25
Boiler pressure, lb./sq. in.	280	280
Tractive effort at 85% boiler pressure, lb.	37,500	31,000
Total weight of engine in working order, tons	94 3/4	86

Fig. 271.—*Southern Railway: Bulleid's lightweight* 4-6-2 *" Battle of Britain " class* (1945).

The " West Country " class locomotives were important as they represented a marked change in the motive power policy of the Southern Railway—not only in design, but in large scale replacement of ageing and obsolescent steam locomotive stock. Hitherto the emphasis had been almost entirely upon electrification, leaving the Chief Mechanical Engineer to maintain existing locomotives until they were made redundant by electrification. With the introduction of the light-weight " West Country " 4-6-2s, embodying all the distinctive features of the heavy express passenger " Merchant Navy " class, an order for no fewer than 70 of the " West Country " engines was authorised. The total strength of the class was eventually 110 locomotives. The general performance of both classes of Bulleid " Pacific " in ordinary service will be discussed in a later chapter of this book; but in view of the special design features incorporated in them, some reference is appropriate at this stage to the series of full dress trials carried out at the Rugby testing station, and on the Settle and Carlisle line of the former Midland Railway, with an engine of the " Merchant Navy " class.

Two factors made the locomotive a difficult one to test. The special valve gear revealed a number of peculiarities in working, and there was always the tendency of the engine, whether on the test plant or on the road, to go into a heavy slip which resulted in buckled side rods. With regard to the valve gear, the published test bulletin has this comment:

" It was found that the actual cut-off bore no definite or consistent relationship to the setting of the reversing gear, not only for the locomotive as a whole but especially for the individual cylinder ends. In the shorter cut-offs particularly, there was a general tendency for the actual mean cut-off to lengthen with increasing speed but not in a smooth or regular manner. The power out-put in the short nominal cut-offs in the upper part of the speed range was found to be greater than that of other locomotives, size for size, and was in some cases more than would be theoretically possible at an actual cut-off equal to the nominal cut-off, even assuming that the cylinders were completely filled up to the point of cut-off with steam at fullsteam chest pressure and that no early release occurred. The true cut-off must have been longer than the nominal. At times quite random changes occurred that appeared to be caused by minute changes of speed or boiler pressure. Some of these random changes were relatively small, though enough to upset the test conditions, but others were of relatively large amount. Whilst changes that occurred over a period of weeks or months might be ascribed to wear of the motion, however small, this could hardly be the case with changes from one day to the next or which occurred, sometimes more than once, in a single test period."

With other locomotives that were subjected to precise testing, either at Rugby or Swindon, it was always a matter of considerable importance to determine the limit of evaporation of the boiler—either the " grate limit " or the " front-end " limit. With the " Merchant Navy " class 4-6-2 it was inexpedient to attempt this because of the proneness of the locomotive to slipping. While usually a severe slip could be tolerated either on the test plant or on the road, with no more than the inconvenience of having to make the particular test again, the frequency with which slipping led to buckled side rods was naturally a matter of some concern. This tendency to bad slipping was ascribed by the test engineers to leakage of lubricating oil from the enclosed oil bath. This inevitably reached the wheel treads and the rollers of the stationary testing plant. Thorough cleaning before every test and wiping of the test plant rollers at frequent intervals were necessary to enable any high powered tests to be completed. Even with this reservation however the locomotive registered some remarkably high outputs of power. On the test plant the locomotive was steamed up to a maximum evaporation of 42,000 lb. per hour of feed water, requiring a coal consumption of 7,000 lb. of Grade 1A coal per hour. On the Settle and Carlisle line the maximum steaming rate was 29,000 lb. per hour.

The conclusions to be reached from the tests were that the " Merchant Navy " as originally designed, was a most effective and capable locomotive, but one that was relatively uneconomical. To quote the report:

" If operated, as they frequently are in service, with the reversing gear in a relatively long cut-off and with the regulator very little open, these locomotives will be still less economical though their performance will then be more reliable from the mechanical point of view and from that of power output. The vibration, associated with full regulator and very short cut-off working, is also avoided.

" In spite of the good draught provided by the multiple jet blast pipe the combustion was never very good, particularly with the lower grade coal. There was, however, no evidence that the thermic syphons had anything to do with this. On the other hand there was some evidence that wider spacing of the grate bars and the provision of a deflector plate and admission of rather more air through the firehole door, with the doors closed, would effect some improvement.

" The actual draughting arrangement could almost certainly be improved by some re-design of the chimney. The actual choke is located about 7 in. above the effective choke where the vacuum is highest."

On the London and North Eastern Railway the death of Sir Nigel Gresley in 1941, and the appointment of Edward Thompson as Chief Mechanical Engineer led to a re-appraisal of the first-line express passenger locomotive stock. In war conditions the Gresley conjugated valve gear for actuating the valves of the middle cylinder in a 3-cylinder engine, suffering from inadequate maintenance, frequently developed so much slogger in the pin joints as to result in serious over-running of the valves and failures of various kinds, and one of the " P2 " 2-8-2 engines of the " Cock o' the North " class was taken as a convenient unit on which to try out a modified arrangement of a 3-cylinder layout. As might be expected the Gresley arrangement, with

Fig. 272.—*L.N.E.R. E. Thompson's* 3-*cylinder* " A2 " *class* 4-6-2, *adapted from Gresley* " P.2 " *class* 2-8-2.

Fig. 273.—*L.N.E.R. The Gresley* " A4 " *in post war guise, with garter blue livery restored, but valances not refitted.*

Fig. 274.—*L.N.E.R. A Thompson* " A2 " *Pacific, showing the unusual wheel spacing* (1944).

Type Designer Wheel dia., ft. in.	" A3 " & " A4 " Gresley 6-8	" A2 " Thompson 6-2	" A2 " Peppercorn 6-2	" A1 " Peppercorn 6-8
	ft. in.	ft. in.	ft. in.	ft. in.
Bogie to leading driver	5-6	8-2	5-7	5-9
Leading to mid-driver	7-3	6-6	6-6	7-3
Mid to rear driver	7-3	6-6	6-6	7-3
Rear driver to trailing truck	9-6	9-6	9-9	9-9

different lengths of connecting rod for the inside and outside cylinders, resulted in certain inequalities of steam distribution, as between the three cylinders, and Thompson laid down the principle that all three cylinders should have connecting rods of equal length. If the drive was to remain divided between two axles this would entail placing the inside cylinder well ahead of the two outside ones; but by rebuilding a " P2 " engine as a 4-6-2 a cylinder layout on this basis could be got in, and a useful trial of the arrangement made. The six engines of the " P2 " class were not the most successful of the Gresley designs from the viewpoint of availability and maintenance costs, and while carrying out the experiment in cylinder layout it was hoped to improve their general efficiency.

The resulting engine looked something of a makeshift (Figs. 272, 274), with the inside cylinders placed just ahead of the leading pair of coupled wheels, and the unusual distance of 8 ft. 2 in. between the leading coupled axle and the rear bogie axle. There was some re-arrangement of the boiler, resulting in a reduction of the total heating surface, while the cylinder diameter was reduced from 21 to 20 in. The nominal tractive effort at 85 per cent. boiler pressure became 40,318 against the original figure of 43,462 lb. and adhesion weight was reduced from 79 to 66 tons. In the severe conditions existing on the East Coast main line in Scotland the capacity of the locomotives was of course somewhat reduced, but the performance of the rebuilt engines was such as to justify the class being chosen as a future L.N.E.R. standard, and in 1946 further engines of the same general type were built new (Fig. 275). These latter had cylinders 19 in. diameter and a boiler pressure of 250 lb. per sq. in. making the nominal tractive effort 40,430 lb. against 40,318 lb. in the conversions from Class " P2 ". In the author's experience these engines, designated Class " A2 ", worked quite effectively, but were subject to a curious defect of detail in the exhaust connection from the front of the outside cylinders. The jointing faces at the extremities of this connecting piece were at right angles to each other; the joints " worked " and leaked, and it was not unusual for the nuts to slacken off the stud fastening at the front end.

In 1947, when further " Pacific " locomotives were authorised, A. H. Peppercorn, who had succeeded Thompson as Chief Mechanical Engineer, produced a modified design on more orthodox lines, moving the outside cylinders forward, and bringing the bogie nearer to the driving wheels (Figs. 276, 277). In so doing the Thompson principle of having all connecting rods of equal length was abandoned, but a locomotive of much neater appearance was the result. The wheel spacing in successive L.N.E.R. " Pacific " design is shown in the table above.

The boiler was common to the Thompson and Peppercorn " Pacifics " and differed from that of the Gresley " A4 " class principally in a reduced distance between tube plates, and in a much larger grate area. The relevant dimensions were as follows:

L.N.E.R. Pacific Boilers

Type	Gresley " A4 "	Thompson/ Peppercorn
Heating surfaces, sq. ft.:		
Tubes	2345.1	2216.07
Firebox	231.2	245.3
Superheater	748.9	679.67
Combined total	3325.2	3141.04
Grate area, sq. ft.	41.25	50
Distance between tube plates, ft. in.	17-11 3/4	17-0

Fig. 275.—*L.N.E.R. A Standard Thompson " A2 ", in post-war standard apple-green livery* (1946).

Fig. 276.—*L.N.E.R. A. H. Peppercorn's* " A1 " 4-6-2 *for express passenger service, built Doncaster*, 1948.

The Thompson and Peppercorn " Pacifics " of the L.N.E.R. represented the attempt of another railway mechanical engineering establishment to adapt their policy to the changing times, and it centred mainly around the elimination of the conjugated valve gear for the inside cylinder, and the provision of a boiler with a grate area of 50 sq. ft. instead of 41 $^1/_4$ sq. ft. The 50 sq. ft. grate had already been used on the L.N.E.R. in the " P2 " class 2-8-2 engines from which the first Thompson " Pacifics " were rebuilt. It cannot be recorded however that either the Thompson or the Peppercorn engines came to supersede those of Sir Nigel Gresley. The large fireboxes were advantageous if a very high rate of steam production was required; but with a return of peacetime conditions, albeit at first in times of great austerity, the very heavy train loads of wartime soon began to reduce to more normal formations. When the time came for appreciable acceleration of service the traffic policy on the East Coast Route became one of running a more frequent service between London and Newcastle than had been operated before the war, and the loading of the trains was reduced accordingly.

Furthermore, with the return of more normal conditions the unreliability of the Gresley conjugated valve gear also began to recede, and one could see the difficult conditions prevalent in the days of theThompson régime as no more than a passing phase. In this later post-war era the large fireboxes of the Thompson and Peppercorn " Pacifics " proved something of a disadvantage, and on comparable duties the newer engines proved heavier coal burners. For this reason, when non stop running was resumed between King's Cross and Edinburgh, and eventually the service was accelerated beyond the best that had been scheduled in pre-war years, these trains were always entrusted to the Gresley " Pacifics " (Fig. 273). Some details of the working of the Anglo-Scottish expresses in the last years of steam are given in a later chapter of this book. The reliability of the " A4 " class in particular was enhanced by the fitting of a modified design of big-end on the inside connecting rod, which greatly reduced the tendency to heat which was sometimes present in these engines in their original condition.

Fig. 277.—*L.N.E.R. The Peppercorn version of* " A2 " *Pacific with normal wheel spacing* (1947).

Fig. 278.—*L.M.S.R. One of the last two Stanier " Pacifics "* (1947) *with roller bearings on all axles, rocker grates and self-cleaning smokeboxes. Engine No.* 46256 Sir William A. Stanier F.R.S.

On the London Midland and Scottish Railway no changes in the basic design of the largest express passenger locomotives took place, and development took the form of improved accessories such as rocker grates, self-cleaning smokeboxes, and the use of roller bearings on all axles. Streamlining was abandoned on the " Duchess " class " Pacifics ", and in course of time all those engines originally equipped thus were modified to a conventional appearance. The last two engines of the " Duchess " class to be built Nos. 6256 and 6257, completed at Crewe in 1947, were in many ways representative of the highest development of the British steam railway locomotive for the heaviest express passenger working, and by a happy inspiration engine No. 6256 was named *Sir William A. Stanier F.R.S.* (Fig. 278). In later years a locomotive of this class under controlled road testing conditions sustained the greatest output of power ever recorded for any length of time continuously with a British passenger locomotive.

In the later years of the war the principles of design applied so successfully to the " Pacific " engines were used in a complete redesign of the " Royal Scot " class 4-6-0, (Fig. 279) making them virtually new engines. The design was based on experience gained with experimental engine No. 6170 built in 1935 as a replacement for the ill-fated high-pressure compound No. 6399 described in Chapter Nine of this book. The engine 6170 (Fig. 280) had a taper boiler and a front-end more in line with the latest Stanier " Pacific " practice than with that of the original " Royal Scots ". The boilers of these varieties of the class were:

Royal Scot Boilers

Variety	Original	Engine 6170	Rebuild of 1943
Heating surfaces, sq. ft.			
Tubes	1892	1669	1655
Firebox	189	195	195
Superheater	399	360	420
Combined total	2480	2224	2270
Grate area, sq. ft.	31.2	31.25	31.25

Fig. 279.—*L.M.S.R. One of the " Royal Scot " class* 4-6-0s *as originally built in* 1927*: the* Lancashire Witch.

Fig. 280.—*L.M.S.R. The first taper-boilered "Scot": No.* 6170 British Legion, *built* 1935, *on the chassis of the high pressure* 4-6-0 *No.* 6399.

The rebuilt engines turned out at Crewe from 1943 (Fig. 281) onwards had double blastpipes and chimneys as per the accompanying drawing (Fig. 282). The motion details were as follows:

Piston valve dia., in.	9
Steam lap, in.	$1\,^{3}/_{8}$
Port width, in.	$2\,^{1}/_{4}$
Exhaust clearance, in.	$^{1}/_{16}$
Average lead, in.	$^{5}/_{16}$
Valve travel in full gear, in.	$6\,^{13}/_{32}$

Apart from the boiler, the following features also contributed to the great success of these engines.

(1) Bolster bogies of the de Glehn type.

(2) Cast steel coupled axleboxes with pressed-in brasses, and a large capacity underkeep with pad lubrication.

(3) The smokebox saddle was cast integral with the inside cylinder, and circular smokebox.

These locomotives proved capable of magnificent performance on the road, which will be referred to in detail in a later chapter. At this stage some early indicator results obtained with engine No. 6138 showed that the engines were capable of a very high output of power on relatively short cut-offs.

Indicator Results: L.M.S. Engine 6138

Cut-off %	Speed m.p.h.	Indicated h.p.
5	62	925
10	00	1070
12	54	1110
15	60	1520
18	62	1670
22	56	1700
26	52	1820
32	44	1840
38	30	1670
46	22	1440
Full Gear	5	420

Fig. 281.—*L.M.S.R. One of the standard taper-boilered "Royal Scot"* 4-6-0s, *first converted from original type in* 1943.

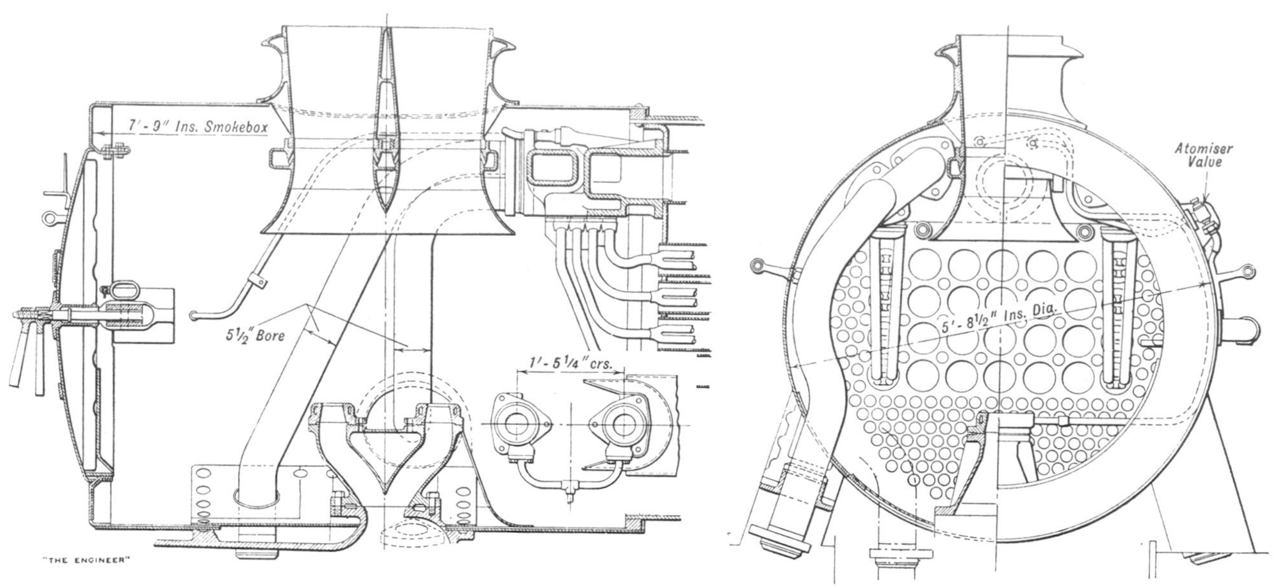

Fig. 282.—*L.M.S.R. Sections of smokebox, taper-boilered " Royal Scot " class* 4-6-0.

The modernised " Royal Scots " provided a classic case of the improvement in performance that could be effected by careful attention to every detail of the front-end, while using Walschaert's valve gear. Nevertheless, the L.M.S.R. under H. G. Ivatt initiated some further interesting experiments with valve gears other than standard on locomotives of the highly successful Stanier Class " 5 " mixed traffic 4-6-0 design, in 1947. At the same time further engines, with the standard arrangement of the Walschaert's gear, were built with roller bearing axles (Fig. 287). The 30 engines built for special observation (Figs. 284, 285) were fitted as follows:

10, Plain bearings, British Caprotti valve gear
10, Roller bearings, British Caprotti valve gear
9, Roller bearings, Walschaert's valve gear
1, Roller bearings, Stephenson's link motion, outside.

The inclusion, among these trial engines of one with the Stephenson link motion was extremely interesting and the engine in question, No. 4767 (Fig. 286), was fitted with a double blastpipe and chimney.

It was not often that devices calculated to save in day-to-day maintenance, and to reduce fuel consumption could be incorporated in an otherwise standard locomotive without some increase in prime cost, or weight; and the roller bearing axles on these Class " 5 " 4-6-0s were a case in point. The driving axles, with their special housings for the roller races constituted a massive ensemble and to provide adequate clearance from the front of the ashpan it was necessary to increase the distance between the driving and trailing pair of coupled wheels by 4 in. This in turn meant lengthening the frames, and resulted in an increase in weight of the engine from 72.1 to 75.3 tons. This application of roller bearings took place just at

Fig. 283.—*L.M.S.R. A " Jubilee " class* 3-*cylinder* 4-6-0 *as rebuilt with the tapered " Royal Scot " boiler.*

Fig. 284.—*L.M.S.R. Class* " 5 " 4-6-0 *Caprotti valve gear* (1948).

Fig. 285.—*L.M.S.R. Class* " 5 " 4-6-0 *with Caprotti valve gear, and double blastpipe and chimney* (1948).

Fig. 286.—*L.M.S.R. Class* " 5 " 4-6-0 *with outside Stephenson's link motion, and double blastpipe and chimney* (1947).

Fig. 287.—*L.M.S.R. Class* " 5 " 4-6-0 *with roller bearings on all axles* (1947).

the time of nationalisation of the railways of Great Britain, and the L.M.S.R. experiments were sufficiently successful for roller bearings to be adopted as standard for the new British Railways locomotives designed later under the direction of R. A. Riddles.

In the new L.M.S.R. locomotives the general layout was carefully arranged to give quick access to all parts likely to need attention. With the valve gear, the cambox was arranged to slide out on to a temporary platform so that the valves could be inspected. Generally this use of poppet valve gear differed from most others previously tried in Britain, in that the standard engine was re-designed where necessary to suit the valve gear, rather than the reverse. Thus the valve gear incorporated all those features that the wide experience of the manufacturers had shown desirable, without imposing the restrictions that were sometimes necessary when poppet valves were applied, with all the appearance of a " last hope " to some existing locomotives of doubtful capacity. To avoid any confusion on the part of drivers who would be called upon to work a Caprotti engine one day, and a Walschaert's engine the next, the reversing gear in the cab was arranged so that the same number of turns was required with each gear, to work from the full forward to the full reverse position.

No figures were ultimately published regarding comparative results obtained with the different forms of valve gear, and it can be generally inferred that the poppet valve gear, even in this series of experiments showed no marked advantage over the standard arrangement of piston valves. This conclusion also applied to the one engine fitted with the Stephenson link motion. In actual working on the road this latter engine displayed exactly the characteristics one would have expected. The layout of the gear closely resembled that of the standard Great Western 2-cylinder 4-6-0s, particularly in the shortness of the eccentric rods. This feature tended to accentuate the extent to which lead was increased as the engine was linked up. On the G.W.R. the amount of lead was adequate for fast and free running at 20 to 25 per cent. cut-off; it was reduced to nothing at all at 40 to 50 per cent., and became *negative* when the cut-off was further increased till it became about $^1/_8$ in. negative in full gear. This setting had the effect of giving enhanced power at slow speed, and very rapid acceleration. These characteristics the L.M.S.R. engine No. 4767 possessed in full measure.

On the Great Western Railway itself, in the last years before nationalisation, to outward appearances there was little development in design practice, and the only new class to appear, the " County " class 2-cylinder 4-6-0 of 1945 (Figs. 288, 289) might, if taken at its face value, have seemed a curious series of departures from the long-established and highly successful precepts of Swindon Works, in that non-standard features were introduced without any obvious reason. Furthermore, the locomotives were not notably successful in traffic in their early days. Seen in retrospect the " County " class can now be set down as one of the " guinea pigs " of locomotive history, but a guinea pig from which the intended outcome never materialised. After C. B. Collett had retired from the post of Chief Mechanical Engineer of the G.W.R. in 1941, and had been succeeded by F. W. Hawksworth, as opportunity presented itself during the war years some consideration was given to post-war express passenger locomotive development. Successful though the " King " class had been in pre-war years the fuel situation, which had affected the Great Western Railway perhaps more than other administrations virtually precluded ideas of further development that would be based on a narrow firebox —at any rate for a high-powered locomotive, and detailed attention was given to the possibilities of a " Pacific ". While developments did not proceed very far it is generally understood that ideas had progressed to the extent of envisaging a 4-cylinder engine with 6 ft. 3 in. coupled wheels and a boiler pressure of 280 lb. per sq. in.

During the war years however work towards a new design intended purely for high-class express passenger

Fig. 288.—*G.W.R. F. W. Hawksworth's " County " class* 4-6-0, (1945).

service was ruled out, but towards the end of the war the authorisation of further 4-6-0 mixed traffic engines presented an opportunity to try out what would have been an outstanding feature of the proposed new " Pacific ", namely a boiler with a working pressure of 280 lb. per sq. in. Unlike Bulleid on the Southern Railway, Hawksworth felt that the increase from the 250 lb. per sq. in. successfully used on the " King " class 4-6-0s was not great enough to justify any departure from the standard methods of boiler construction used at Swindon, and for the mixed traffic 4-6-0 it was originally intended to use the " Castle" boiler, suitably modified to take the higher pressure. Unfortunately the weight came out too heavy, but the wartime building at Swindon of the Stanier type 2-8-0 had provided flanging blocks of a size that

Fig. 289.—*G.W.R. The first " County " class* 4-6-0 *of* 1945, *No.* 1000 County of Middlesex *and the only one originally to have a double blastpipe and chimney.*

permitted of the building of a boiler intermediate in size between the " Castle " and the famous Churchward standard No. 1 used on the " Hall " class. So, without any appreciable tooling the 280 lb. boiler was produced, of a weight suitable for the new mixed traffic 4-6-0. The only detail part that was new was the 6 ft. 3 in. driving wheel, foreshadowed for the intended express passenger " Pacific ". Thus emerged the " County " class mixed traffic 4-6-0 of 1945, and the first engine of the class was further distinguished by a double blast-pipe and chimney.

The basic dimensions of these engines were:

Cylinders (2) dia., in.	18 1/2
Cylinder stroke, in.	30
Coupled wheel dia., ft. in.	6-3
Heating surfaces, sq. ft.:	
Tubes	1545
Firebox	169
Superheater	265
Combined total	1979
Grate area, sq. ft.	28.84
Boiler pressure, lb./sq. in.	280
Nom. tractive effort at 85% boiler pressure, lb.	32,580

Two points of interest concerned the boilers of these engines. Although Hawksworth had begun to move away from Swindon traditions in a batch of " Hall " class engines constructed in 1944 (Fig. 290), using a somewhat higher degree of superheat, the " Counties " had relatively small superheaters. The second point is to note the general correspondence in boiler dimensions to those of the Stanier " 8F " 2-8-0s, eighty of which were built at Swindon during the war. These latter engines had a grate area of 28.65 sq. ft.; a firebox heating surface of 171 sq. ft., and superheaters providing 245 sq. ft. In the L.M.S.R. engines the boiler pressure was 225 lb. per sq. in.

Fig. 290.—*G.W.R. Improved " Hall " class* 4-6-0 *with modified frames and bogie, and higher degree of superheat* (1944).

Chapter Fifteen

Detail Features: New Standard Locomotives

Self-cleaning smokebox : oil-firing : grates and ashpans : tenders : roller bearings.

Even before the end of World War II the need to consider any features of detail design that would reduce the amount of day-to-day maintenance on locomotives, and assist towards the making of longer mileages between routine examinations and repairs, was becoming urgent. Labour was scarce, and in a changing age fewer young men were being attracted towards locomotive operating. Towards this end one device that was widely accepted was the so-called "Self-cleaning smokebox". For many years certain of the privately-owned British railways, even before the grouping, had fitted devices to locomotives to enable some clearance of smokebox ash to be made at the discretion of the driver. On the London and North Western Railway, for example, a very effective ash blower was a standard feature of Crewe-built locomotives for many years. Modern practice was to install equipment in which the process was continuous and requiring no action whatever on the part of the driver.

The accompanying diagram (Fig. 291) shows a typical layout of a large locomotive smokebox, on which the arrows indicate the direction of the exhaust products of combustion. Where no device for clearing smokebox ash is fitted the larger cinders, drawn through the tubes by the draught accumulate in the smokebox, because they are too large to be ejected by the blast through the chimney. The purpose of the arrangement shown herewith is so to entrain the exhaust gases as to make them pass through a wire mesh screen at the front of the smokebox. The effect is to break up the larger cinders, in their contact with the wire screen, until they are of such a size that they can be ejected satisfactorily. This simple arrangement worked very well, and experience with locomotives so equipped was such that no removal of smokebox ash at running sheds was necessary between successive boiler washouts. Small plates with letters SC were fitted to the smokebox doors of locomotives fitted with self-cleaning screens, to indicate to shed staff that no attention was necessary in ordinary day-to-day servicing.

The use of self-cleaning equipment proved so attractive in post-war British operating conditions that it was fitted to certain existing locomotive classes engaged in mixed-traffic, common user duties, where a high availability was very desirable; but this led to some deterioration in performance. The insertion of the baffle plates and screens, together with the wire mesh naturally provided an obstruction to the flow of the exhaust gases—as indeed it was intended to do, to break up the larger cinders; in so doing however the obstructions had an adverse effect upon the draughting, with the result that the steaming of the locomotives was affected. An interesting example of this, and the way it was overcome, was provided by the "V2" class 2-6-2 mixed traffic engine of the L.N.E.R. introduced in June 1936 by Sir Nigel Gresley, and of which a total of 184 locomotives was eventually built. The class as a whole had a long record of successful performance, but to assist in their more intensive utilisation in heavy mixed traffic working self-cleaning plates were fitted into the smokebox from 1946 onwards, and this caused a deterioration in steaming capacity to a point far below the previous high standards. Some trial adjustments of the chimney and blastpipe proportions were made, including a reduction in the diameter of the blastpipe orifice from $5\,^1/_4$ to 5 in.—a retrograde step. But these modifications had no appreciable effect, and it was then decided to send

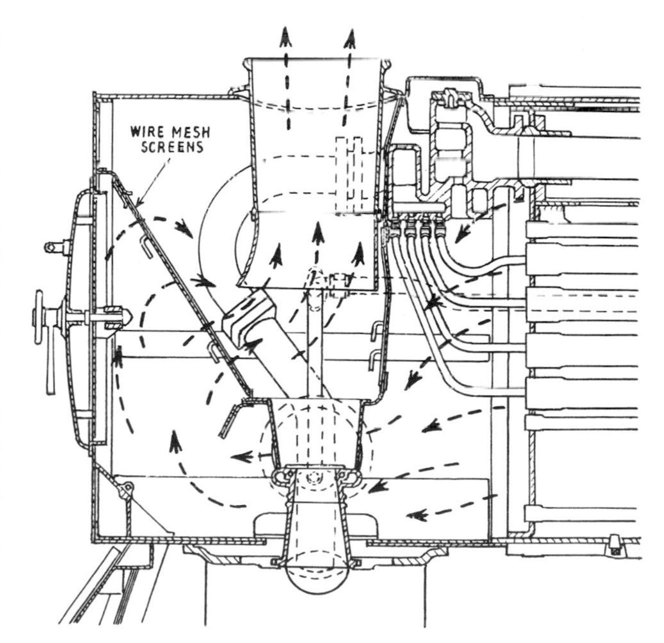

Fig. 291.—*Diagram of self-cleaning smokebox, as applied on the South African Railways.*

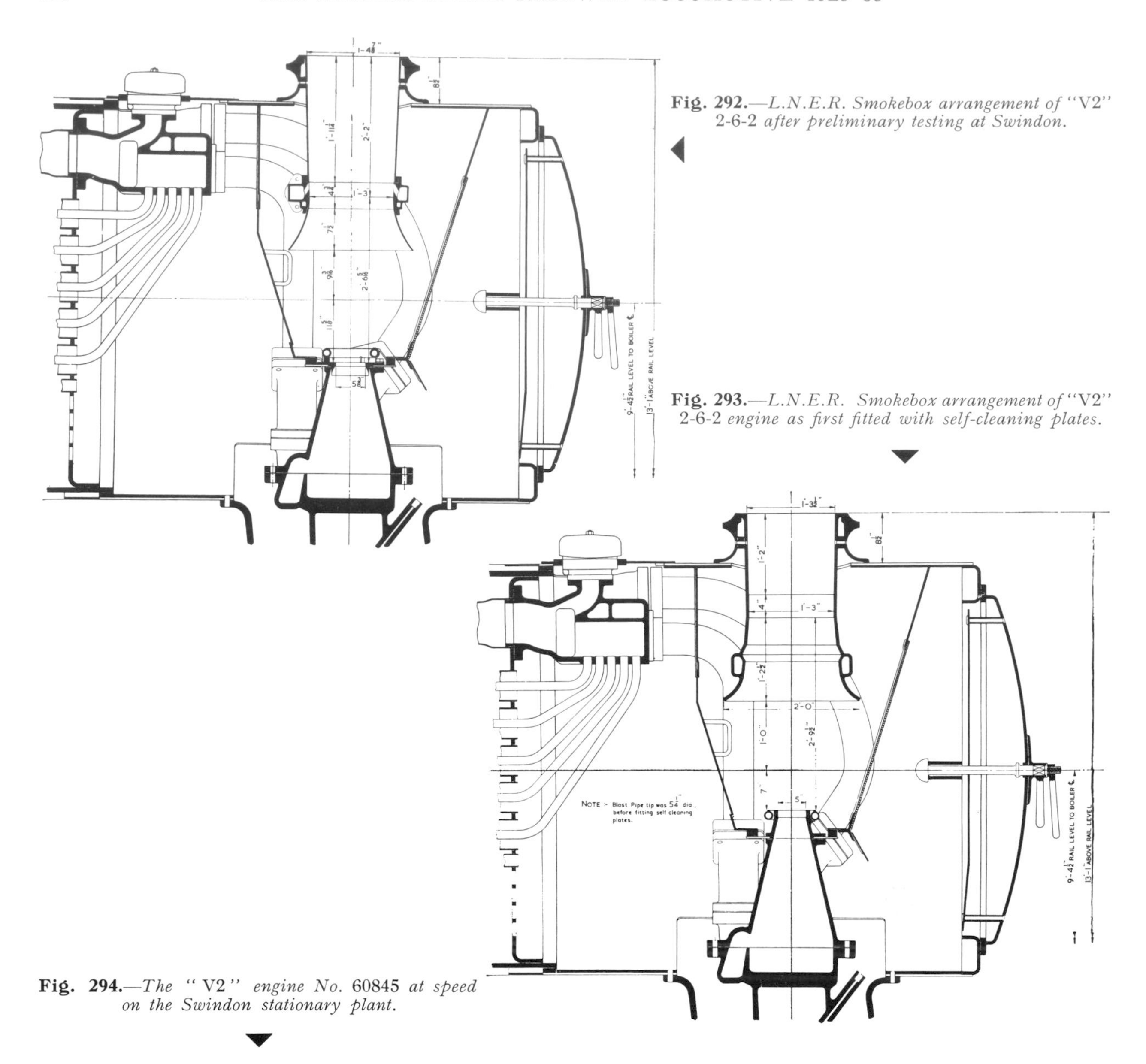

Fig. 292.—*L.N.E.R. Smokebox arrangement of "V2" 2-6-2 after preliminary testing at Swindon.*

Fig. 293.—*L.N.E.R. Smokebox arrangement of "V2" 2-6-2 engine as first fitted with self-cleaning plates.*

Fig. 294.—*The "V2" engine No.* 60845 *at speed on the Swindon stationary plant.*

a representative engine of the class to the Swindon stationary testing plant for examination, and modification if found necessary.

The accompanying diagram (Fig. 293) shows a cross-section of the smokebox of this engine as received, with the reduced diameter of blastpipe orifice. First tests on the plant found the steaming to be very poor. The air supply was insufficient to maintain continuously an evaporation higher than 14,000 lb. per hr. When the self-cleaning plates were removed, and the steaming was much improved, the evaporation could be sustained at a maximum of 24,000 lb. per hr. But the exhaust was very discoloured at the best of times; there was much black smoke at others, and in any case the reduction of the diameter of the blastpipe orifice as compared with the original probably meant that the cone of exhaust steam was not entirely filling

Fig. 295.— *G.W.R. Oil-fired "Hall" class* 4-6-0 *No.* 5955 Garth Hall.

Fig. 296.—*G.W.R. Oil-fired* 2-8-0 *No.* 2872.

Fig. 297.—*G.W.R. " Castle " class* 4-6-0 *No.* 5091 Cleeve Abbey *equipped for oil firing, and fitted with small tender.*

the chimney. From the accompanying drawing it will be seen that originally the chimney itself was very nearly parallel.

Some very careful studies into the theory of blastpipe and chimney design had been undertaken at Swindon, and the chimney was altered to convergent-divergent form as shown in the second smokebox cross-sectional drawing (Fig. 292). To enable the divergent portion, which tapers at the rate of 1 in 14 to be made 2 ft. 2 in. long the choke was lowered, as can be seen from a comparison of two cross-sectional drawings. The top of the blastpipe had to be lowered considerably to bring it into the correct position in relation to the chimney choke, and a new blastpipe had to be made to accomodate all these parts in their existing positions. With this arrangement and self-cleaning plates in position the boiler was found capable of a sustained rate of evaporation of 30,000 lb. per hr. while using Grade 2B coal from Blidworth colliery (Fig. 294). This experience with the " V2 " engine was extremely interesting as showing how the performance of a locomotive can be seriously impaired by the addition of some " improved " auxiliary unless all the factors involved are carefully reconsidered.

At earlier times in the history of the British steam railway locomotive there had been occasions when the normally good grades of locomotive coal were not available, and performance had suffered in consequence. But these occasions had been no more than passing phases, and even the period of the prolonged coal strike in 1926, serious though it was at the time, appears as no more than a brief interlude when the history of steam traction in Great Britain is viewed in retrospect. But in later stages of World War II and some time afterwards the deterioration in the quality of coal available, even for first class express passenger work was such as to cause important alternatives to be considered. At earlier times of emergency the conversion of locomotives to oil-firing had been carried out as a temporary measure, but in 1946 the policy initiated by the Great Western Railway was conceived on a long-term basis. It was intended that certain well-defined geographical areas, and groups of services should be changed over completely to oil-firing. Cornwall was one such area, and the underlying principle in making such a choice was that not only would the use of coal be reduced, but that the long hauls from the collieries to Plymouth and depots beyond would be avoided.

The decision to convert the stud of main line freight engines working in South Wales might in the ordinary way have seemed a strange one, seeing that the line ran through an area adjacent to one of the principal sources of Great Western locomotive fuel. But immediately after the war, under Government direction, a large tonnage of the best Welsh coals was set aside for export only, and ironically enough much of the local workings in the coalfield area was dependent upon low grade imported coal. The author had the extraordinary experience, in 1947, of travelling in a local train on the Taff Vale main line on which a 0-6-2 tank engine of otherwise excellent reputation, was steaming so badly on imported coal, that the ejector was failing repeatedly, and the train stopped intermediately in consequence because the brakes were leaking on.

The Great Western Railway adapted engines of three main classes for oil-firing: the " 28XX " class heavy mineral 2-8-0 (Fig. 296); the " Castle " class express passenger 4-6-0 (Fig. 297) and the " Hall " class mixed traffic 4-6-0 (Fig. 295). A number of " Hall " class engines so equipped were drafted to Laira shed, Plymouth for working in Cornwall, and

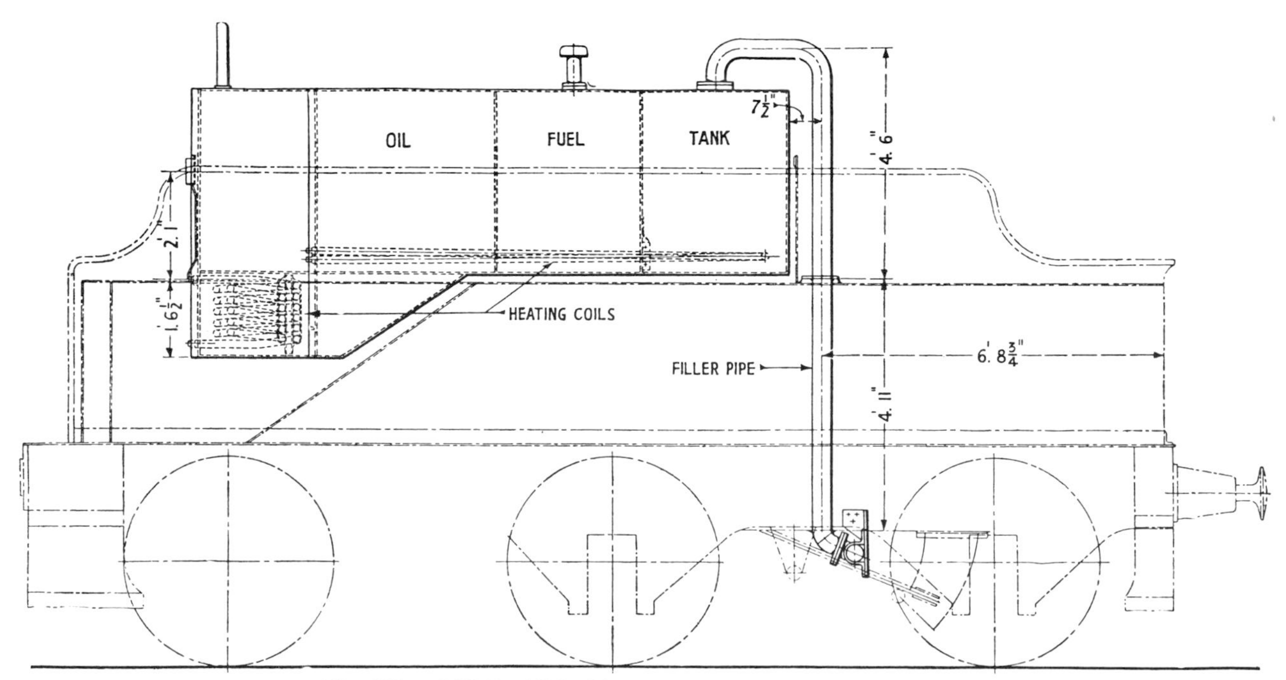

Fig. 298.—*G.W.R. Oil fired locomotives: arrangement of tender.*

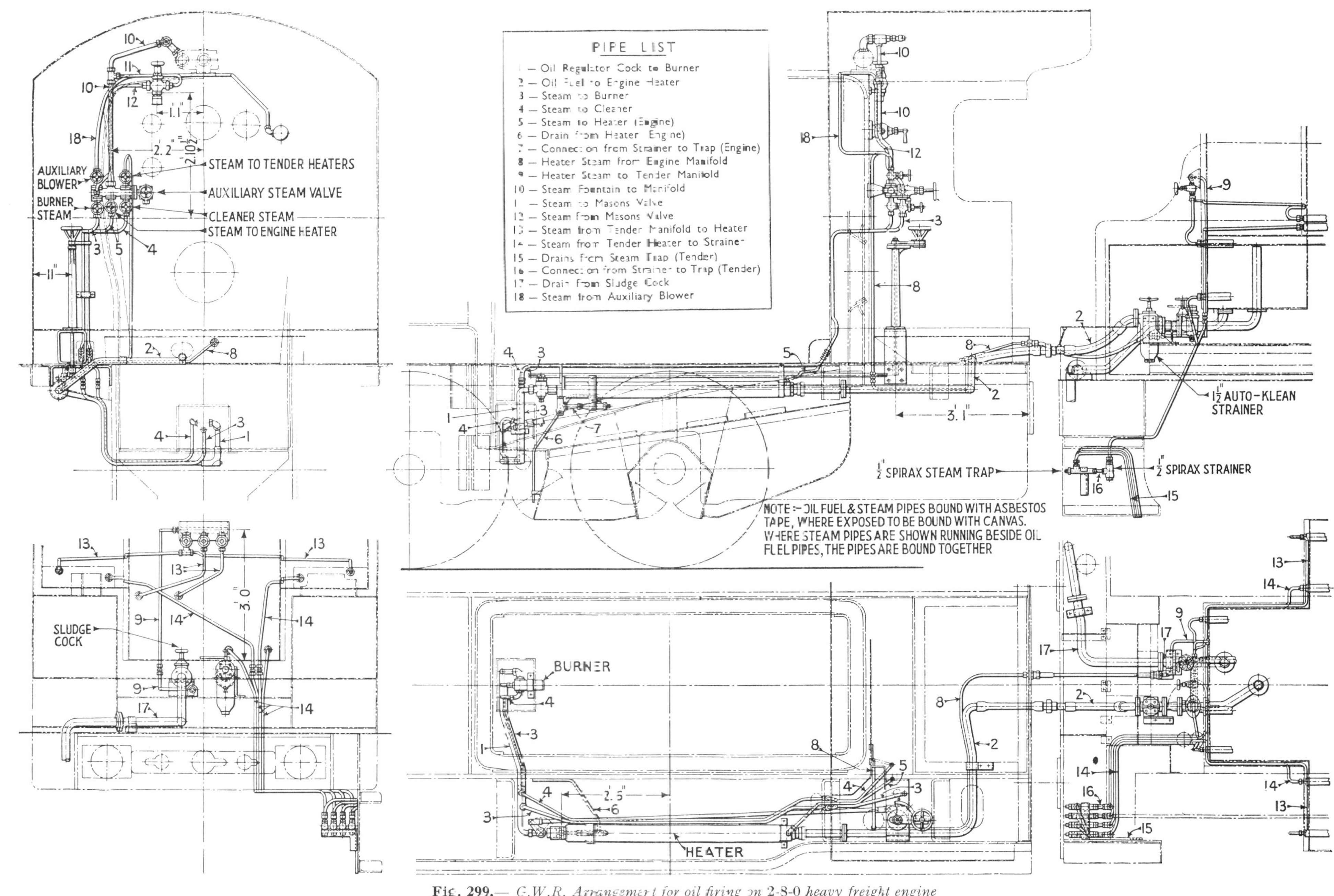

Fig. 299.— *G.W.R. Arrangement for oil firing on 2-8-0 heavy freight engine*

G.W.R. Oil-fired " Castle " Working

Engine No.	5039	5079	5079
Engine Name	*Rhuddlan Castle*	*Lysander*	*Lysander*
Load, gross trailing tons	375	365 to Truro 325 to Penzance	365
Train	9.5 a.m. ex-Paddington	Down " Cornish Riviera "	Up " Cornish Riviera "
Length of trip, miles	118.3	79.5	79.5
Total time of trip, min.	138 1/4	135 1/2	150 1/2
Water consumption, gall.	3600	2900	2700
Oil consumption, gall.	340	380	370
Water per train mile, gall.	30.4	36.5	34.0
Oil per train mile, gall.	2.88	4.78	4.65
Evaporation: gall. of water/gall. of oil	10.5	7.6	7.3
Ratio A: percentage of total time in which engine was steamed, %	85	56	47
Evaporation: Ratio A	0.124	0.135	0.155
Overall speed of trip, m.p.h.	51.4	35.2	31.7

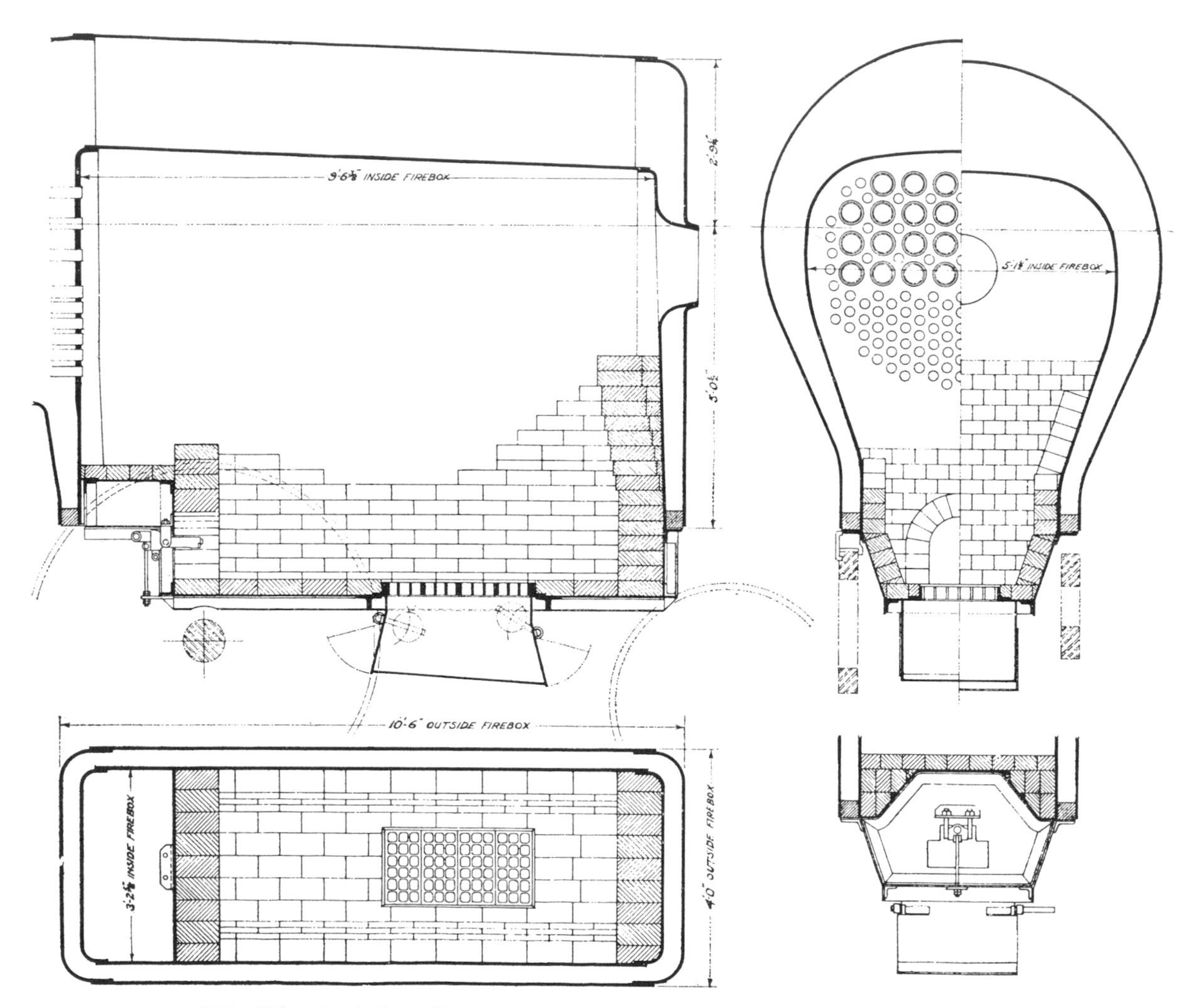

Fig. 301.—*Iraqi State Railways: Firepan and brick lining, Pacific locomotives.*

Experience in Cornwall, also on "Castle" class locomotives showed, as would be expected, that a very much heavier oil consumption per train mile was involved, partly on account of the heavy gradients encountered, and even more so because of the relatively low proportion of the total journey time during which the engines were being steamed. These figures, tabulated herewith, alongside corresponding ones for the Bristol route provide ample explanation in themselves why the Great Western management decided to make Cornwall an area to be wholly served by oil-fired locomotives. The scope for saving coal was clearly greater there than in relatively easier country.

At the time the above journeys in Cornwall were made engine No. 5079 was regularly making two return trips from Plymouth to Penzance and back every day, covering a daily mileage of 318. The steaming of the engine, especially on the heavy gradients of 1 in 60 and 1 in 70, was very reliable, while the reversing gear had frequently to be advanced to more than 40 per cent. cut-off, with full regulator.

The success of the Great Western project on a limited scale, led the Government of the day to embark on a nation-wide scheme for oil-firing, to alleviate in some way the acute shortage of coal that was such a source of difficulty at that time. The scheme was announced in the summer of 1946, and the Minister of Transport authorised the conversion of no fewer than 1,217 main line locomotives, with an estimated saving of 1,000,000 tons of coal per annum. Unfortunately the necessary foreign exchange to purchase oil fuel was not eventually forthcoming, and the scheme was halted in its early stages. At this period in time however it is interesting to recall the extent of the conversions proposed, and the classes of locomotives concerned.

Proposed Conversion to Oil Firing—1946

Railway	Engine		Number to be converted
	Type	Class	
G.W.R.	2-8-0	"28XX"	63
	4-6-0	"Hall"	84
	4-6-0	"Castle"	25
	Beyer-Garratt		33
L.M.S.R.	0-8-0	Class 7	175
	2-8-0	Class 8	266
	2-8-0	S.D.J.R.	11
L.N.E.R.	2-8-0	"O4"	112
	2-8-0	"O2"	16
	2-8-0	"O1"	39
	2-8-0	"WD"	111
	2-6-0	"K3"	91
	0-8-0	"Q6"	46
	0-6-0	"J39"	35
Southern	4-6-2	"West Country"	20
	4-6-0	"N15" and "H15"	16
	2-6-0	"N" and "U"	34
	4-4-0	"D15"	10
	4-4-0	"L11" and "T9"	30

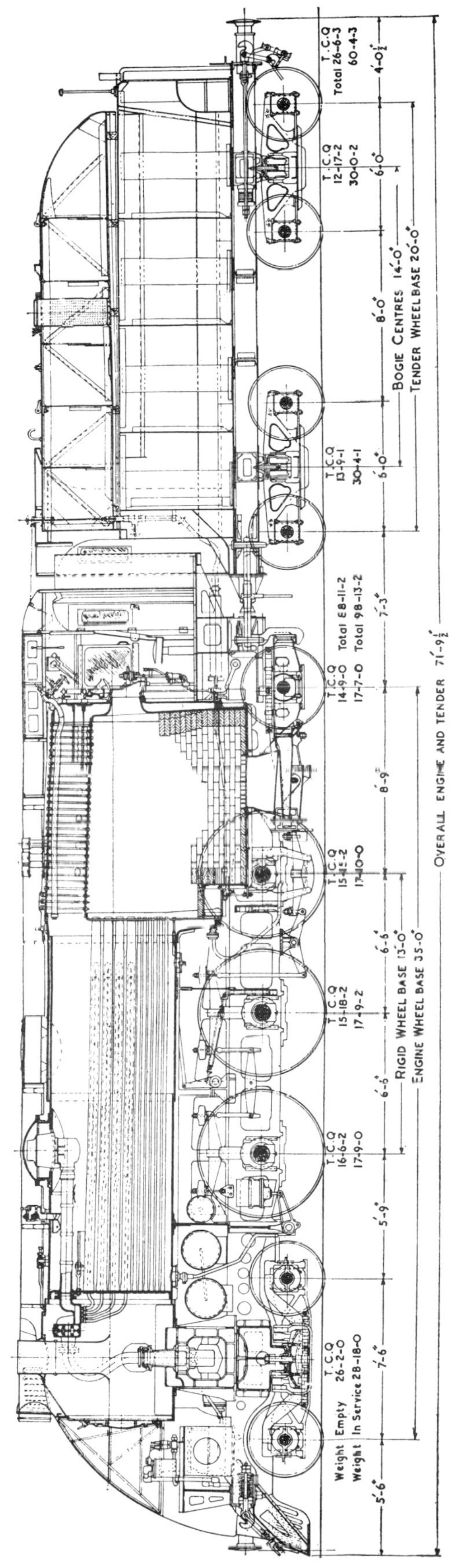

Fig. 302.—*Iraqi State Railways: arrangement of oil-fired 4-6-2 locomotive built by R. Stephenson and Hawthorns Ltd.* (1941).

Before the project was halted some progress had been made on the Southern Railway, where it was intended to work the majority of duties centred upon Fratton shed with oil-fired locomotives. Some footplate experiences on passenger train services between Portsmouth and Salisbury showed how successfully the basic arrangement developed on the Great Western Railway could be adapted to relatively small and old engines like the Drummond " L11 " and " T9 " 4-4-0s of the former London and South Western Railway. It will be noted from the foregoing table that with the exception of the Bulleid " West Country " 4-6-2s of the Southern all the engine classes proposed for conversion had narrow fireboxes. In this connection it is interesting to compare the cleverly adapted firebox of the G.W.R. " Castle " class with the firebox of a locomotive of comparable power, designed specifically for oil-firing, namely the Iraqi State Railways " PC Pacifics ", built in 1941 by Robert Stephenson and Hawthorns Ltd. and having a grate area of 31.2 sq. ft., and a nominal tractive effort at 86 per cent. boiler pressure of 31,080 lb. (Figs. 301, 302).

Following the end of the large oil-firing project the British steam locomotive was forced back entirely upon coal, and while there was a continuous fall in the average quality of fuel available one of the great difficulties experienced by the running departments was the variation in the quality—very often from day-to-day. It was a situation that the railway authorities themselves could do little to control. One might have locomotives loaded with reasonably good Welsh steam coal one day, and a mixture of Kitchen Nuts and ovoids the next. The author had the experience in Scotland on two successive days of running with a very heavy train on the L.N.E.R. north of Edinburgh on which bad coal caused a loss of 57 min. in a four-hour run one day, and on the next, with the same engine an excellent run on the same duty. In Great Britain devices for dealing with the effects of inferior fuel were rather slow in being adopted because of the good coal that was nearly always available before World War II; but from the time of nationalisation the rocking grate was generally incorporated in all new designs. This arrangement provided means for tilting the firebars backwards and forwards through a small angle, so as to disturb the firebed and cause accumulations of ash to fall into the ashpan instead of staying on the grate and fusing into clinker. Again the general adoption of hopper type ashpans greatly reduced the labour required at sheds for the removal of ash from locomotives during servicing operations.

A factor contributing much to the availability of locomotives, and the mileages between successive shoppings is of course bearing performance, and the period under review saw the adoption of roller bearings on an extensive scale. Purely from the viewpoint of reducing rolling resistance of locomotives and coaching stock the roller bearing did not offer any great advantage over a well-designed and well lubricated plain bearing; indeed G. J. Churchward of the Great Western Railway once staged a demonstration to show that the reverse was the case, when he was being pressed to adopt roller bearings. But in the greatly changed social and economic conditions prevailing after World War II maintenance as it was practised previously was not everywhere possible, and the roller bearing came into its own as a means of minimising the incidence of hot boxes and of securing increased mileages between repairs for locomotives engaged in fast and onerous duty. From an historical viewpoint it is interesting to recall that Sir Nigel Gresley was the first user of the self-aligning ball bearing as far back as the year 1916 on the return cranks of the Walschaerts valve gear applied to certain locomotives of the Great Northern Railway. The arrangement is shown in the accompanying drawing (Fig. 303).

Prior to nationalisation an interesting application of roller bearings was on the last two L.M.S.R. " Pacifics " of the " Duchess " class built in 1949. These locomotives, the *Sir William A. Stanier F.R.S.* and the *City of Salford* were completely roller-borne, and they attained mileages well in excess of the normal permitted maxima before shopping. Both spherical and taper roller bearings were used and the accompanying drawings shows the design of the self-aligning axle box on the inside crank axle (Fig. 304). This presented a special problem owing to the limited length of 10 in. available, and the combination of high axle loading of 22 $^1/_2$ tons and maximum speeds of over 90 m.p.h. The self-aligning type of bearing used was ideally suited to its application on a built-up crank axle, because it ensured an efficient distribution of the load when flexing of the axle took place. Furthermore the underslung suspension provided a good arrangement in conjunction with the self-aligning bearings in permitting the boxes to align with the horn blocks, when the spring deflections differed.

The Cannon type axle box, as illustrated herewith (Fig. 305), has also been extensively used, with

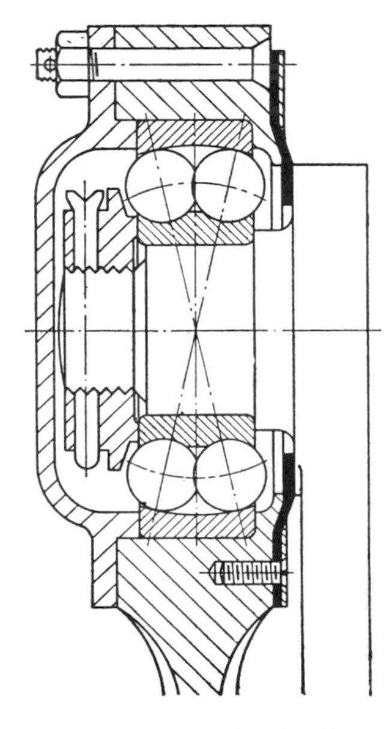

Fig. 303.—*Gresley's first use of self-aligning ball bearing, Great Northern Railway,* 1916, *on return cranks, Walschaerts valve gear.*

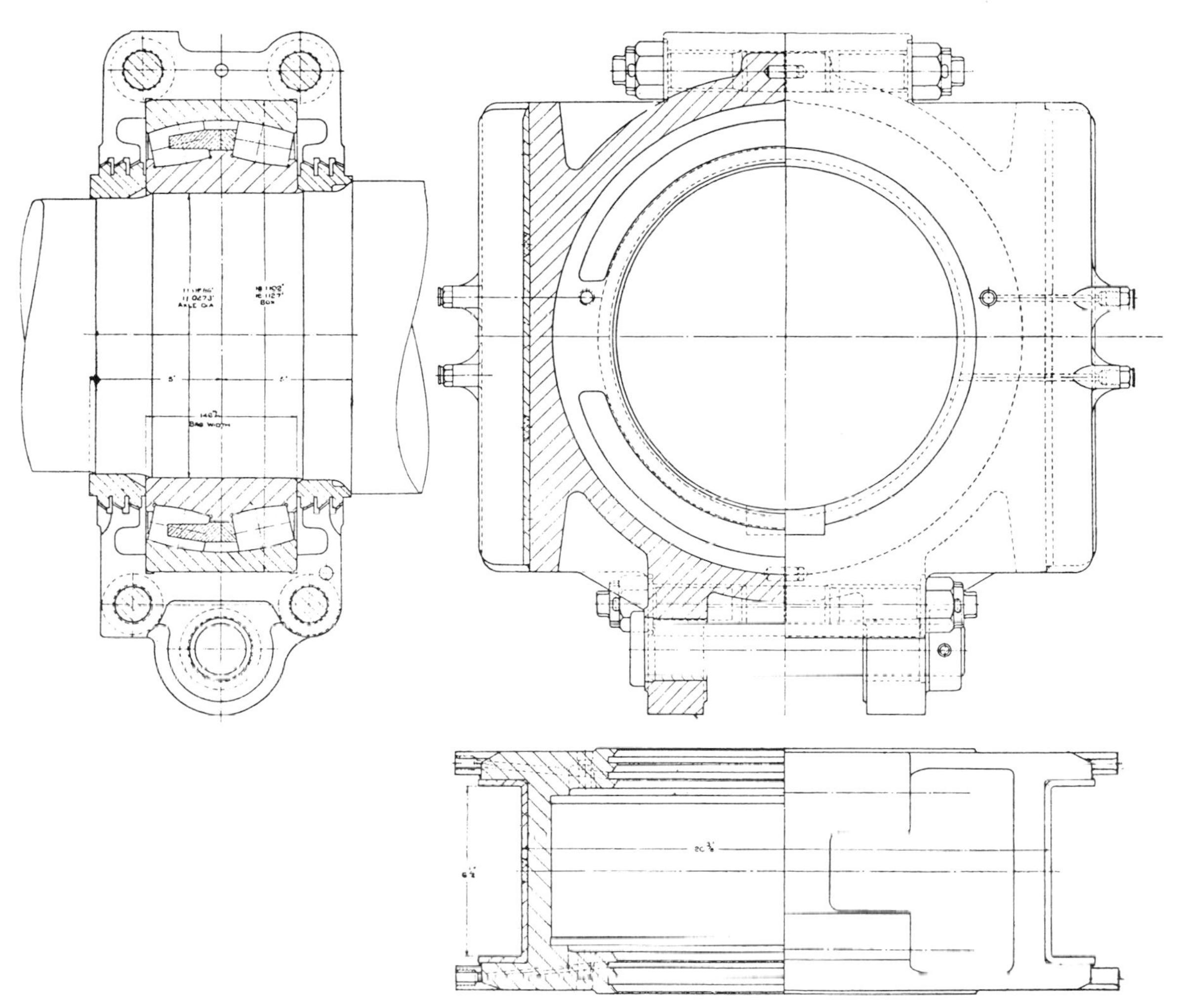

Fig. 304.—*Self-aligning axlebox on inside crank axle of British Railways locomotives* Sir William Stanier F.R.S. *and* City of Salford.

variations in the form of roller bearings incorporated. The example illustrated is one used on the " Ja " class 4-8-2 mixed traffic locomotive of the New Zealand Government Railways, and built by the North British Locomotive Company. In some applications a solid housing is used, and in others, to facilitate inspection, it is split. The bearings themselves, which are a heavy press fit on the axle, are mounted in a cast steel tubular housing, either solid or split according to the desired application. With this tubular design there is less danger of extraneous loads being applied to the bearings than with the conventional journal type axle box, and the assembly follows any irregularities in the track without disturbing the alignment of the bearings. An interesting example of roller bearings applied to tender axles is afforded by the arrangement shown in Fig. 306, used on the 8-wheeled tender of Sir Nigel Gresley's famous streamlined " Pacific " engine " Silver Link ".

A very successful arrangement of roller bearings to all axles was made on five L.N.E.R. " Pacific " express locomotives of the Peppercorn " A1 " type. These engines ran some exceptionally high mileages between general repairs. At the same time important improvements in detail design, using conventional methods,

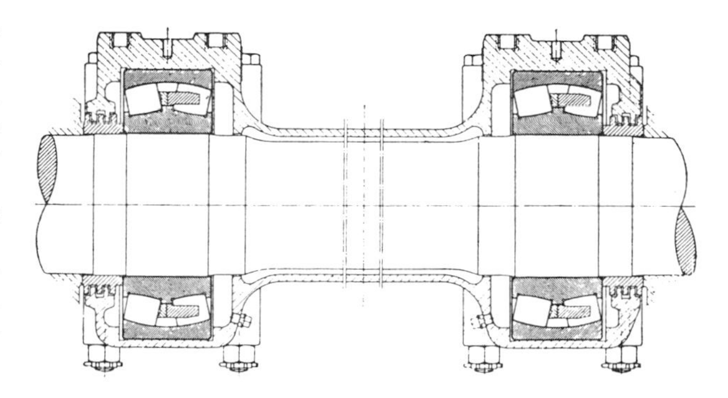

Fig. 305.—*Cannon type axle box: New Zealand Class* " Ja " 4-8-2 *engine.*

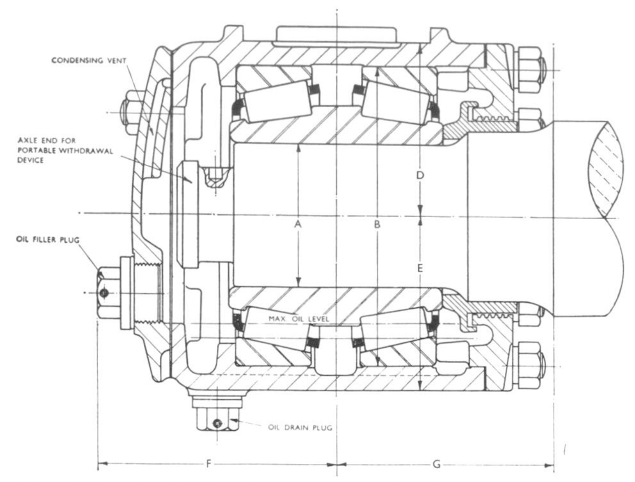

Fig. 306.—*L.N.E.R. Arrangement of roller bearings on 8-wheeled tender of streamlined* 4-6-2 *locomotive Class* " A4 ".

have led to a great improvement in availability, and to increased mileage between repairs. Even with a very successful locomotive such as the Stanier Class " 5 " 4-6-0 a relatively simple change, such as the use of manganese steel liners instead of white metal faces on the coupled axle box slides increase the mileage between the repairs by more than 30 per cent. Figures relating to British Railways as a whole indicate the extent to which availability has been increased and the number of engines undergoing or awaiting repairs reduced. These figures were taken before the new national standard designs had begun to make any appreciable impact.

Year	Total No. of steam locos. owned	Under, or awaiting repair	
		No. of locos.	% of total
1923	23,894	4038	16.9
1933	20,823	1676	8.1
1943	19,938	1288	6.5
1952	19,133	960	5.0

In concluding this chapter a brief reference is needed to developments in the design of tenders. With the increased length of run demanded by the economics of modern operating methods the tenders of main line locomotives were designed to carry more coal. The tender bodies mostly of all-welded construction were built out to the limits of the loading gauge, and their sides extended upwards practically to the maximum (Fig. 307). On most routes ample supplies of water were available, either from track troughs or from columns at intermediate stations, and although some tenders, like those of the L.M.S.R. " Duchess " class " Pacifics " were designed to carry 10 tons of coal the water capacity was not greatly increased and, except on the L.N.E.R., they were carried on six wheels. The steam operated coal-pusher, which proved such a boon to the firemen on long non-stop runs, was used only on the L.M.S.R.

Examples are illustrated of the London and North Eastern Railway programme of standardisation initiated by Edward Thompson in which the former practice of Sir Nigel Gresley in using 3-cylinder locomotives was reversed, except for the largest main line passenger and mixed "Pacifics". All the locomotives illustrated had two cylinders, 20 in. diameter by

Fig. 307.—*G.W.R. New type of straight-sided tender applied to locomotives built since* 1945.

Fig. 308.—*L.M.S.R. H. G. Ivatt's Class* " 4 " 2-6-0 *with double-blastpipe and chimney; self cleaning smokebox:* (1947).

26 in. stroke, and three out of the four were created by rebuilding existing classes, namely:

New class	Type	Rebuilt from	Illustration Fig. No.
" K1 "	2-6-0	Gresley 3-cylinder " K4 "	311, 312
" K5 "	2-6-0	Gresley 3-cylinder " K3 "	313
" O1 "	2-8-0	Robinson ex-G.C.R. " O4 "	315

Many locomotives of the " K1 " and " O1 " were subsequently at work, but the " K5 " class was not multiplied beyond the original rebuilt engine No. 206. The " L1 " 2-6-4 tank (Fig. 313) was a new design, though built up entirely from standard parts. There was also the " B2 " 4-6-0, a 2-cylinder rebuild of the 3-cylinder " Sandringham " class. This is not illustrated.

Fig. 309.—*L.M.S.R. One of the new Class* " 4 " 2-6-0s *in passenger service, Somerset and Dorset line.*

Fig. 310.—*L.N.E.R.* 2-6-0 *Class* " K1 ", *rebuilt* 1945 *by E. Thompson, from Gresley* 3-*cylinder* " K4 " 2-6-0.

Fig. 311.—*British Railways* 2-6-0 *Class* " K1 " (1949) *built by North British Locomotive Co. Ltd.*

Fig. 312.—*L.N.E.R.* 2-*cylinder* 2-6-0 *Class* " K5 " *rebuilt* 1945 *by E. Thompson from Gresley* 3-*cylinder* " K3 " 2-6-0.

Fig. 313.—*L.N.E.R. Thompson Class* " L1 " *standard* 2-6-4 *tank engine* (1945).

Fig. 314.—*L.N.E.R. Thompson Class* " O1 " 2-8-0, *rebuilt* 1944 *from Robinson original G.C.R. type.*

L.N.E.R. Thompson standard cylinder types

Class Type	" K1 " 2-6-0	" K5 " 2-6-0	" O1 " 2-8-0	" L1 " 2-6-4T
Cylinders, dia. × stroke, in.	20 × 26	20 × 26	20 × 26	20 × 26
Coupled wheel dia., ft. in.	5-2	5-8	4-8	5-2
Heating surfaces, sq. ft.:				
Tubes	1240	1719	1508	1198
Firebox	168	182	168	138.5
Superheater elements	300	407	344	284
Combined total	1708	2308	2020	1620.5
Grate area, sq. ft.	27.9	28	27.9	24.74
Boiler pressure, lb./sq. in.	225	225	225	225
Max. axle load, tons	19.7	20.65	17.0	20.0
Total engine wt. in working order, tons	66.85	71.25	73.6	89.45
Total engine and tender wt. in working order, tons	111.05	123.25	123.6	—
Nom. tractive effort at 85% working pressure, lb.	32.080	29.250	35.518	32.080

Fig. 315.—*L.M.S.R. H. G. Ivatt's Class* " 2 " *light branch* 2-6-0 (1946).

Chapter Sixteen The Interchange Trials of 1948

At the time of nationalisation of the British railways R. A. Riddles, formerly Vice-President, London Midland and Scottish Railway, was appointed a member of the Railway Executive, with responsibility for mechanical and electrical engineering. Conditions of national austerity prevailed, and one of the most important tasks set to the new administration was to reduce operating costs of running the railways by standardisation of equipment, and finding means of operating the traffic with fewer locomotives than previously. A continuation of existing trends was envisaged in the form of longer locomotive workings, shorter turn-round times at running sheds, and improved workshop methods whereby locomotives spent shorter times in Main Works when under general overhaul. Because of the national economic situation the policy adopted from the beginning of 1948 was to continue using coal-fired locomotives, despite the very marked trends towards diesel traction in many parts of the world. No foreign exchange was needed to purchase locomotive coal, and in first cost steam provided the highest tractive effort per pound sterling—an important consideration in 1948.

An extensive series of interchange trials was at once decided upon, as a means of producing, as quickly as possible, indications of the most desirable features of existing locomotives to incorporate in the design of future steam locomotives. Although these trials were not intended to be a contest between locomotives of similar capacity from the former railway companies, there is no doubt that there was much friendly rivalry developed in the course of the trials, and in certain cases the results were markedly influenced by local circumstances and the personality of individual enginemen. The engineers of the Railway Executive realised from the outset that indications of desirable features of design would be of a very broad nature, as the trials were carried out on service trains under normal operating conditions, without any special preparation of the locomotives, and it was agreed that the units selected for test should have run between 15,000 and 20,000 miles since last general overhaul, and should be taken straight from the regular links. The fuel was to be uniform throughout for each locomotive in the three broad groups:

1. Express Passenger and mixed traffic:	Grade 1A South Kirkby hards
2. Goods:	Grade 2B Blidworth
3. All tests in Scotland:	Grade 2B Comrie

The use of " hard " coals reacted unfavourably on the locomotives of Great Western design, the fireboxes and grates of which were designed to use soft Welsh coal; and when certain tests were repeated in order to obtain a truer comparison of the general performance of the locomotives concerned the use of Welsh coal showed a reduction in coal consumption per d.h.p. hour of 6 $^1/_2$ per cent. on the " King " class engine. On the other hand the comparison between the runs with " hard " and " soft " coal made by engines of the " Hall " class was not a true one, because circumstances at the time of the main series of Interchange Trials led to a somewhat inferior performance on the part of the " Hall " class engine actually engaged in the tests between Bristol and Plymouth. Although it had no more than an indirect bearing upon locomotive engineering this was perhaps the most extreme case in the whole series of trials where personalities entered in, and had a seriously adverse effect on performance. The circumstances were these: it was decided that the locomotives engaged on this particular group of tests should be based upon Bristol, and the workings involved a double-home turn to Plymouth. For many years previously however the G.W.R. men at Bristol had not worked " lodging turns ". Volunteers for the duty were called for, and the only response came from men of somewhat limited main line experience, who had in any case to learn the difficult road west to Newton Abbot specially for the job. The result was a relatively poor set of runs. When the tests were repeated, with Welsh coal, the engine was worked by a fully experienced Plymouth crew, and the result showed an improvement of 18 per cent. in the coal consumption per d.h.p. hour—far greater than could be attributed merely to the coal.

The locomotives tested were not as representative as the Railway Executive originally desired. Because of structure gauge clearance limitations the Great Western locomotives were prohibited from running on any part of the London Midland Region, and the " King " class 4-6-0 was prohibited between Waterloo and Exeter, on the Southern Region. This meant that the only " foreign " running made by the " King " was between King's Cross and Leeds; and the " Hall " was prohibited from working the St. Pancras-Manchester, and Perth-Inverness tests, on both of which it should have shown up well in the heavy hill-climbing involved. It was originally desired that the former L.N.E.R. should have been represented by one of the latest " Pacifics " of Thompson or Peppercorn design. At the time of the trials however, none of these new varieties had been sufficiently proved to withstand the rigours of long mileages on " foreign " lines, and so the

Fig. 316.—*G.W.R. High-speed test run* (1947) *in connection with automatic train control trials.*

Fig. 317.—*G.W.R. The Interchange trials of* 1925*: L.N.E.R. " Pacific " engine No.* 4474 *on Cornish Riviera Express near Reading.*

Region	Route and trains	Locomotives		Illustration Fig. No.
		Region	Class	
Western	1.30 p.m. Paddington–Plymouth	W.R.	" King "	—
	8.30 a.m. Plymouth–Paddington	E.R.	" A4 "	319, 321
		L.M.R.	" Duchess "	323
		L.M.R.	" Rebuilt Scot "	329
		S.R.	" Merchant Navy "	318, 324
Eastern	1.10 p.m. King's Cross–Leeds	W.R.	" King "	328
	7.50 a.m. Leeds–King's Cross	E.R.	" A4 "	—
		L.M.R.	" Duchess "	—
		L.M.R.	" Rebuilt Scot "	325
		S.R.	" Merchant Navy "	327
London Midland	10 a.m. Euston—Carlisle	E.R.	" A4 "	—
	12.55 p.m. Carlisle–Euston	L.M.R.	" Duchess "	—
		L.M.R.	" Rebuilt Scot "	—
		S.R.	" Merchant Navy "	—
Southern	10.50 a.m. Waterloo–Exeter	E.R.	" A4 "	—
		L.M.R.	" Duchess "	—
	12.37 p.m. Exeter–Waterloo	L.M.R.	" Rebuilt Scot "	—
		S.R.	" Merchant Navy "	—
Western	1.45 p.m. Bristol–Plymouth	W.R.	" Hall "	—
	1.35 p.m. Plymouth–Bristol	E.R.	" B1 "	327
		L.M.R.	Class " 5 "	—
		S.R.	" West Country "	—
Eastern	10 a.m. Marylebone–Manchester (London Rd.)	W.R.	" Hall "	322
		E.R.	" B1 "	—
		L.M.R.	Class " 5 "	326
		S.R.	" West Country "	—
London Midland	10.15 a.m. St. Pancras–Manchester (Central)	E.R.	" B1 "	320, 331
		L.M.R.	Class " 5 "	—
		S.R.	" West Country "	—
Scottish Region	4 p.m. Perth–Inverness	E.R.	" B1 "	—
		L.M.R.	Class " 5 "	—
	8.20 a.m. Inverness–Perth	S.R.	" West Country "	—

L.N.E.R. was represented by the Gresley " A4 " " Pacifics ". Here again personalities entered in to produce something of an unrepresentative result. Presumably for prestige purposes high authority decided that the engine to be used was the world record-holder *Mallard*. At the time she was definitely not one of the best of the stud, and in ordinary working from King's Cross shed was apt to run hot. In vain the local men pleaded against her selection, and sure enough she failed twice with a heated middle big-end, once on the Western Region and once on the Southern. These unfortunate events tended to reinforce the views of those who contended that the Gresley front-end layout was unsuitable for modern conditions.

At the time of the trials the booked times of the express passenger trains generally were little faster than those of the worst war period, and there was little chance for locomotives to display any capacity they had for sustained steaming at high power output. In many cases too, the trains most convenient for test running were in themselves not the fastest then operating over the routes in question. On the other hand the loads were augmented specially for test purposes, and a train like the London Midland 12.55 p.m. from Carlisle to Euston, with various intermediate stops, became a difficult proposition with a load of 500 tons. Leaving out of consideration for the moment the trials with the 2-8-0 and 2-10-0 freight engines which gave perhaps the least conclusive results of any, the trials with express passenger trains were made as shown in the table above.

In the past it had frequently been contended that locomotives required to be designed to suit the conditions existing on their respective roads, and that a visiting engine from another railway was not likely to do so well as the local product. In the grouping era, on the L.M.S. in particular, this time-honoured contention had been largely exploded by the outstanding success of the Stanier Class " 5 " 4-6-0, used in passenger and freight service over the entire system; and the Interchange Trials of 1948 amply confirmed the experience of the L.M.S.R. This was particularly the case with the Bulleid " West Country " class 4-6-2. It was only to be expected that this class of engine would do well over the undulating main lines of the former Great Central and Midland Railways, south of Nottingham and Derby respectively; but an engine of this class also did exceptionally well on the severe gradients of the Highland main line. It could be argued that the gradients on that route are no more severe than some existing on Southern Region lines west of Exeter, and particularly on the Ilfracombe

Fig. 318.—*The* 1948 *Interchange Trials:* "*Merchant Navy*" *class* 4-6-2 *on W.R. near Reading.*

Fig. 319.—1948 *Interchange trials: the Gresley* 4-6-2 Mallard *on Paddington–Plymouth express near Reading.*

branch; but on these Southern lines neither the loads nor the train schedules called for such sustained output of power as that needed between Blair Atholl and Dalnaspidal on the Highland line. The agreed load for the 4 p.m. express from Perth to Inverness was 350 tons. This required the service of an assistant engine on the climb of Struan bank; by attaching this in rear the dynamometer car recorded only the performance of the train engine on the long ascent, and some fine work was certainly registered by all the various locomotives involved.

During the entire series of trials none of the locomotives were indicated. Assessment of performance was entirely by the drawbar records. Three dynamometer cars were used, namely those of the former G.W.R., L.N.E.R., and L.M.S.R. In the case of the last-mentioned the car used was that originally built by the Lancashire and Yorkshire Railway, and not the new one which in 1948 had been recently introduced for working in conjunction with the new Mobile Test Units of the L.M.S.R. equipped with electrical braking apparatus for obtaining constant speed testing regardless of the gradients of the line. Quite apart from the intervention of personalities the tests carried out were subject to all the incidental events connected with the working of service trains: variations of weather conditions; traffic delays; the need for observation of all temporary and permanent speed restrictions. The results therefore constitute an overall picture of typical British locomotive working at that particular period, on a wide variety of service trains, on routes extending from London to Plymouth, London to Leeds, Manchester, and Carlisle, and between Perth and Inverness. While they cannot be regarded as a close scientific assessment of the relative merits of the classes of locomotive involved, by reason of the various extraneous influences that affected performance, the overall result is very important in the comparison it affords to the standards of performance discussed in Chapter Two of this book, representing the overall picture as it existed in the year 1925.

The following tables have been compiled to show the summary figures of coal consumption of the express passenger and mixed traffic locomotives involved. In the case of the Western Region "King" and "Hall" classes the results of the subsequent tests with Welsh coal are also included—both the actual consumption and equivalent figures, adjusted for the difference in calorific value of the Welsh and South Kirkby coals used. Figures are also included for a "King" class locomotive then experimentally fitted with a very much larger superheater than hitherto used on that class of engine.

Express Passenger Locomotive
Coal consumption summaries—1948

Engine	Route	Coal per d.h.p. hr., lb.
W.R. "King" (Standard)	Eastern Region	3.39
	Western (hard coal)	3.74
	,, (Welsh coal)	3.33 (actual)
	,, ,, ,,	3.50 (S. Kirkby equiv.)
W.R. (high Superheater)	,, ,, ,,	3.10
E.R. "A4"	Western Region	3.19
	Eastern Region	2.92
	L.M. Region	3.00
	Southern Region	3.20
L.M.R. "Duchess"	Western Region	3.24
	Eastern Region	3.04
	L.M. Region	3.07
	Southern Region	3.17
L.M.R. "Rebuilt Scot"	Western Region	3.64
	Eastern Region	3.26
	L.M. Region	3.37
	Southern Region	3.24
S.R. "Merchant Navy"	Western Region	3.61
	Eastern Region	3.73
	L.M. Region	3.57
	Southern Region	3.52

Mixed Traffic Locomotives
Coal consumption summaries—1948

Engine	Route	Coal per d.h.p. hr., lb.
W.R. "Hall"	Eastern Region	3.84
	Western (Hard coal)	4.11
	Western (Welsh coal)	3.22 (actual)
	,, ,, ,,	3.38 (S. Kirkby equiv.)
E.R. "B1"	Western Region	3.96
	Eastern Region	3.32
	L.M. Region	3.34
	Scottish Region	4.01
L.M.R. Stanier Class "5"	Western Region	3.39
	Eastern Region	3.29
	L.M. Region	3.71
	Scottish Region	3.9
S.R. "West Country"	Western Region	4.28
	Eastern Region	3.90
	L.M. Region	3.80
	Scottish Region	4.77

One's immediate comment upon these results studied in connection with those described in Chapter

Southern Region: "West Country" 4-6-2s Some High Power Outputs

Region	Location	Gradient	Speed	Recorded		Equivalent		Cut-off	Boiler Press.,	Steam Chest
		1 in	m.p.h.	Pull-tons	d.h.p.	Pull-tons	d.h.p.	%	p.s.i.	p.s.i.
W.R.	Cullompton	155	63	3.9	1469	4.16	1565	30	270	165
W.R.	Stoke Canon	217	57	4.4	1500	5.04	1715	25	250	195
E.R.	Chalfont Road	104	43	5.16	1325	6.40	1639	25	270	235
E.R.	Annesley	132	46	5.74	1574	7.15	1962	30	245	225
E.R.	Whetstone	176	67.8	4.12	1667	4.97	2010	27	260	240
Sc.R.	Struan Bank	70	36.3	5.08	1100	6.95	1506	25	265	240
Sc.R.	Slochd summit	60	35.8	5.42	1158	7.78	1663	35	265	225
Sc.R.	Drumochter	80	46.2	5.64	1555	6.94	1912	30	245	205
Sc.R.	Struan bank	80	35.8	5.40	1154	7.42	1585	30	270	240
Sc.R.	Slochd summit	60	37.2	5.42	1202	7.8	1730	30	265	245
Sc.R.	Drumochter	80	47.0	5.40	1515	6.95	1950	35	260	235

Fig. 320.—*L.M.S. The Eastern Region* " B1 " 4-6-0 *leaving St. Pancras on a dynamometer car test run to Manchester.*

Two, is that none of the express passenger engines involved in the 1948 trials exceeded 3.74 lb. per d.h.p. hour, whereas in 1925, with the exception of the Churchward and Collett engines of the Great Western Railway it was only an occasional and isolated Midland compound that showed a basic coal consumption of less than 4 lb. per d.h.p. hour. Even the " Merchant Navy " class of the Southern, which in 1948 were generally considered to be heavy on coal, did not exceed 3.73 lb. per d.h.p. hour in any group of tests. Among the mixed traffic locomotives the reasons for the relatively poor results from the " Hall " class 4-6-0 locomotive on its own road have already been explained, but the comparatively high figures from the " West Country " 4-6-2 of the Southern could be explained partly on account of the exhuberance of her crew in running ahead of time on most of the test runs, and in Scotland particularly making some exceptional large uphill efforts.

Taking the results as a whole the figures for basic coal consumption form a remarkable tribute to the working efficiency of the principal locomotive types on all four of the British main railways of pre-grouping days, and the uniformity of results as between the five express passenger classes concerned was extraordinary, having regard to the fact that the tests were all made on service trains. The Great Western " King ", as the oldest express passenger design involved, was perhaps handicapped by having design features that could have been considered obsolescent in 1948; but nevertheless with the simple modification of high degree superheating an engine of this class made an extremely good showing in the tests carried out later in the same year. Another interesting comparison could be made between the mixed traffic 4-6-0s of the G.W.R., L.N.E.R. and L.M.S.R. There is no doubt that relative to the size of the locomotives concerned the 1.35 p.m. express from Plymouth to Bristol was one of the hardest turns worked in the entire series of Interchange Trials, with a load of about 490 tons gross behind the tenders from Exeter to Bristol. The disparity in coal consumption between the L.N.E.R. " B1 " 4-6-0 and the Stanier Class " 5 " 4-6-0 can to some extent be accounted for by the way in which the respective drivers worked the test trains. But a comparison between the two types that could not be made as a result of the Interchange Trials of this character would have been between the economic performance of the two different types of boiler. On the L.M.S.R. the tapered barrel, and a firebox with the distinctive Churchward shaping had been proved superior in its maintenance costs to the more conventional type with straight barrel and a Belpaire firebox of Derby design. But the L.N.E.R. " B1 " engine had the very simple form of round-topped firebox favoured by Doncaster; it steamed very freely, but no comparative figures as to its maintenance costs have ever been published.

Some of the most interesting individual performances took place during the trials of the mixed traffic locomotives on the Highland line, and the following details are taken from the official report:

Struan-Dalnaspidal Bank: 1 *in* 70

Engine	Speed	Duration of effort	Average d.h.p.	Average d.h.p.	Gain on booked time
	m.p.h.	min.	actual	equiv.	min.
E.R. " B1 " ..	31 1/2	18	800	1130	8
L.M.R. " 5 " ..	27 1/2	22	600	895	6
S.R. " West Country " ..	36 1/4	16 1/2	1115	1545	11 1/2

Fig. 321.—*The world-record holder No.* 22 Mallard *leaving Paddington for Plymouth.*

The visiting engines were being worked hard, evidently taking no chances on this formidable bank, while the local man on the L.M.S. Class " 5 " 4-6-0, from his experience could take things less strenuously.

The details of engine performance contained in the official report represent the only test results that have been published relating to the Bulleid " West Country Pacifics ", and some extracts are included here, as showing the remarkable feats of weight haulage and hill climbing of which those locomotives

Fig. 322.—*W.R. " Hall " class* 4-6-0 *No.* 6990 *at Marylebone, after test run from Manchester via the G.C.R. line.*

Fig. 323.—*The L.M.R. Stanier " Pacific "* City of Bradford *passing Old Oak Common with a Western Region express from Plymouth.*

Fig. 324.—*The Bulleid " Pacific " No.* 35019 *prior to leaving Paddington on* 1.30 *p.m. express for Plymouth.*

Fig. 325.—*" Royal Scot " class* 4-6-0 *No.* 46162 *on dynamometer car test run King's Cross to Leeds near Barnet.*

London Midland Region: " Rebuilt Royal Scot " 4-6-0s
Some High Power Outputs

Region	Location	Gradient 1 in	Speed m.p.h.	Recorded		Equivalent		Cut-off %	Boiler Press. p.s.i.	Regulator position
				Pull tons	d.h.p.	Pull tons	d.h.p.			
W.R.	Hele	306	59.5	4.05	1440	4.34	1540	25	248	1/2
W.R.	Lavington	222	61	3.85	1402	4.47	1630	25	235	1/2
W.R.	Tiverton Jc. . .	155	59	3.75	1321	4.79	1685	25	225	3/4 1st valve
L.M.R.	Oxenholme	111	52.8	3.9	1215	4.6	1442	22	240	1/3
S.R.	Seaton Jc.	80	57	3.7	1259	4.38	1493	22	228	Full 1st valve
S.R.	Chard Jc.	140	47	4.65	1305	5.64	1582	30	225	1/4
S.R.	Crewkerne	200	57.5	4.5	1548	5.19	1782	30	242	1/4
S.R.	Sherborne	80	26.0	7.9	1228	9.63	1495	45	245	1/2

were capable. It is significant of the vigour with which the locomotive was being driven that the highest value of equivalent d.h.p. mentioned in the report should be with a mixed-traffic, and not an express passenger engine. The latter, relative to their tractive power, were not being so fully extended. The maximum equivalent d.h.p. values for each class mentioned in the report were:

Region	Engine	Max. Equiv. d.h.p.	Speed m.p.h.
Western	" King "	1480	34
Eastern	" A4 "	1728	35.8
London Midland	" Duchess "	1865	44
London Midland	" Rebuilt Scot "	1782	57.5
Southern	" Merchant Navy "	1929	49.8
Western	" Hall "	1287	32
Eastern	" B1 "	1341	42.5
L.M.R.	Class " 5 "	1398	56
Southern	" West Country "	2010	67.8

Another locomotive of which the Interchange Trials provided some interesting performance data was the rebuilt " Royal Scot ", which did a good deal of excellent work. The running was not always consistent, and there were times when it seemed the driver was working to save coal, and was not keeping strict point-to-point times. But the maximum efforts of this locomotive fully justified its inclusion with the most powerful express passenger classes. On most routes over which tests were made this engine took the same loads as the " King " and the " Pacifics " which were of Class " 8 " capacity, as against Class " 7 " for the " Scot ". A selection of notable individual performances have been tabulated at the end of this chapter.

In view of the references made in Chapter fourteen to the indicator trials with these engines it is interesting to observe from recordings taken during the Interchange Trials that the two different " Royal Scot " engines used, Nos. 46154 and 46162, were driven with relatively small regulator openings and cut-offs of not less than 22 per cent. The report records certain items of performance on the London Midland line at cut-offs down to 17 per cent.; but in actual working of the locomotives one of the great advantages of a valve and valve-gear layout that gives ample port openings is that the technique used by the driver is not critical, within certain limits, and if working with a fully opened regulator, and very short cut-offs tends to result in knocking, a narrower regulator opening and longer cut-offs can be used without appreciably affecting the economy.

Fig. 326.—*L.M.R. Stanier Class " 5 " 4-6-0 on the G.C.R. line near Northwood.*

Fig. 327.—" *Merchant Navy* " *class* 4-6-2 *No.* 35017 *preparing to leave King's Cross for Leeds.*

The freight engine trials were interesting in that examples of the Austerity 2-8-0 and 2-10-0, built for war service abroad, were included. The test routes were:

Western Region:	Acton and Severn Tunnel Junc.
Eastern Region:	Hornsey and Peterborough
L.M.:	Toton yard and Cricklewood
Western and Southern.	Bristol and Eastleigh

The coal consumption summaries were as follows:

Engine	Route	Coal: lb. per d.h.p. hr.
W.R. " 28XX "	Western	3.54
	Eastern	3.25
	Western & Southern	3.43
E.R. " O1 "	Western	3.37
	Eastern	3.25
	London Midland . .	3.31
	Western & Southern	3.63
L.M.R. " 8F "	Western	3.81
	Eastern	3.17
	London Midland . .	3.48
	Western & Southern	3.58
Austerity 2-8-0	Western	4.02
	Eastern	3.56
	London Midland . .	3.55
	Western & Southern	4.11
Austerity 2-10-0	Western	3.59
	Eastern	3.09
	London Midland . .	3.65
	Western & Southern	3.66

Again the close uniformity of the results will be apparent, and as in the case of the passenger and mixed traffic locomotives could have presented the engineers of the newly-formed Railway Executive with some difficult decisions to make when it came to settling the principles of design to be followed in future British standard locomotives. But while the Interchange Trials showed that in ability to work the traffic, and in the coal consumption figures there was very little to choose between the engines of the former privately-owned railway companies, there were, in certain cases, major points of design that would need far more close analysis before they could be either adopted or rejected in future practice, such as the fully-enclosed motion and chain-driven valve gear of the Bulleid " Pacifics "; the question of round-topped *versus* Belpaire fireboxes, as exemplified by the differing practices of the L.M.S.R. and of the L.N.E.R.—both on " Pacific " engines and 4-6-0s. The natural development of Churchward's historic practice on the Great Western was to be seen in Stanier's work on the L.M.S.R., so that while the actual representatives of the G.W.R. engaged in the Interchange Trials were themselves of designs dating back more than 20 years, the natural line of development from them, particularly in respect of the boiler design was to be seen in the "Duchess" class "Pacifics", the rebuilt "Royal Scots " and the mixed traffic Class " 5 " 4-6-0.

Although the results were inconclusive in that no design, or group of designs emerged as superior to the rest, the complete set of trials as viewed in retrospect, provide a most comprehensive, and historic picture of the running characteristics of British locomotives as a

Table III —

Engine class		W.R. " 2800 " class, 2-8-0			E.R. " O1 " class, 2-8-0			
Engine No.		3803	3803	3803	63773	63773	63789	63789
Route		Acton Severn Tunnel Jc.	Ferme Park New England	Bristol Eastleigh	Acton Severn Tunnel Jc.	Ferme Park New England	Brent Toton	Bristol Eastleigh
Train miles	Down	118.9	74.1	77.25	118.9	74.1	127.0	76.35
	Up	116.75	73.9	75.6	116.95	73.4	127.8	75.6
Average speed, m.p.h.	Down	21.94	22.9	17.6	17.21	23.8	21.2	18.6
		20.31	24.2	18.1	20.27	23.2	21.4	17.1
	Up	18.02	20.0	20.5	18.24	18.0	14.6	19.8
		17.01	23.7	20.6	18.23	17.9	17.4	19.7
Average d.h.p. (under power)	Down	416	415	477	392	546	488	448
		454	607	463	516	521	536	405
	Up	345	605	401	419	474	534	397
		332	628	416	377	417	536	425
Coal per train mile, lb.	Down	54.04	51.5	73.48	58.1	61.7	68.1	65.72
		64.13	72.3	61.87	59.0	64.8	73.0	61.49
	Up	57.84	73.2	51.15	53.7	62.9	84.0	55.68
		55.35	74.0	52.18	55.9	60.5	73.0	65.77
Coal per d.h.p. hr., lb.	Down	3.64	3.09	3.83	3.78	3.11	3.32	3.81
		3.57	3.05	3.40	3.14	3.26	3.24	3.81
	Up	3.52	3.35	3.22	3.17	3.19	3.29	3.35
		3.43	3.50	3.27	3.43	3.46	3.37	3.54
Evaporation, lb. water/lb. coal (actual)	Down	7.38	8.72	8.05	7.40	8.20	7.63	7.79
		6.93	7.25	8.79	7.49	7.70	7.35	7.85
	Up	7.48	7.45	9.43	7.79	7.87	7.71	7.72
		7.93	7.35	8.77	7.49	8.03	7.46	7.50

Fig. 328.—*Trials on Western Region,* 1955*:* a *Stanier* 4-6-2 *No.* 46237 City of Bristol, *on the Cornish Riviera Express near Reading West.*

Freight Engines

L.M.R. " 8F " class, 2-8-0				" Austerity ", 2-8-0				" Austerity ", 2-10-0			
48189	48189	48189*	48189	77000	63169	63169	77000	73774	73774	73776	73774
Acton Severn Tunnel Jc.	Ferme Park New England	Brent Toton	Bristol Eastleigh	Acton Severn Tunnel Jc.	Ferme Park New England	Brent Toton	Bristol Eastleigh	Acton Severn Tunnel Jc.	Ferme Park New England	Brent Toton	Bristol Eastleigh
118.8 116.25	74.1 73.6	126.8 127.5	76.5 75.5	118.9 114.0†	74.1 73.4	127.1 128.0	76.4 75.6	119.3 116.3	74.1 73.4	126.35 127.85	76.65 75.58
20.41 20.90 17.91 17.22	26.5 26.2 18.7 20.1	22.4 21.5 14.4 15.8	17.1 16.9 22.3 21.0	19.76 20.57 17.24 16.06	26.2 26.1 17.3 16.3	19.7 22.7 17.4 15.6	19.2 19.1 21.6 19.4	20.3 19.7 17.4 18.1	27.2 25.7 22.0 19.2	21.6 20.2 14.9 15.1	18.3 18.79 20.87 22.21
465 390 349 343	505 515 478 514	575 596 498 518	463 471 523 455	405 426 331 341	512 425 436 429	510 510 561 498	422 435 449 434	415 405 392 380	590 486 576 535	544 505 498 490	486 527 528 499
61.8 54.8 59.3 60.3	51.2 52.3 66.7 62.3	78.19 79.00 91.83 79.70	71.99 67.30 65.04 61.93	65.0 59.5 67.5 75.1	56.8 47.1 72.8 79.0	80.2 69.1 84.4 82.4	72.97 79.38 59.34 75.65	55.4 53.8 61.9 55.9	56.6 50.3 61.2 72.7	78.12 70.41 87.12 86.45	75.93 70.86 69.72 59.81
3.86 3.80 3.71 3.86	2.98 2.91 3.46 3.29	3.41 3.14 3.76 3.61	3.97 3.69 3.33 3.32	4.01 3.76 4.16 4.14	3.09 3.07 3.83 4.13	3.46 3.46 3.6 3.67	4.45 4.68 3.26 4.03	3.75 3.59 3.56 3.47	2.80 2.80 3.24 3.49	3.50 3.33 3.74 4.01	4.27 3.68 3.27 3.40
7.06 7.94 7.59 7.44	8.43 8.27 7.46 7.30	8.03 7.95 8.08 7.59	7.47 7.88 7.62 7.89	7.23 7.60 7.71 7.81	7.85 7.98 7.23 7.32	7.81 7.46 7.45 7.70	8.10 6.82 8.73 7.80	7.85 7.62 7.83 8.01	8.08 8.32 7.91 7.23	7.68 8.00 7.49 7.66	8.36 8.77 8.86 9.07

* Engine 48400 used on first down and first up test on this route.

† On the second up journey test terminated at Iver.

Fig. 329.—" *Royal Scot* " *class* 4-6-0 *No.* 46162 *on Plymouth-Paddington express near Reading.*

Fig. 330.—*W.R. " King " class* 4-6-0 *on test with a King's Cross–Leeds express near New Barnet.*

Fig. 331.—*St. Pancras–Manchester express leaving Elstree Tunnel hauled by* " B1 " *class* 4-6-0 *No.* 61251.

whole at that time. Purely from the viewpoint of locomotive engineering interest it is perhaps a matter for regret that certain Southern Region locomotives of slightly earlier vintage could not have been tested, such as the "Lord Nelson" 4-6-0s with the Bulleid modifications, and also the "Schools" class 4-4-0s. The reasons why they were not included can be appreciated, but in the event no comprehensive thermodynamic trials with these two excellent designs were ever undertaken. Again it was unfortunate that the Peppercorn "A1 Pacific" appeared just too late to be included. When the Interchange Trials were first proposed the running of a Great Western "Castle" class engine was also considered. The work of this famous design is perhaps more fully documented than any other in British practice, and it was felt that the "King" sufficiently covered the development of the Churchward 4-cylinder 4-6-0 engine, and would be able to compete on an equal loading basis with the "Pacifics" of the other regions. With these few reservations the Interchange Trials can be considered to have covered the best features of modern British steam locomotive practice, and are most important historically on that account.

Chapter Seventeen The British Standard Locomotives

In a survey of British steam locomotive development from the opening of the Stockton and Darlington Railway in 1925 until the year 1954 when the last British locomotive designed purely for express passenger service was completed at Crewe one sees first a process of expansion, and extreme diversification in design and constructional practice, and then from 1900 onwards a gradual contraction. At the great locomotive building centres of the home railways the aim was clearly tending towards using fewer different locomotive classes to handle the traffic, in which tendencies the London and North Western and the Great Western Railways were the acknowledged leaders. The grouping of 1923 brought still further contraction and concentration in new design effort, and finally, with nationalisation in 1948, policy for the entire country became vested in the single authority of the Railway Executive, and in the gathering twilight of steam it became possible to gather together all the varied and tangled threads of development over more than 120 years of intense experience and produce a new range of standard locomotives, which would be a synthesis of all that was best.

And yet it cannot be said that the designers had a completely free hand in this respect. In 1948 much of the traffic flowing on British Railways was being worked by locomotives designed for operating con-conditions far different from those familiar enough in the years of post-war austerity. The standard locomotives of the Great Western Railway, and the distinguished range of 3-cylinder designs with conjugated valve gear built for the L.N.E.R. in Sir Nigel Gresley's time will immediately come to mind. There is no doubt that the thermodynamic performance of the low-superheat "King" and "Castle" 4-6-0s in the years 1930–9 was some of the finest this country has ever witnessed; but it was achieved with the aid of fine-quality Welsh coal, and the kind of meticulously accurate firing that would keep the gauge needle between the limits of 245 and 250 lb. per sq. in. throughout from Paddington to Exeter without once blowing off. Similarly the day-to-day work of the Gresley "A4 Pacifics" on the L.N.E.R. high-speed streamlined trains has probably never been surpassed in its particular class. Yet in 1948 neither the "King" nor the "A4" was acceptable as a future national standard.

Immediately after nationalisation a Locomotive Standards Committee was set up to consider the standardisation of detail for the future. In some cases it was easy to agree which was the best among regional alternatives, but in others superiority one way or the other was not so clear cut. For new designs the use of two cylinders only was decided on for the following reasons:

(a) to attain the ultimate in simplicity and accessibility.
(b) the split inside big-end can be a source of trouble.
(c) a built-up crank axle is expensive, both in first cost and maintenance.
(d) other things being equal, four exhausts per revolution promote better steaming than do six or eight.

In preparing the new designs, the following eight points were laid down as representing firm guiding principles:

1. Maximum steam raising capacity.
2. Simplicity, with the least number of working parts and all readily visible and accessible.
3. Each new engine class so proportioned as to give the widest range of mixed traffic working.
4. High standard of bearing performance, either by use of roller bearings, or plain bearing with manganese steel liners for the horns.
5. Simplified shed preparation, by wide use of mechanical lubricators, and grease lubrication.
6. More rapid disposal; self-cleaning smokeboxes; rocking grates; self-emptying ashpans.
7. High factors of adhesion—to minimise slipping.
8. High thermal efficiency, through large grate areas, high degree of superheat, long-lap valve gear.

By the early months of 1949 the work of various inter-regional committees, together with the results of the 1948 Interchange Trials described in the previous chapter had provided enough data for work to be started on seven British standard designs. The work had necessarily to be divided between the headquarters drawing offices of the previous four main line companies, namely Brighton, Derby, Doncaster and Swindon; but instead of allocating a complete locomotive to one office each of the four was made responsible for preparation of standard drawings for particular components in respect of all the engine types, thus:

Brighton: Brakes and sanding gear.
Derby: Bogies and trucks: tenders, wheels, tyres, axles and spring gear.
Doncaster: Coupling and connecting rods; valve gear and cylinder details.
Swindon: Boiler and smokebox details; steam fittings.

Fig. 332.—*The first British standard* 4-6-2, *No.* 70000 Britannia *at Crewe.*

In addition each office was designated " parent " for one or more of the standard designs, with responsibility for making the general arrangement drawings, though the locomotives concerned would not necessarily be built in the works associated with that drawing office. By this procedure, and under the guiding influence of E. S. Cox, Executive Officer, Design, of the Railway Executive, a notable uniformity in style and detail design was achieved. The six classes undertaken at first were:

Power Class	Type	Parent Office	Illustration Fig. No.
7 MT	4-6-2	Derby	332, 347
6 MT	4-6-2	Derby	336
5 MT	4-6-0	Doncaster	337
4 MT	4-6-0	Brighton	340
4 P	2-6-4T	Brighton	342
3 P	2-6-2T	Swindon	—

No new design for heavy express passenger work was at first undertaken, though the Class " 7 " 4-6-2 (Fig. 332) with a contemplated tractive effort of about 30,000 lb. was intended to work on similar duties to those then being undertaken by the ex-G.W.R. " Castles ", the ex-L.M.S. Rebuilt " Royal Scots ", the Gresley " V2 " 2-6-2, and the Bulleid " West Country " class 4-6-2.

A student of British locomotive history viewing in retrospect the gradual build-up of design policy on the former privately-owned railway companies might feel that the decision to build only two-cylinder engines even for the largest types, was a retrograde one. All four of the regional equivalents of the new Class " 7 " 4-6-2 were multi-cylindered machines. But in contrast to the latest practice in Continental Europe the practice in the U.S.A. and Canada for many years previously had been to use two-cylinder simples. This, in the case of new British standard designs did not bring any disadvantages in the effect of locomotives on the track. In 1928 the Report of the Bridge Stress Committee had recommended that on all new locomotives designed from that time onwards, at a speed of 5 rev. per sec. the axle hammer blow should not exceed one-fourth of the static load, with a maximum of five tons and that the " whole engine " blow should not exceed 12 $^1/_2$ tons. In the new British standard designs the proportion of reciprocating weights balanced were limited to 40 per cent. on the larger engines, and 50 per cent. on the smaller. This, combined with the use of light-weight materials for the motion parts enabled the figures for hammer blow as set out in the accompanying table to be obtained, which were very satisfactory from the viewpoint of the track.

Typical British Standard Locomotives
Hammer Blow at 5 *r.p.s.*

Engine Class	4-6-2 Class 7	4-6-2 Class 6	4-6-0 Class 5	2-6-2T Class 3
Speed, at 5 r.p.s., m.p.h.	66	66	66	56
Max. coupled axle wt., tons	20 $^1/_4$	18 $^1/_2$	19 $^1/_2$	16 $^1/_4$
Reciprocating masses per cyl., lb.	845.3	831.3	825.8	736.8
Percentage of reciprocating wt. balanced, %	40	40	50	50
Hammer blow, tons:				
per wheel	2.12	2.12	2.59	2.17
per axle	2.55	2.55	3.11	2.58
per rail	5.50	5.50	6.72	5.93
whole engine	6.60	6.60	8.15	7.04

Certain detail features of the new designs are worthy of special note. In the boilers the well known

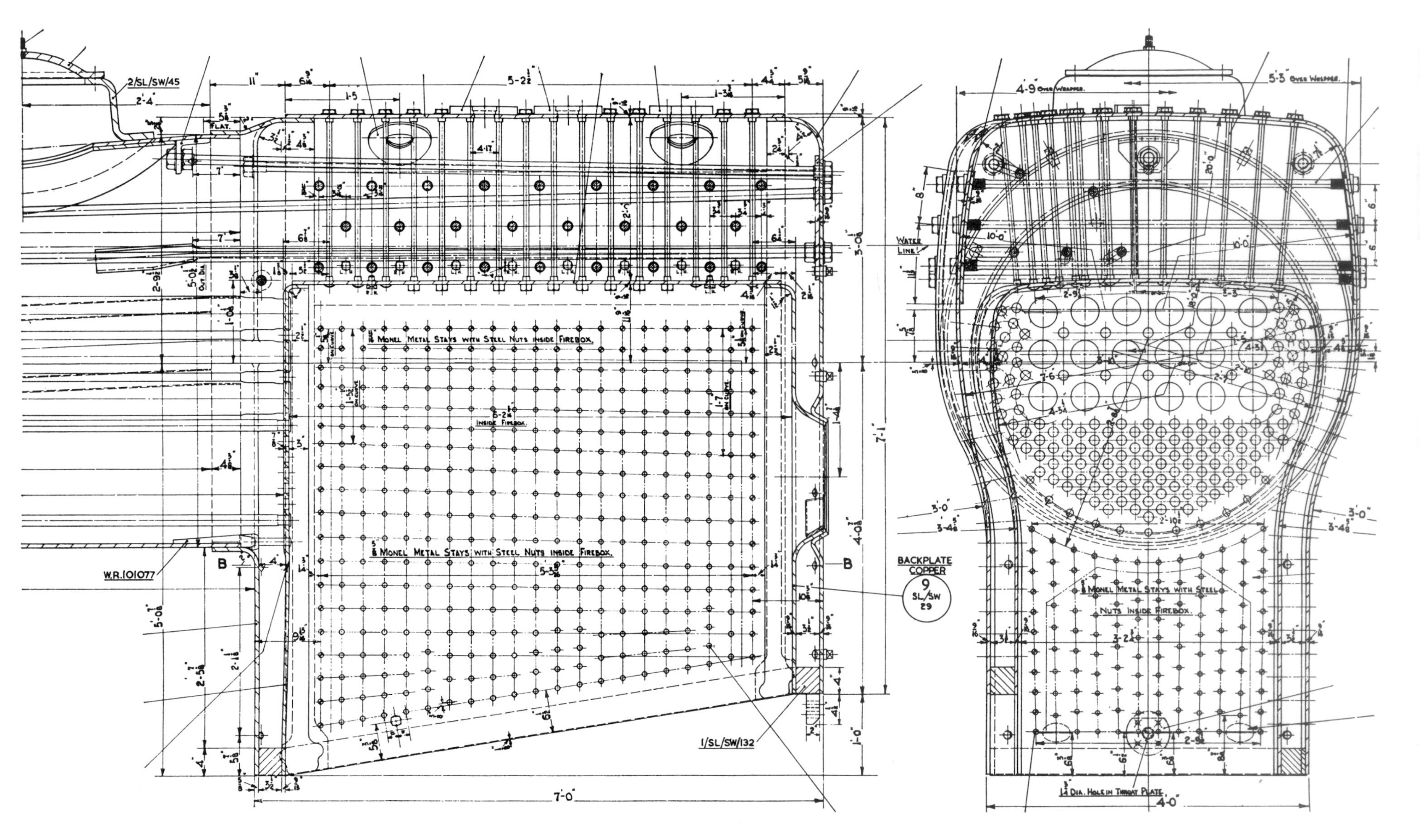

Fig. 333.—*Arrangement of narrow firebox, as applied to B.R. standard Class " 3 " 2-6-2 tank engines.*

yardstick of performance, namely the free gas area through the tubes in relation to grate area was kept above 14 per cent. The figures of the new locomotives relative to certain very successful regional classes are shown in the accompanying table; but it is of particular note that a ratio of 16.2 per cent. was achieved in the case of the Class " 7 " 4-6-2, because it is not always an easy matter in design to achieve such a ratio with a boiler having a wide firebox and combustion chamber. In view of contemporary Continental practice, and the success that had attended the fitting of double blastpipes and chimneys to certain L.M.S.R. and L.N.E.R. locomotives, not to mention the Bulleid 5-nozzle blastpipes on the Southern, it was a little surprising that the Class " 7 " 4-6-2 had only a single blastpipe (Fig. 334). But it had to be borne in mind that a great part of the work of mixed traffic locomotives on British Railways was performed at moderate, rather than maximum output. In a comprehensive series of tests at both Rugby and Swindon stationary test plants the effectiveness of double blastpipes, and other improved methods of drafting was proved beyond question at maximum output; but the performance was less efficient at low outputs, and the general view was that the locomotives were not so versatile over the full range of duties.

Values of Free Gas Area

Region	Engine Class	Total free gas area sq. ft.	Grate area sq. ft.	Gas Area as % of grate area
Western	" King "	5.24	34.3	15.3
	" Hall "	4.33	27.07	16.0
Southern	" West Country "	5.17	38.25	13.5
Eastern	" A1 "	6.24	50.0	12.5
	" A4 "	6.24	41.25	15.1
	" V2 "	6.24	41.25	15.1
London Midland	" Duchess "	8.89	50.0	13.8
	Rebuilt Scot	5.09	31.25	16.3
	Class " 5 " 4-6-0	4.75	28.65	16.6
British Standard	4-6-2 " 7 "	6.79	41.8	16.2
	4-6-0 " 5 "	4.55	28.65	15.9
	2-6-2T " 3 "	3.08	20.35	15.2

The boilers and fireboxes, in keeping with the designs generally, were very simple in their conception. The narrow-firebox engines had boilers and fireboxes of the well-tried Swindon type, as modernised by Stanier in his latest L.M.S.R. engines, while the Class " 6 " and Class " 7 " 4-6-2 engines had a straightforward wide firebox, without arch tubes or thermic syphons. Details of typical wide and narrow British standard fireboxes are shown in the accompanying drawings (Figs. 333 and 335). The original arrangement of the steam collector on the Class " 7 " boiler gave serious trouble, by allowing water to be carried over to the cylinders in sufficient amounts to cause broken pistons. The regulator was in the smokebox, which meant that there was an open steam pipe from the dome. A combination of circumstances led to water being carried over, and as a result a change was made in the design of the steam collector as shown in the accompanying drawings (Fig. 336), which duly cured the trouble. In view of the manner in which design work over the standard range of locomotives was co-ordinated it is interesting to record the extent to which components were standardised over the first six types, and the features of previous railway practice that were adopted as a national standard. These are covered in the accompanying schedule. A further table gives dimensional details of the first six standard classes.

Components either identical with, or following former railway companies' practice

G.W.R.

Mechanical lubrication, atomiser gear	Brick arch
Cylinder relief valve	Air Valve
Water gauge	Pressure and vacuum gauges
Steam valves	Live steam injector
Whistle for smaller engine	Fusible plug
Pipe unions in smaller sizes	Firedoor (for engines allocated to Western Region)
Smokebox door	Tender and tank water lever gear

L.M.S.

Bogie complete	Pony trucks
Plain bearing axleboxes*	Manganese liners
Hornblocks and horn stays (other than 4-6-2)	2-bar crosshead and slidebar (engines with leading pony truck)
Engine spring suspension	Valve heads
Piston head and fastening	Connecting rod big end, and side rod bushes and lubrication*
Valve spindle guides	Brake cylinder
Connecting rod, little end lubrication (2 bar crosshead)*	Ashpan hopper, doors and gear
Carriage warming valve	Firedoor (other than engines allocated to Western Region)
Water pick-up gear	
Rocking grate gear	

* These were originally G.W.R. designs brought to the L.M.S. by Sir William Stanier and in some cases slightly modified.

L.N.E.R.

3-bar slidebars and crosshead	Gudgeon pin and little end lubrication (3-bar arrangement)
Piston rod packing	Brakeblocks
Whistle (for larger engines)	Drawhook
Nameplates	

Southern

Frames central with welded horns (4-6-2)	Three compartment ashpan (4-6-2)
Trailing truck (4-6-2)	Engine and tender drawgear
Washout plugs	Top feed valve
Tender spring suspension	Tyre fastening

Four of the new classes, while incorporating to a major extent the standardised design details and components were largely adaptations of well-tried existing designs. For example the new " BR5 " mixed traffic 4-6-0 (Fig. 338) was based on the Stanier Class " 5 " 4-6-0 of the L.M.S.R. and the " BR4 " 2-6-4 tanks (Fig. 344) was similarly derived from L.M.S.R. practice. The " BR3" 2-6-2 tank was derived from the Great Western " 45XX " class 2-6-2, while the " BR4 " 4-6-0 (Fig. 341) was a lighter version of the " BR5 ". Principal interest, of course, was centred upon the " BR7 " 4-6-2 mixed traffic engine, the now well-known " Britannia " class, which has given sterling service on many regions of British Railways, once certain teething

British Standard Locomotives

Power Class Type	7 4-6-2	6 4-6-2	5 4-6-0	4 4-6-0	4 2-6-4T	3 2-6-2T
Cylinders, dia. × stroke, in.	20 × 28	$19^1/_2$ × 28	19 × 28	18 × 28	18 × 28	$17^1/_2$ × 26
Coupled wheel dia., ft. in.	6-2	6-2	6-2	5-8	5-8	5-3
Piston Valves, in.:						
diameter	11	11	11	10	10	10
steam lap	$1\ ^{11}/_{16}$	$1\ ^{11}/_{16}$	$1\ ^{11}/_{16}$	$1\ ^{11}/_{16}$	$1\ ^1/_2$	$1\ ^1/_2$
lead	$^1/_4$	$^1/_4$	$^1/_4$	$^1/_4$	$^1/_4$	$^1/_4$
Exhaust clearance, in.	Nil	Nil	Nil	Nil	Nil	Nil
Max. travel in full fore gear, in.	$7\ ^{47}/_{64}$	$7\ ^{47}/_{64}$	$7\ ^{47}/_{64}$	$7\ ^{11}/_{32}$	$6\ ^{37}/_{64}$	$6\ ^{29}/_{64}$
Heating surfaces, sq. ft.:						
Large tubes	978	857	497	366	344	262
Small tubes	1286	1021	982	935	879	671
Firebox	210	195	171	143	143	118
Superheater	704	615	369	265	246	190
Combined total	3178	2688	2019	1709	1612	1241
Grate area, sq. ft.	42	36	28.65	26.7	26.7	20.35
Boiler pressure, lb./sq. in.	250	225	225	225	225	200
Nom. tractive effort at 85% boiler pressure, lb.	32,150	27,520	26,120	25,100	25,100	21,490

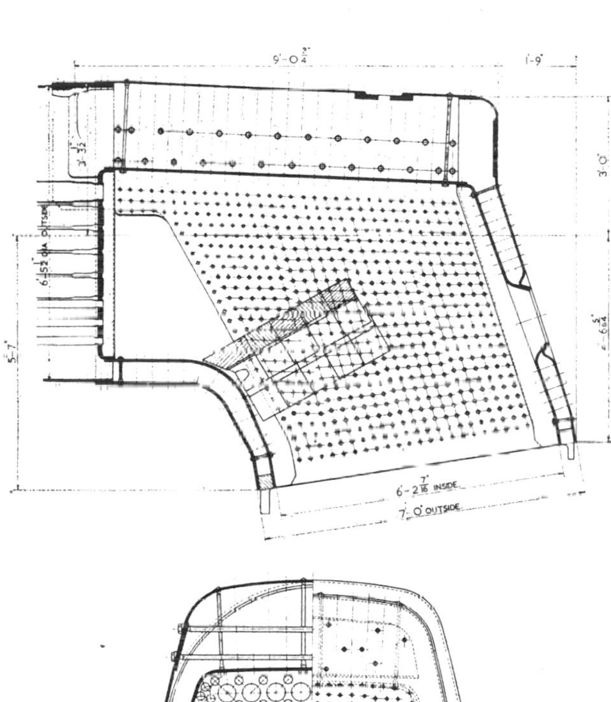

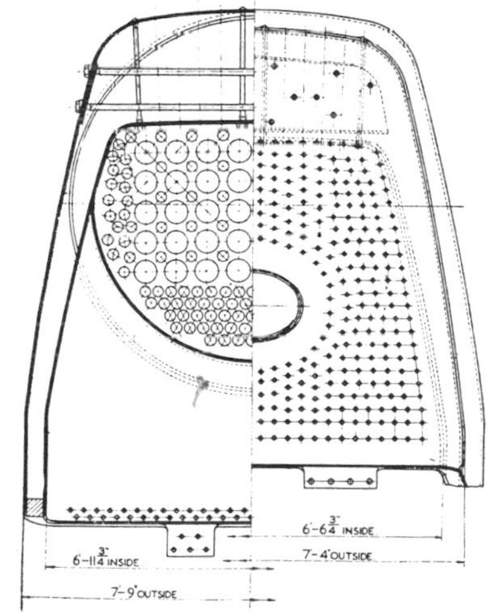

Fig. 335.—*Arrangement of wide firebox on B.R. 4-6-2 locomotives.*

troubles had been overcome. The breaking of pistons due to water carry-over began when the first engine was only a few weeks out of the shops, and the trouble was speedily corrected. But a more serious trouble arose from coupled wheels shifting on the axles, and a series of incidents culminated in the failure of engine No. 70004 on the up Golden Arrow Pullman boat train at Headcorn, which naturally attracted a great deal of public attention. All the 25 locomotives of the "BR7" class then in service were withdrawn until a full diagnosis of the causes had been made.

The trouble was traced to the method of fitting the cannon-type solid roller bearing axle boxes. At the time the first engines of the class were built at Crewe the shop equipment was such that the quartering of crank pin holes and the balancing of wheel sets could not be carried out after the roller bearings were in position on the axles. It was, therefore, the practice first to press the wheels on to a special axle, using a relatively light load. After fitting the tyres, and carrying out the work of balancing and quartering the wheels were pressed off the special axle and pressed on to the final axle, to which the cannon-type roller bearings had already been fitted. Another feature of design may also have contributed to the conditions that led to shifted wheels. In common with standard Great Western and L.M.S. practice for large express locomotives the axles had been bored out to have a $4\ ^1/_2$ in. diameter hole from end to end. This feature was originally introduced at Swindon on the "Abbey" series of "Stars", in 1922, not primarily to reduce weight but in connection with the heat treatment of axles. With a solid axle the heat treatment did not penetrate to the centre, but by boring it out the coolant affected the metal throughout the thickness of the axle. On the G.W.R. the reduction in weight was regarded as a bonus. On the "Britannia" however, with the increased weight on the axle due to the cannon-boxes the engineers of the Railway Executive were glad enough to have the saving of 2 cwt. per axle, which the $4^1/_2$ in. diameter central hole provided.

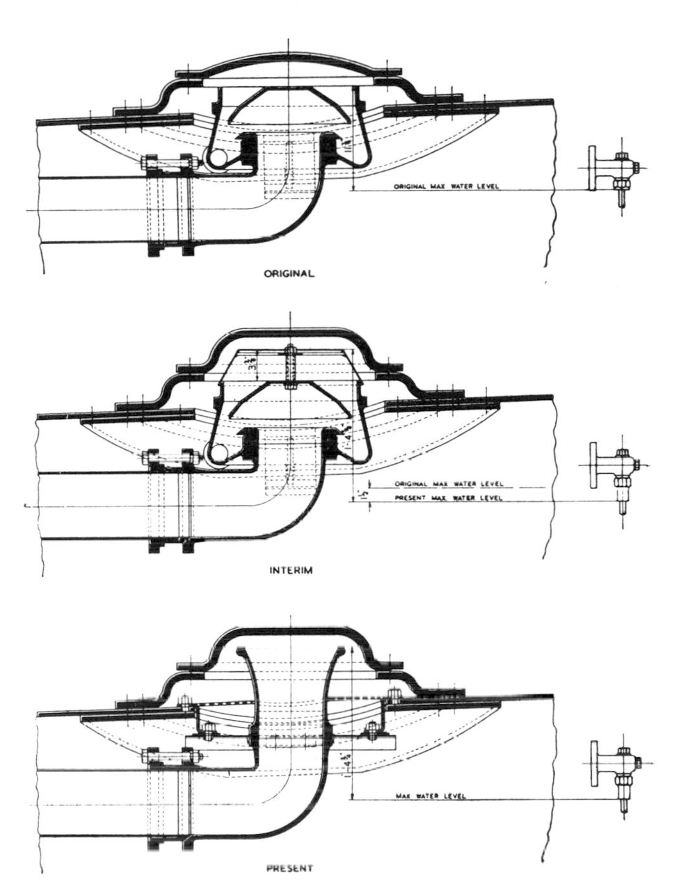

Fig. 336.—*Arrangement of steam collector on B.R. Class 7 4-6-2 locomotives.*

Taken together, the double process involved in the fitting of the wheels, and the possible loss in pressure between the wheel boss and the axle as a result of squeezing the hollow ring when pressing on probably combined to produce the marginal conditions of fit that resulted in complete success in some cases, and failure in others. Of the 25 engines of the "Britannia" class originally built seven became failures with shifted wheels; but among those which had shown no sign of trouble at the time the whole class was withdrawn there was the remarkable case of engine No. 70005 which was sent new to Rugby testing station. There it was put through an extremely severe series of trials at maximum power, and was afterwards sent to the Settle and Carlisle line for controlled road tests. In the course of these latter trials there were times when heavy slipping occurred, causing vibration which could have been expected to shake the wheels loose, in view of what happened elsewhere. But this engine gave not the slightest trouble, and evidently all the variables in the degrees of fit of the various members must have been on the "plus", rather than the "minus" side. As a result of the failure of the seven engines all members of the class were re-wheeled, to improved methods of assembly on the axles, and the hollow axles were plugged for the length of the wheel fit.

Once these troubles were overcome the "Britannia" 4-6-2s became excellent engines, powerful, free-steaming and light on maintenance. Like other two-cylinder locomotives balanced to provide minimum hammer blow their action was naturally not so smooth to those riding on them as that of the three-cylinder and four-cylinder express passenger locomotives of regional origin. But the "Britannia" class was well suited to the conditions it was designed to meet, and if this last British example of a powerful locomotive intended for express passenger as well as mixed traffic duties was a hard-slogging work-horse rather than a highly bred "racer", it was an ample sign of the times. It is true that three years later an express passenger 4-6-2 was designed and built, the three-cylinder No. 71000 with Caprotti valve gear; but this locomotive was an isolated prototype, which was never fully developed, because of the decision to cease building steam locomotives. As will be told in a succeeding chapter of this book, No. 71000 did some work that was remarkable in its class; but among locomotives that took an appreciable share in the hauling of the heavy express passenger traffic of this country the "Britannia" was the last of the line.

From the six original British Standard designs, the first examples of which appeared in 1951-52, four other small types were introduced in 1953-54, all

Fig. 337.—*British standard Class "6" 4-6-2 "Clan" class* (1952).

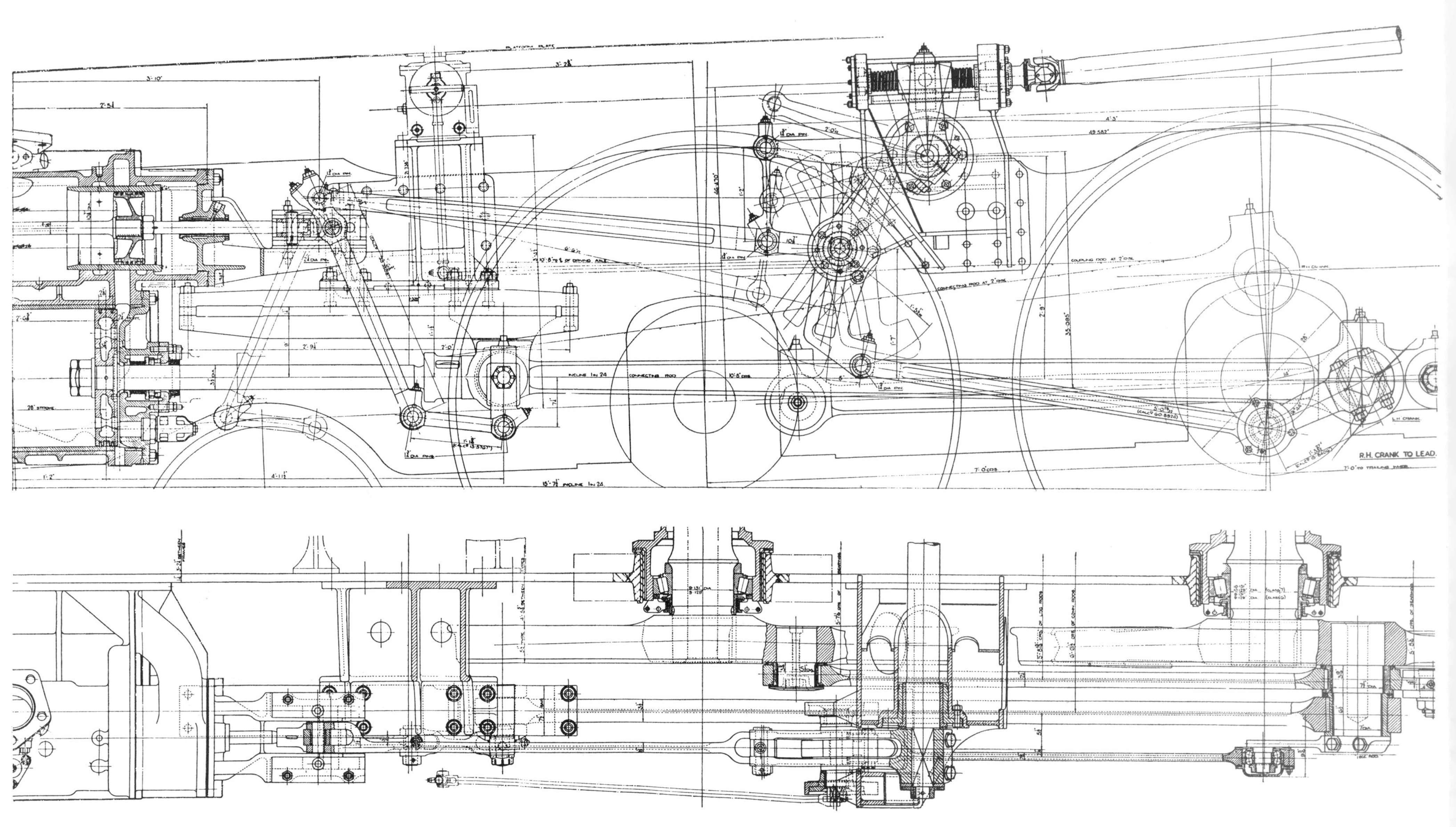

Fig. 343.—*B.R. Class "7" 4-6-2 engine: layout of motion.*

Fig. 344.—*B.R. standard class* " 4 " 2-6-4 *passenger tank engine* (1951).

Fig. 345.—*One of the later engines of the "Britannia" class* (*B.R.*, 7) *No.* 70030 William Wordsworth.

Fig. 346.—*B.R. Class* " 3 " 2-6-0 (1954).

Fig. 347.—*B.R. Class "4" 2-6-0 engine: introduced* 1953 *Engine No.* 76114 *illustrated was the last steam locomotive built at Doncaster.*

Fig. 348.—*B.R. Class "2" 2-6-0* (1953) *developed from Ivatt standard L.M.S.R. type.*

maximum performance of various standard classes when they were in the most run-down condition in which they would be permitted to continue in traffic.

Maximum Steaming Rates
British Standard Locomotives

Class	Absolute maximum at draughting limit	Maximum sustained in run-down condition	D.h.p. at max. steam rate
	lb. per hr.	lb.	
BR7 4-6-2	31,410	24,000	1450
BR5 4-6-0	24,000	18,000	1100
BR4 4-6-0	19,600	14,700	820
BR4 2-6-0	17,000	12,700	750
BR2 2-6-0	11,200	8,400	350

The one and only British standard express passenger No. 71000 (Fig. 349) was built in replacement of the London Midland Region "Pacific" No. 46202, which was destroyed in the disastrous double-collision at Harrow in October, 1952. This latter engine was in itself a replacement for the Stanier 4-6-2 Turbomotive which had borne the same number. Opportunity was taken of the need to replace the engine destroyed at Harrow to design and build a prototype Class "8" "Pacific". The aim was to have a locomotive of approximately 40,000 lb. tractive effort, and capable of a continuous high output of power. Thus while the boiler barrel and the tube heating surfaces were the same as those of the "Britannia" the "BR8" had a larger firebox, and the smokebox was equipped with a double blast-pipe and chimney. The comparative boiler proportions between the "BR7" and "BR8" were thus, as follows:

British Standard Pacifics

Class	"BR7"	"BR8"
Heating surfaces, sq. ft.:		
Tubes	2264	2264
Firebox	210	226
Superheater	718	691
Combined total	3192	3181
Grate area, sq. ft.	42	48.6

The accompanying drawing (Fig. 350) shows a longitudinal cross-section of the smokebox, with the very simple arrangement of double-blastpipe. The three cylinders were 18 in. diameter by 26 in. stroke, but although the engine was designed primarily for first class express passenger working the coupled wheel diameter was kept at 6 ft. 2 in. as in the "Britannia". Experience with these latter engines had shown that they would run freely up to 90 m.p.h.;

Fig. 349.—*B.R. Class "8" 3-cylinder express passenger locomotive No.* 71000 Duke of Gloucester, *with Caprotti valve gear: built at Crewe* 1954.

so there appeared no reason to depart from the standard diameter in the Class "8" "Pacific".

The general details of construction followed the standard practices established in the original six designs that appeared in 1951-52; but this locomotive was the only one in the entire British Standard range to have an inside cylinder. A feature that was new on British Railways was the method of locking the inside big-end cotter. This took the form of a serrated locking plate, on which the serrations were phased on both sides of the plate in such a way that different positions of the serrations relative to the serrations on the back edge of the big-end cotter could be obtained. This allowed an adjustment of $^{1}/_{32}$ in. in the cotter position, and as the taper on the cotter was 1 in 16 an adjustment of approximately 0·002 in. on the big end bearing was obtained.

The steam distribution was by British Caprotti valve gear. The separate inlet and exhaust poppet valves at each end of the cylinder were actuated by rotary cams. Machined from solid steel forgings, the valves were housed in cages to form complete units. The cages expanded under influence of temperature at the same rate as the valves, thus avoiding differential expansions and promoting maximum steam tightness; this was further safeguarded by the

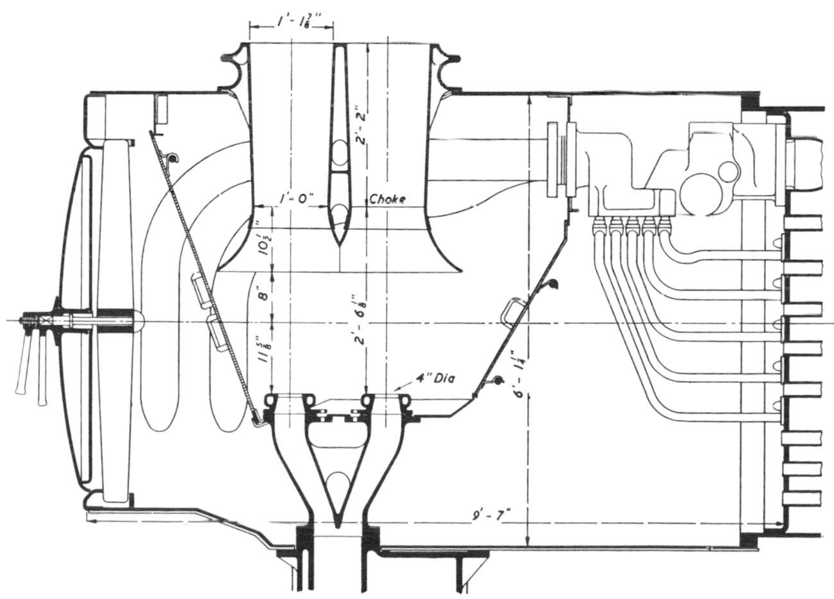

Fig. 350.—*B.R. Class "8"* 4-6-2*: diagram of smokebox showing draughting arrangements.*

position and design of the valve seats, the bottom being flat and the top having an angle which was a function of the distance between them. The valves were designed to give good port openings, free steam flow, and adequate strength with minimum weight. The main engine regulator incorporated a special valve which automatically admitted saturated steam through an actuation pipe to the bottom of the valve spindles before the main steam supply was admitted to the cylinders. The actuation steam lifted the valves into the working position and provided the closing force during running, although the major force on the inlet valves was the steam chest pressure acting on the section of the spindle above the valve. The valves were thereby springless, and worked without lubrication.

The camshaft assembly is illustrated in Fig. 352.

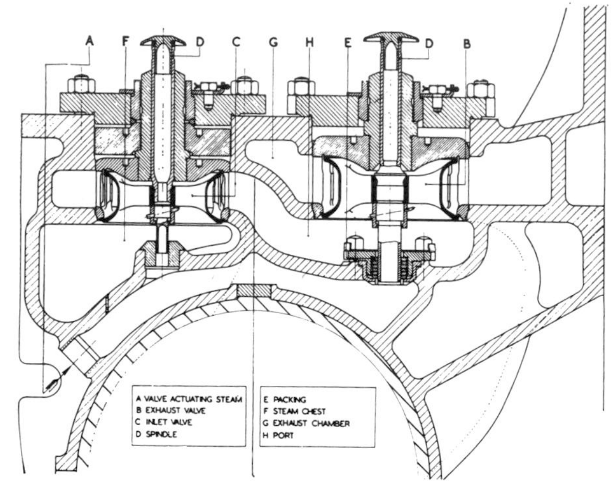

Fig. 351.—*B.R. Class* " 8 " 4-6-2 *drawing showing layout of valves.*

The cam profiles were designed for low opening and closing velocities with smooth acceleration in order to avoid roller bounce. Two inlet cams were mounted on a bush at one end of the camshaft and an exhaust cam assembly at the other, these assemblies being shown on the accompanying diagrams (Fig. 353). Between the assemblies, two scrolls were fitted to the square threaded portion of the camshaft. Each scroll was connected to an inlet cam by two cam-rods in a manner which allowed relative angular movement to be imparted to the inner and outer cams. The scrolls were housed in collars which were connected to the reversing crankshaft. The exhaust cams were directly coupled to cam-rods, the system being identical in principle to that of the inlet cams.

As may be seen in the photograph on page 229 the drive for the cam boxes on the outside cylinders was taken from the intermediate coupled wheels by means of worm gears mounted on return cranks. Tubular shafts transmitted the drive. The drive for the outside cambox was taken from an extension on the worm shaft on the left hand cambox through a right-angle bevel gear box. From each of the three reversing gearboxes, a transmission tube connected with a cambox, enabling the reversing of the engine to be done by advancing or retarding the angular position of the cams relative to the camshafts. Any cut-off desired was obtained by the angular adjustment of the inlet cams relative to each other. The designed valve events are shown in the accompanying diagram (Fig. 354).

Although applied to only one British locomotive in this particular form the valve gear has been described and illustrated in some detail because it was very successful in service and gave an excellent cylinder performance. Examples are illustrated in Fig. 353 of indicator cards taken at 75 m.p.h., the results from which were:

Engine 71000: *speed* 75 *m.p.h.*

Cut-off %	Steam rate lb./hr.	I.H.P.
5	19,000	1,446
7	21,740	1,712
10	23,380	1,837
15	28,200	2,280
17	30,000	2,310

The maximum capacity of the locomotive was probably not determined. During the full-dress trials carried out at Swindon in 1955 the water-coal relation displayed characteristics curiously unlike those of the " Britannia ". With the latter engine it was the front-end limit that marked the upper limit of capacity; with engine No. 71000 it was the grate limit, and at that limit the evaporation was less than that of the " Britannia ".

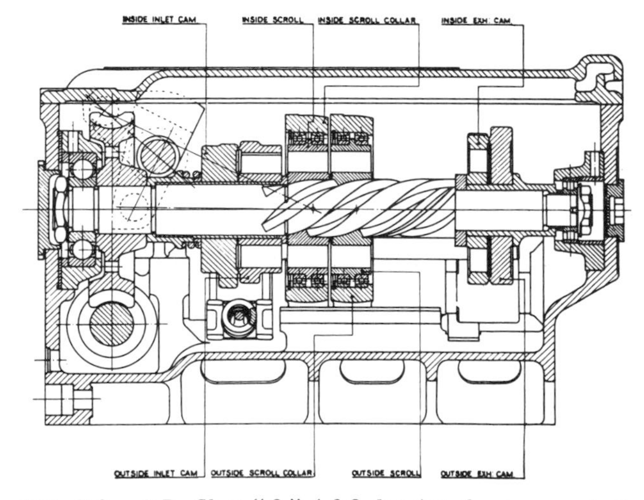

Fig. 352.—*B.R. Class* " 8 " 4-6-2 *drawing showing camshaft.*

Water-Coal Ratios: B.R. Pacifics

Class	BR7	BR8
Coal Rate (Blidworth) lb. per hr.	Evaporation steam to cylinders lb. per hr.	lb. per hr.
1000	8,500	9,000
2000	15,000	15,800
3000	21,200	21,200
4000	26,500	25,500
5000	31,300	28,200
5066	31,410†	—
6000	—	29,800
6850*	—	30,250

* Front end limit of " BR7 Pacific "   † Grate limit of " BR8 Pacific "

Clearly the optimum performance of the " BR8 " engine was not being obtained, because despite the use of a larger grate, and the improved draughting that could be expected from the double blastpipe the steaming was inferior to that of the " Britannia ". On the Swindon stationary plant repeated attempts were made, using two firemen, to raise the rate of evaporation. By the time these tests were made however the decision had been taken to discontinue the use of steam traction, and the experimental work obviously necessary to obtain the correct balance between the boiler and the draughting arrangements was unfortunately not carried out. Engine No. 71000 was allocated to Crewe North shed, and worked in the links with Stanier " Pacifics " of the " Duchess " and " Princess Royal " classes. As would be expected she proved a relatively heavy coal burner, and became an unpopular engine on that account.

The last of the British Standard locomotives to be described was in many ways the most remarkable of them all, and it was perhaps appropriate that the last steam locomotive to be built for service on the home railways should have been of this class—the " BR9 " 2-10-0 (Fig. 355). At one time consideration was given to building this class as a 2-8-2; but the success of the " Austerity " 2-10-0 described in Chapter 13 of this book largely influenced the decision to use the 2-10-0 wheel arrangement. The boiler

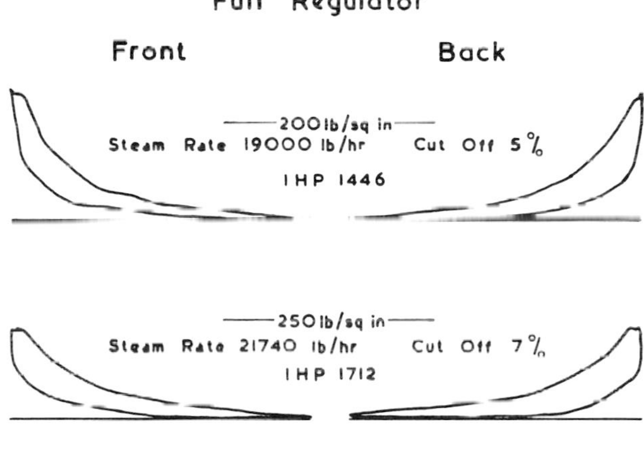

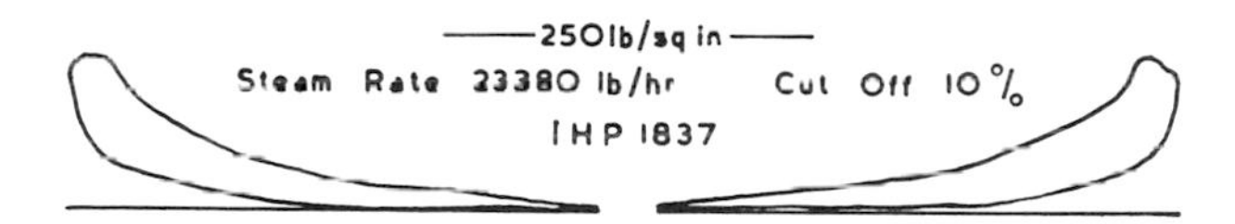

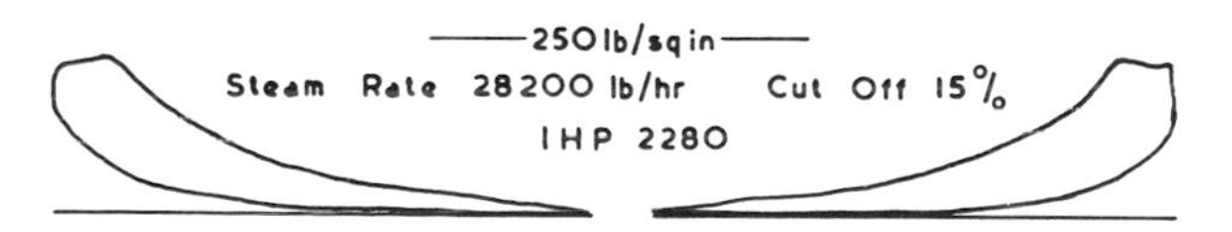

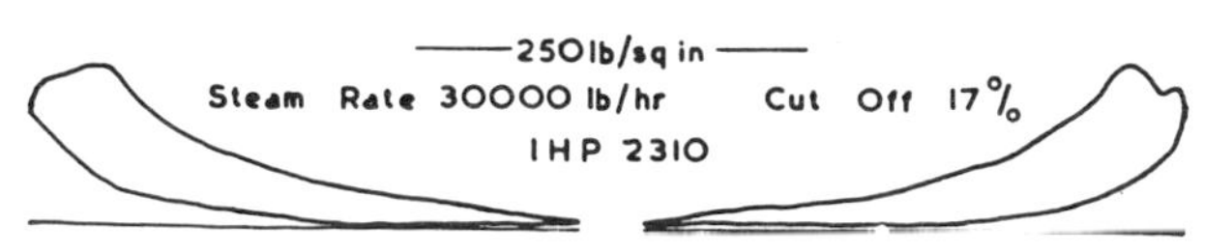

Fig. 353.—*B.R. Class " 8 " 4-6-2: examples of indicator cards taken at 75 m.p.h.*

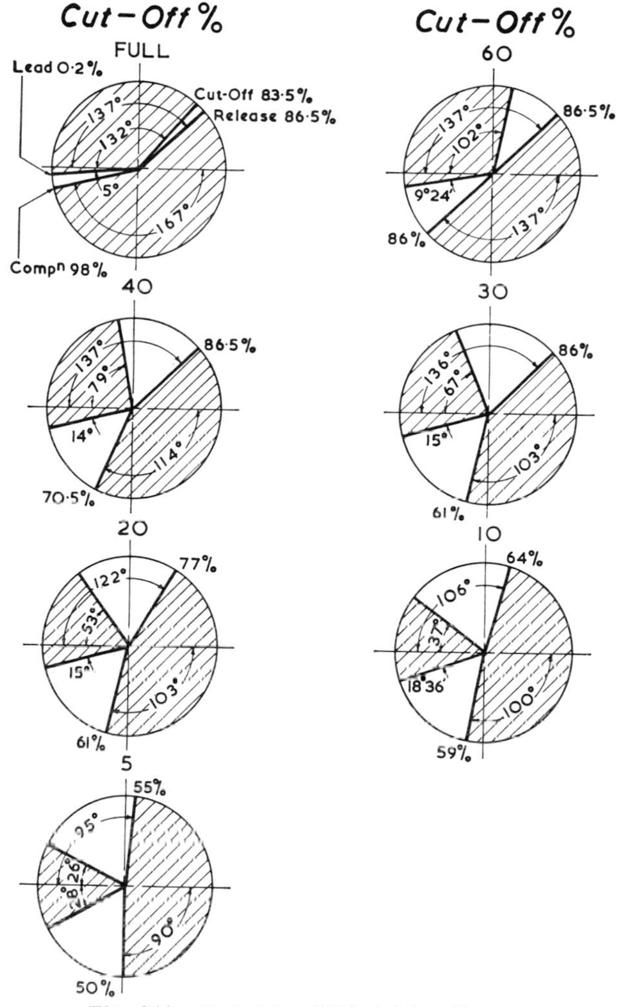

Fig. 354. *B.R. Class " 8 " 4-6-2. diagram showing designed valve events.*

Fig. 355.—*The British standard* 2-10-0 *Class* " 9F " *built Crewe* 1954.

was similar to that of the " Britannia " but shorter, being 15 ft. 3 in. between the tube plates, against 17 ft. 0 in. As the locomotive was intended for mixed traffic as well as heavy freight duties the coupled wheel diameter was made 5 ft. 0 in., and this entailed the use of a shallower firebox, than on the "Pacifics", and a grate horizontal at the back and sloping in front. The heating surfaces were: tubes 1,836 sq. ft.; firebox 179 sq. ft.; superheater 535 sq. ft.; combined total 2,550 sq. ft. The grate area was 40·2 sq. ft.

The 2-10-0 locomotives were equipped with all the devices adopted to reduce maintenance and shed work on the earlier standard locomotives, including rocker grates, hopper ashpans, and self-cleaning smokeboxes; but plain bearing axleboxes were employed throughout. The locomotives of this class proved extremely free-running, and so steady at speed that when occasions arose of their being used on main line express trains, in emergency, the remarkable maximum speed of 90 m.p.h. was attained on two fully authenticated journeys. One of these involved working the " Flying Scotsman " from Grantham to King's Cross with a heavy train, at an average speed of 58 m.p.h. start-to-stop. In view of this kind of performance the piston valve dimensions and the designed valve events are of particular interest.

B R 9 2-10-0 *Locomotives*
Cylinders and Valve Dimensions

Cylinders, dia., in.	20
Cylinders stroke, in.	28
Piston valves:	
nominal dia., in.	11
steam lap, in.	1 $^{11}/_{16}$
lead, in.	$^{1}/_{4}$
Exhaust clearance	Nil
Max. cut-off in forward gear, %	78
Max. travel in full fore gear, in. . .	7 $^{7}/_{8}$

The designed valve events were as given in table below only the figures for forward gear being given, as it is these which clearly had a bearing upon the speedworthiness of the locomotives. Later engines of this class were fitted with double blastpipes, and chimneys, and a further engine was fitted experimentally with a Giesl oblong ejector (Fig. 415).

" *B R* 9 " 2-10-0 *Locomotive*
Designed Valve Events—Walschaerts Valve Gear

Nominal Cut-off %	Travel of Valve	Lead, in.		Opening to steam in.		Steam Cut-off %		Release %		Compression %	
		F.P.	B.P.	F.P.	B.P.	F.P.	B.P.	F.P.	B.P.	F.P.	B.P.
					Forward Gear						
78	7.875	0.25	0.25	2.19	2.31	80.5	75.6	94.0	91.8	91.8	94.0
75	7.31	,,	,,	1.92	2.02	77.6	72.3	93.0	90.5	90.5	93.0
70	6.66	,,	,,	1.59	1.69	72.8	67.8	91.3	88.6	88.6	91.3
60	5.67	,,	,,	1.11	1.19	62.6	57.4	87.7	84.2	84.2	87.7
50	5.06	,,	,,	0.81	0.875	51.8	48.3	83.8	80.2	80.2	83.3
40	4.61	,,	,,	0.59	0.64	40.6	39.3	79.2	75.8	75.8	79.2
30	4.36	,,	,,	0.44	0.47	29.8	30.2	74.8	71.1	71.1	74.8
25	4.19	,,	,,	0.39	0.42	24.5	25.6	70.6	68.6	68.6	70.6
20	4.06	,,	,,	0.33	0.36	19.5	20.6	67.2	65.5	65.5	67.2
15	3.97	,,	,,	0.30	0.30	14.4	15.4	62.2	61.7	61.7	62.2
Mid Gear	3.875	,,	,,	0.25	0.25	6.8	6.9	50.3	52.0	52.0	50.3

NOTE.—F.P. = Front Port B.P. = Back Port

Chapter Eighteen Some Results of Scientific Testing

In the long history of the British steam railway locomotive there have been many instances where a potentially sound, and amply proportioned locomotive has not come up to expectations. At one time it became almost a *cliché* to suggest that the steam locomotive, for all its longevity, remained one of the least understood of machines, and in many cases the running of indicator trials did little to elucidate matters—rather the reverse! Even in the period covered in this book, when a 100 years of experience had been amassed, it was still possible to organise a carefully planned series of dynamometer car trials on a particular pair of service trains; to work the same loads for a week on end, in relatively constant weather conditions, and finish up with a series of basic coal consumption figures, from the same engine and the same driver and fireman ranging from 2.89 and 3.44 d.h.p. hour. Such a variation, in modern parlance, just does not make sense.

It is perhaps ironical, yet typical of the rule-of-thumb methods that characterised much steam locomotive designing in earlier years that it was not until the very last years of its life that testing methods were adopted which really laid bare the performance characteristics of the machine. And although such methods came so late in the day the work done on the stationary plants at Rugby and Swindon did eventually provide a mass of interesting and reliable data on which the capacity of many well-known locomotives may be ultimately assessed. Fortunately the British Railways authorities decided to publish the test bulletins in full, so that the actual results, and the penetrating comments often accompanying them can be studied by those interested in the performance of locomotives. However absorbing the details of design and construction may be it is the end-product—the work of the locomotive on the road, and the economy with which that work is done—that is the ultimate criterion by which design is judged.

Test bulletins have been published by British Railways as follows:

NON-STANDARD LOCOMOTIVES

1. Western Region: " Hall " class 4-6-0
2. Eastern and North Eastern: " B1 " class 4-6-0
3. L.M. Region: Class " 4 " 2-6-0
4. W. D. Austerity 2-8-0 and 2-10-0 freight engines
5. ex-L.N.E.R. "V2 " 3-cylinder 2-6-2
6. Southern Region: " Merchant Navy " class 4-6-2

STANDARD LOCOMOTIVES

1. " BR4 " Mixed Traffic 4-6-0
2. " BR5 " Mixed Traffic 4-6-0
3. " BR7 " Mixed Traffic 4-6-2
4. " BR8 " 3-cylinder Express passenger 4-6-2
5. " BR9 " 2-10-0 heavy freight engine

In addition to the above equally comprehensive trials have been conducted with other classes of locomotive; and although the complete results have not been made public some of the most significant detail has been given in certain papers to engineering institutions, and in articles in the technical press. These further tests, included such locomotives as the ex-G.W.R. " King " class 4-cylinder 4-6-0; the Stanier "Duchess" class "Pacifics" of the L.M.S.R.; the Great Western " Castles "; the rebuilt " Merchant Navy " class 4-6-2s of the Southern Region, and the ex-G.W.R. light 4-6-0s of the " Manor " class. Reference has already been made in a previous chapter to some features of performance recorded with the ex-L.N.E.R. " V2 " class 2-6-2, and with the " BR8 " standard 3-cylinder express passenger 4-6-2.

In all these tests carried out in post-nationalisation days insistence was placed on complete reconciliation of the results. The accompanying diagram (Fig. 356)

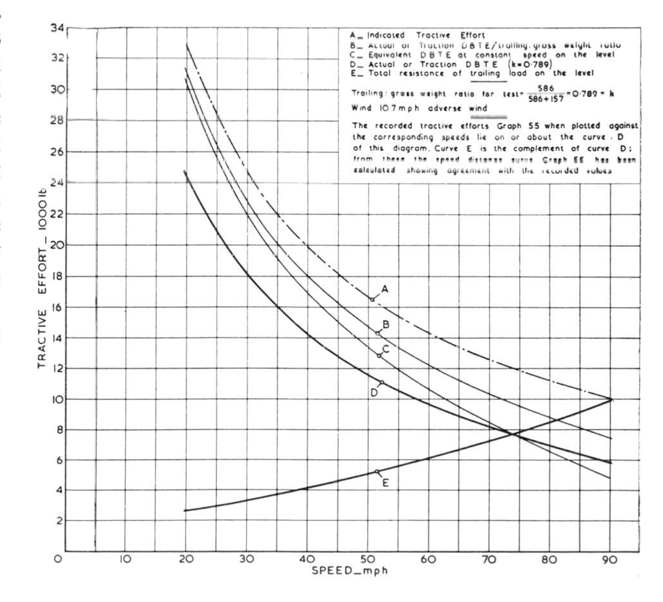

Fig. 356.—*Controlled Road Test with engine* 71000. *Set of characteristic curves.*

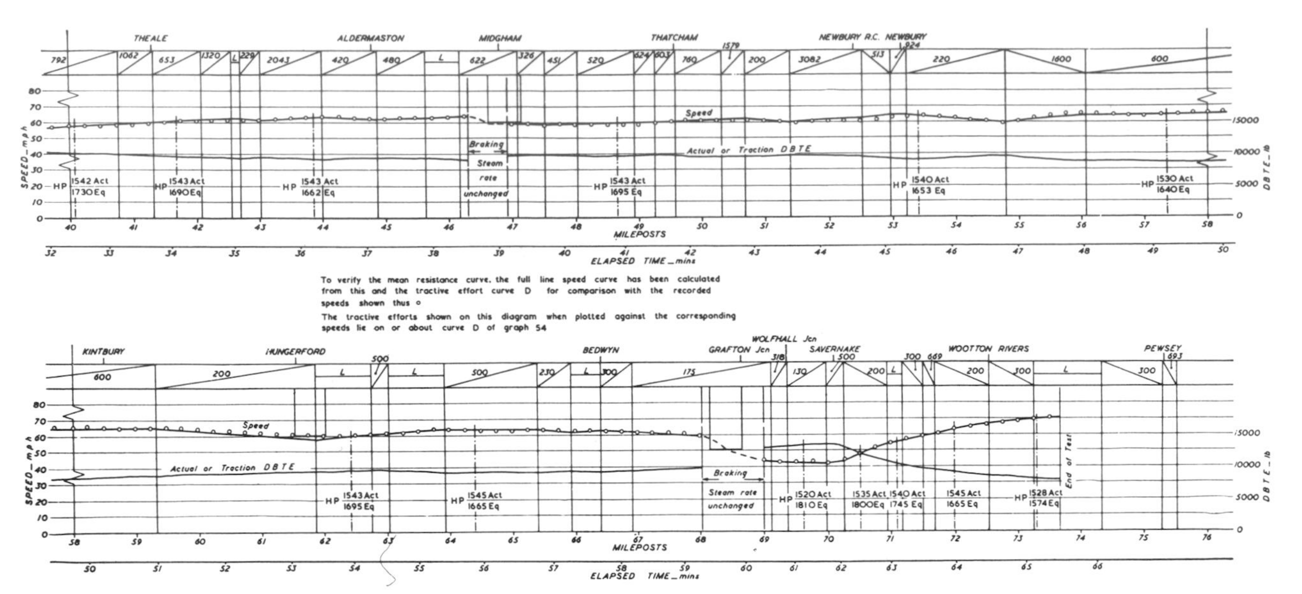

Fig. 357.—*Reconciliation and check of test run with engine* 71000, *at high steam rate.*

shows a series of characteristic curves plotted from values of indicated tractive effort, and actual drawbar tractive effort from a controlled road test of the "BR8" 4-6-2 locomotive No. 71000, when the steam rate was 30,000 lb. per hour, and the corresponding coal rate was 4,850 lb. per hour. These values were recorded at varying speeds during a run from Swindon to Westbury via Reading West, and from these two basic graphs A and B were plotted two more C, that of equivalent drawbar tractive effort at constant speed on level track, obtained by making allowance for the gradient on which the train was running at the speeds corresponding to the various points on curve B, and D representing the total resistance of the trailing load on level track. Curve D is the complement of curve B. From these two curves it is possible to construct a speed-distance curve for the entire test run, at varying speeds. This was done by plotting the calculated speed at intervals of $^1/_4$ mile en route.

The reconciliation of the test results is to be seen in the graphical representation of a portion of the test run shown in the accompanying diagram (Fig. 357), when a trailing load of 586 tons was being hauled up the Kennet Valley section of the West of England main line at a sustained steam rate of 30,000 lb. per hour. The calculated values of speed are seen to be very close to the actual "trace" of the speed recorded by the apparatus in the dynamometer car. There were certain sections of this run where brakes had to be applied in order to observe permanent speed restrictions, as at Midgham and at Grafton Junction. Such reductions were made without changing the steam rate, by steaming the engine against the additional load temporarily provided by the brakes. The gradients of the line, and the duration of the various inclinations are shown in the upper part of this diagram. Two further diagrams (Figs. 358, 359) show the excellent drawbar tractive effort and d.h.p. characteristics of this locomotive. These diagrams are typical, so far as the comprehensive information they contain, of all characteristic diagrams established for locomotives tested on the stationary plants at Swindon and Rugby.

The results obtained with the British standard locomotives in general, represent what is probably the best all round performance recorded in the various

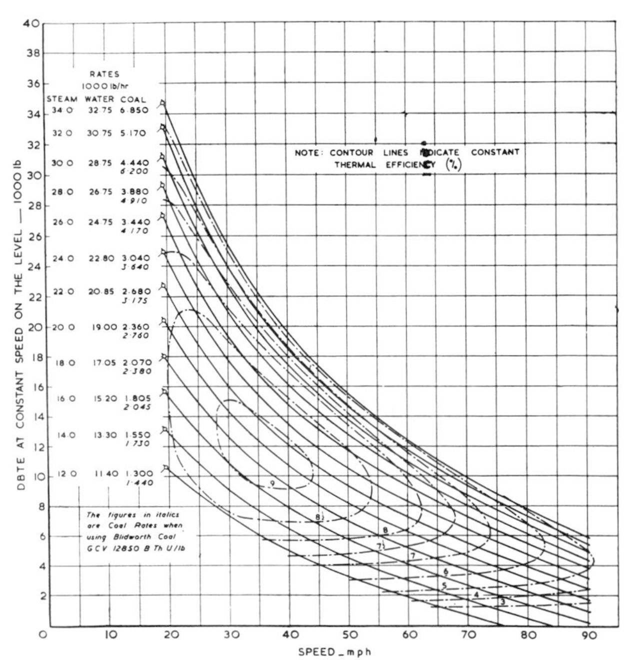

Fig. 358.—*Engine* 71000: *Characteristic curves of equivalent drawbar tractive effort.*

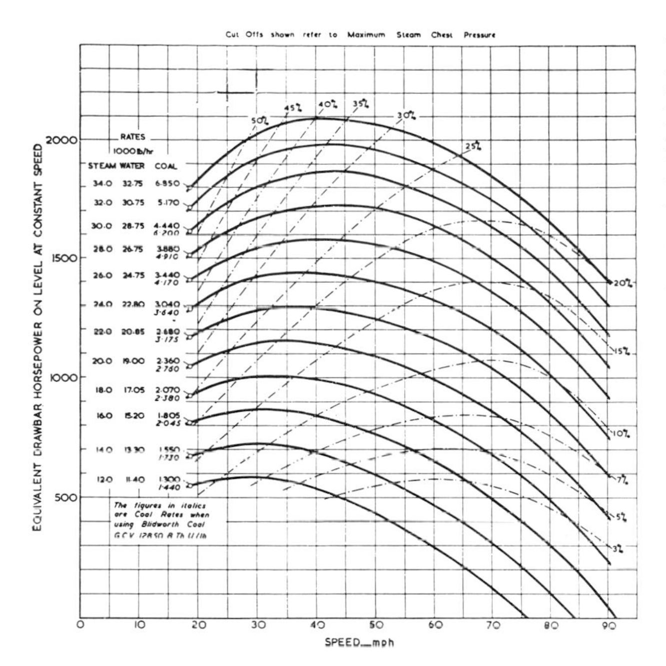

Fig. 359. *Engine* 71000: *Drawbar horsepower characteristics.*

power classes concerned. The exception is the Class "8" express passenger 4-6-2 which for reasons not fully established had a boiler performance considerably below expectations. Nevertheless certain details of the performance are tabulated collectively with other locomotives of the British standard range. For comparative purposes the details tabulated herewith relate to the performance using South Kirkby coal, except as noted below.

Power Output at Maximum Steaming Rate 50 m.p.h.

Power Class	Type	Steam rate lb. per hr.	Coal Rate lb./hr.	D.h.p. on level at constant speed	Coal per d.h.p. hr. lb.
4	4-6-0	19,600	2835*	1090	2.6
5	4-6-0	26,310	3880	1420	2.72
7	4-6-2	31,410	4260	1790	2.4
8	4-6-2	34,000	6850	2075	
9	2-10-0	30,000	4530†	1715	2.7

* Bedwas coal † Blidworth coal

The foregoing steaming rates were obtained under specially regulated test conditions, and in the majority of instances were sustained at firing rates far beyond what could be expected for any length of time in regular service. The maximum firing rate considered practicable in sustained performance on service trains was 3,000 lb. per hour, and the following table has been prepared to show what could be expected from the standard "BR5", "BR7", "BR8" and "BR9" engines in such conditions, with various classes of fuels.

Performance at 50 m.p.h.: 3000 lb. per hour

Power Class	Type	Coal	Calorific value B.Th.U./lb.	D.H.P. on level at constant speed	Coal per D.H.P. hr. lb.
5	4-6-0	South Kirkby Blidworth	13700 12600	1250 1110	2.45 2.55
7	4-6-2	South Kirkby Blidworth	13800 12600	1410 1200	2.13 2.5
8	4-6-2	South Kirkby Blidworth	13550 12850	1400 1200	2.14 2.5
9	2-10-0	Blidworth	12800	1270	2.4

* Maximum output reached at firing rate of 2010 lb. per hour

The translation of the drawbar horse power figures registered on test into actual trailing tons hauled at various speeds is necessary so that the significance of the test results can be appreciated from the viewpoint of hauling revenue-hauling trains, and in this connection it is interesting to compare the performance of the British standard locomotives with the recorded tests results of certain designs of the former private companies.

Fig. 360.—*G.W.R. "Hall" Class 4-6-0 in immediate post-war livery as engaged in dynamometer car trials.*

Fig. 361.—*L.N.E.R. " B1 " Class 4-6-0, as engaged in controlled road tests.*

Class 4 Locomotives: haulage of 300-ton train on level track: Blidworth coal

Class	Speed m.p.h.	D.h.p. involved	Steam rate lb./hr.	Coal rate lb./hr.
L.M.S. 2-6-0	60 70	408 570	11,000 15,000	1440 2255
G.W.R. " Manor " 4-6-0	60 70	408 570	12,700 18,000	1520 2700
" BR4 " 4-6-0	60 70	408 570	11,700 15,900	1500 2300

The L.M.S.R. Class " 4 " 2-6-0 was adapted with very little alteration to become a British standard, but no details of its subsequent test performance were published. It was generally considered to be one of the most useful medium-powered engines in the British Railways standard range.

Class 5 Locomotives: Haulage of 400-ton trains on level track: Blidworth Coal

Class	Speed m.p.h.	D.h.p. involved	Steam rate lb./hr.	Coal rate lb./hr.
G.W.R. " Hall " 4-6-0	60 70	545 775	15,000 21,500	1865 4000
L.N.E.R. " B1 " 4-6-0	60 70	545 775	13,000 17,000	1550 2450
" BR5 " 4-6-0	60 70	545 775	13,000 16,900	1650 2220

The Great Western " Hall " (Fig. 360) class engine suffers in the comparison from its poor boiler performance on Blidworth coal. With Grade 1a Welsh coal with a calorific value of 14,330 B.Th.U. per lb. the required steam rates would have required coal rates of 1,530 and 2,490 lb. per hour at speeds of 60 and 70 m.p.h. respectively. On the other hand the " B1 " (Fig. 361) and " BR5 " engines fired on South Kirkby coal show reductions in coal rates proportionate to the higher calorific value of the fuel. The Great Western engine would have needed a re-arrangement of its firebox and air spaces, and some change in the draughting in order to burn " hard " coal economically.

Some of the finest performances ever registered with locomotives of the basic Churchward 4-cylinder 4-6-0 design were made in the year 1953–55, when a series of dynamometer car test runs were made with " King " class locomotives prior to the restoration, in contemporary conditions of fuel supply, of the pre-war schedules of the Bristolian and Cornish Riviera Expresses: 1 $^{3}/_{4}$ hours between Paddington and Bristol in the first case, and 4 hours between Paddington and Plymouth in the latter. As originally built the " King " class engines had the Churchward jumper ring on the blastpipe; but on examination of the performance of these locomotives with a view to giving a greater margin of power in the conditions existing after World War II the draughting was modified, and a plain blastpipe substituted. Whereas engines of the original design were not required to steam continuously at rates of more than 25,000 to 26,000 lb. per hour, on the heaviest pre-war duties, the locomotives with the modified blastpipe had a " front-end " limit of no less than 33,600 lb. per hour (Fig. 362).

What a steam rate of 30,000 lb per hour sustained for any length of time could involve in actual haulage capacity on the road was shown in a pair of controlled tests with the dynamometer car and a special train, when an enormous load of 25 bogie vehicles, having a trailing tare weight of 796 tons was hauled from Reading to Stoke Gifford sidings and back by the

Fig. 362.—*W.R. " King " Class* 4-6-0 *with modified draughting on high-power tests at Swindon.*

4-cylinder 4-6-0 engine No. 6001 *King Edward VII* (Fig. 363). On each run the steam rate was constantly maintained at 30,000 lb. hour, at a coal consumption of approximately 4,000 lb. per hour. The coal consumption per d.h.p. hour was approximately 3.3 lb. a very good figure in such heavy conditions of working. It must be noted that the average speeds during the duration of the tests, at constant speed were as follows:

Test	Westbound	Eastbound
Distance of test, miles	59.5	60.9
Duration of test, min. . .	61 1/2	60 1/2
Average speed, m.p.h.	58	60 1/4

Fig. 363.—*W.R. " King " Class* 4-6-0 *on* 25-*coach test train near Hullavington.*

The slight difference reflects the marginal difference in gradient, which slightly favoured the eastbound journey. The most pronounced gradients are the distances of approximately 10 miles rising at 1 in 300 from both east and west to the summit point at Badminton station. The ability of the locomotive in such conditions of steaming to sustain average speeds of around 60 m.p.h. with this very long and heavy train over the entire test length was remarkable.

The running conditions on this particular day were ideal, with a light variable wind and a dry rail, and the performance characteristics of the locomotive, confirming the results previously obtained on the stationary testing plant at Swindon, were such as to permit of the 1 3/4 hour "Bristolian" schedule being restored in full confidence (Fig. 364). Indeed, with a load of 232 tons it was soon found possible to provide entirely reliable service with the "Castle" class engines, instead of "Kings". When it came to restoration of the 4 hour schedule of the Cornish Riviera Express between Paddington and Plymouth, some doubts were expressed as to whether the "King" class engines, even with modified draughting, would be adequate, and in the spring of 1955 with the load made up in each direction to the maximum required by the traffic department a series of dynamometer car trials was carried out. Being made on service trains temporarily running to accelerated schedules, these tests were not "controlled" as on the special runs previously referred to. But the manometer gauge used for registering the steam rate was in operation in the dynamometer car, and on sections where heavy work was in progress it was interesting to see how nearly the various drivers concerned did work to an approximately constant rate of steam usage.

The loads required by the traffic department involved double heading over the very severe gradients on the South Devon line, and to provide the necessary time for stopping and attaching the assistant engines at Newton Abbot it was necessary to run the 193.7 miles from Paddington in 189 minutes. On certain of the tests the locomotives were worked hard throughout to ascertain how much time was in reserve, with maximum load, on the proposed new schedules, while on one eastbound journey a strong adverse wind provided a serious handicap to punctual running. The following data was recorded on a fast westbound journey when engine No. 6013 covered the 173.5 miles from Paddington to passing Exeter in 159 1/4 minutes, and reached Newton Abbot in a net time of 177 1/4 minutes.

Load to Heywood Rd. Junc. tare, tons ..	460
Load to Newton Abbot tare, tons	393
Average net steam produced by boiler, lb./hr.	26,950
Firing rate, lb./hr.	3245
Average d.h.p.	
Under power	978
Actual	828
Coal per d.h.p. hr., lb.:	
Inclusive of auxiliaries	3.32
Exclusive of auxiliaries	3.21
Average speed, m.p.h.:	
Actual	62 1/2
Net	65 1/2

Fig. 364.—*W.R. "King" Class* 4-6-0 *on high-speed test run passing Hullavington at* 86 *m.p.h.*

This was an excellent piece of sustained hard work, and it was extremely interesting to find that the relationship between the steam rate and the coal rate on this service train without any specially controlled conditions showed a very close correspondence to the figures obtained for the " King " class engines on the stationary plant thus:

Test	Stationary Plant	Cornish Riviera Express
Steam rate, lb./hr.	26,000	26,950
Coal rate, lb./hr.	3,200	3,246

On one of the up journeys a strong adverse wind was encountered throughout, the effect of which is shown in the following comparison.

Wind and Rail Resistance of Coaching Stock

Date of test		March 10 1955	March 11 1955
Load, tons tare		460	393 $^1/_2$
Wind:			
Speed, m.p.h.		5	7 $^1/_2$
Direction:		45 deg. to tail	45 deg. to head
Resistance at	m.p.h.		
drawbar, lb.	50	3200	3200
,,	60	3900	3900
,,	70	4700	4600
,,	80	5600	5500

From this it is evident that the 12 coach train of March 11, weighing 393 $^1/_2$ tons, had almost exactly the same tractive resistance as the 14-coach train of March 10. On the eastbound journey the effort sustained between Exeter and Reading was probably without parallel for a run on a service train in Great Britain. The firing rate averaged no less than 4,210 lb. per hour, which is far higher than the maximum of 3,000 generally accepted for service conditions. In this instance however the inspector riding on the footplate gave considerable assistance to the fireman in getting the coal forward. On this journey, in comparison with the special test runs with 25-coach trains when the steam rate was 30,000 lb. for just over one hour, the steam rate averaged 29,835 lb. per hour throughout from Exeter to Reading. The overall results on this remarkable trip were as follows:

Exeter–Reading

Engine: " King " class No.	6013
Load: tare trailing, tons	393 $^1/_2$
Distance, miles	138 $^1/_2$
Running time, min.	127.6
Average speed, m.p.h.	64.7
Time under power, min.	117.3
Total coal consumed, lb.	8420
Firing rate, lb./hr.	4210
Net steam rate, lb./hr.	29,835
Average d.h.p.	1020
Coal per d.h.p. hr., lb.:	
Inclusive of auxiliaries	4.13
Exclusive of auxiliaries	4.00

The check and reconciliation of results on a particularly heavy section of the run is shown in the

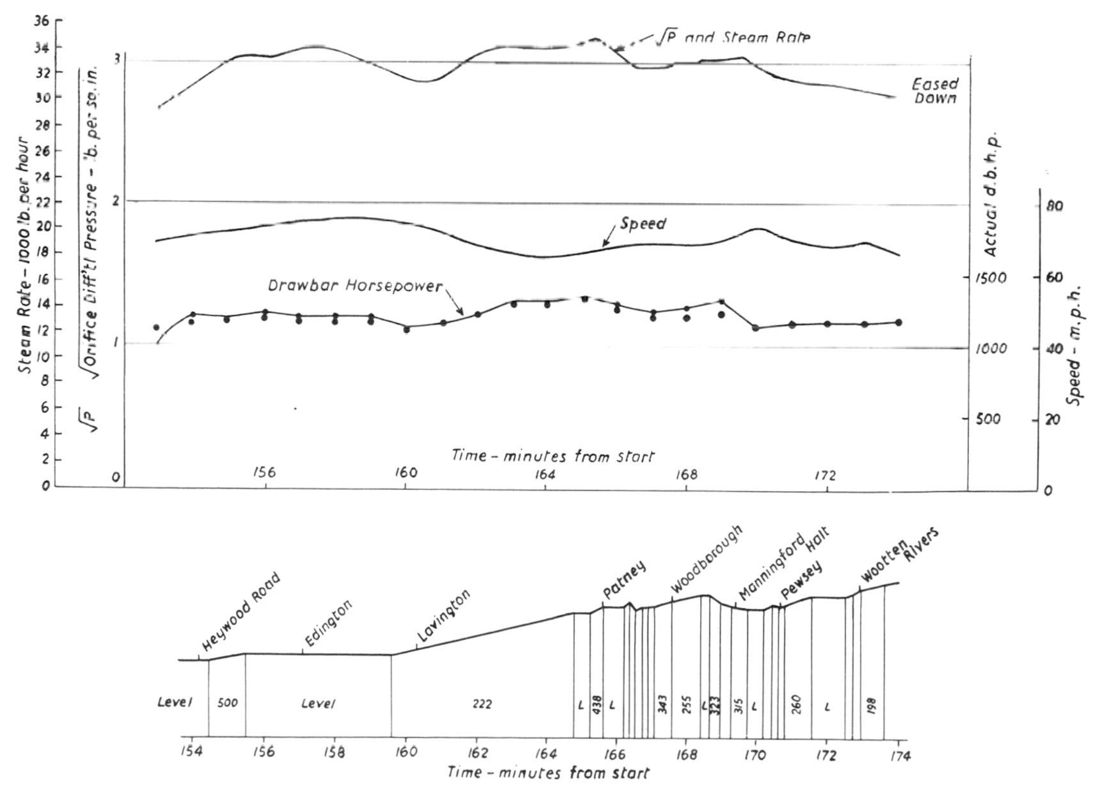

Fig. 365.—*W.R. " King " Class: portion of high-power test run, Exeter to Paddington. Check and reconciliation of results.*

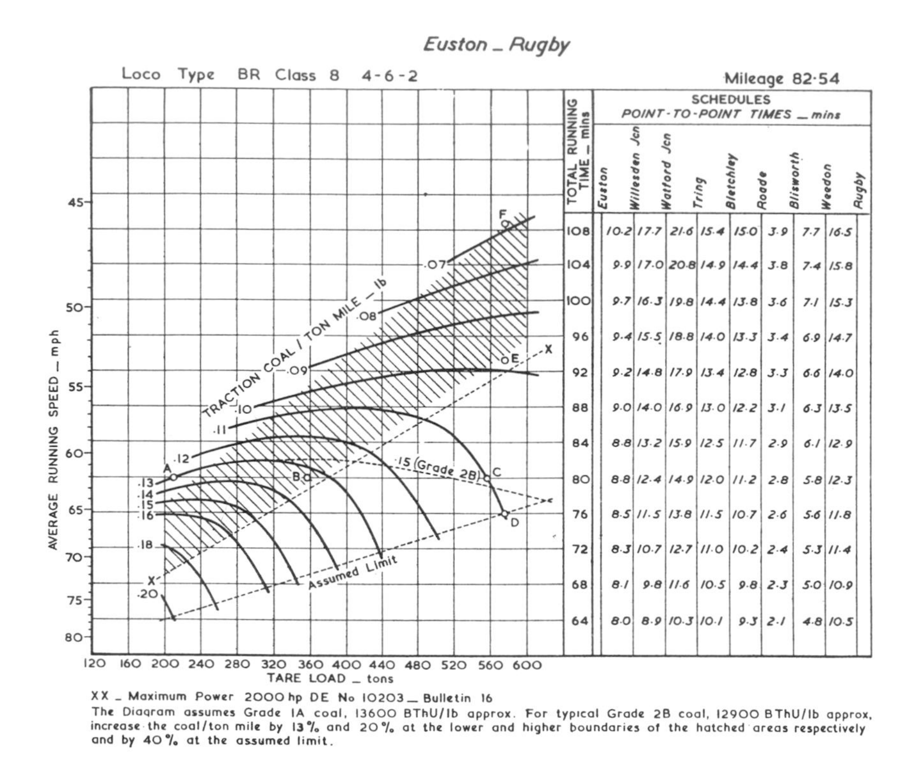

Total running time – mins	Euston – Willesden Jcn	Willesden Jcn – Watford Jcn	Watford Jcn – Tring	Tring – Bletchley	Bletchley – Roade	Roade – Blisworth	Blisworth – Weedon	Weedon – Rugby
108	10·2	17·7	21·6	15·4	15·0	3·9	7·7	16·5
104	9·9	17·0	20·8	14·9	14·4	3·8	7·4	15·8
100	9·7	16·3	19·8	14·4	13·8	3·6	7·1	15·3
96	9·4	15·5	18·8	14·0	13·3	3·4	6·9	14·7
92	9·2	14·8	17·9	13·4	12·8	3·3	6·6	14·0
88	9·0	14·0	16·9	13·0	12·2	3·1	6·3	13·5
84	8·8	13·2	15·9	12·5	11·7	2·9	6·1	12·9
80	8·8	12·4	14·9	12·0	11·2	2·8	5·8	12·3
76	8·5	11·5	13·8	11·5	10·7	2·6	5·6	11·8
72	8·3	10·7	12·7	11·0	10·2	2·4	5·3	11·4
68	8·1	9·8	11·6	10·5	9·8	2·3	5·0	10·9
64	8·0	8·9	10·3	10·1	9·3	2·1	4·8	10·5

Fig. 366.—*Performance of* " BR8 " 4-6-2 *engine No.* 71000 *applied to Euston-Rugby route.*

accompanying diagram (Fig. 365) which relates to the section of line between Westbury and Savernake summit. Here the engine was being steamed virtually to its limit, at an average rate of about 33,000 lb. per hour. In this case the check was made by calculating the d.h.p. at one minute intervals, based only on the speed and the steam rate, and comparing these with the actual values of the d.h.p. as registered at the same points in the dynamometer car. As will be seen from the diagram the correspondence is very close.

Although the evaporation-coal ratio in the case of the " King " class engines as thus modified was very good, even when pressed almost to the limit the cylinder performance, related to the design of 1927 vintage has been considerably improved upon in later British locomotives of the Class " 8 " power capacity. In this connection some work associated with the tests of the " BR8 " 3-cylinder " Pacific " No. 71000 at Swindon is of interest. The accurate assessment of locomotive performance made possible by modern work on the stationary plants, and the subsequent verification of results in controlled road tests with the dynamometer car, has enabled the testing staffs to prepare realistic schedules for express passenger and through goods trains for any route of which the physical characteristics and speed restrictions are known. One result of the comprehensive tests carried out at Swindon on the " BR8 " " Pacific " was the preparation of a series of performance charts for this engine over the main line of the Region on which it was to be used, namely the London Midland.

The chart for the section between Euston and Rugby is reproduced in Fig. 366. The diagram is one in which tare load on the horizontal scale is plotted against overall time on the right hand vertical scale. Horizontal projection across the diagram to the left hand vertical scale indicates the average train speed. Thus an 80 min. timing corresponds to an average speed of 62 m.p.h. and the schedule which gives minimum coal per ton mile for the timing is set out on its right. In practice the point-to-point times would be rounded off to the nearest half-minute as this is the minimum practical time period. The vertical ordinates of all loads from 200 tons to over 600 tons cut the 80 min. line within the boundaries of the diagram and above the " Assumed Limit " line. This range of loading therefore falls within the capabilities of the locomotive for the 80 min. timing but for practical purposes the range must be limited at its higher end. The vertical ordinate for 320 tons cuts the assumed limit line at a timing 12 min. less than 80 min. This is the nominal margin in a locomotive with 320 tons. It means that if the train left Euston 12 min. late it could reach Rugby on time, if the road were clear, by working at its assumed maximum rate wherever possible; or leaving Euston on time, and whilst running to time, met a 6 min. delay halfway in the journey. But for 600 tons the nominal margin is only 3 min. on the whole distance which is insufficient to preserve the time-table under practical operating conditions, besides making a high physical demand on the fireman. The practical nominal margin would not be less than 6 to 8 min. which would limit the load to 460-500 tons for the 80 min. timing.

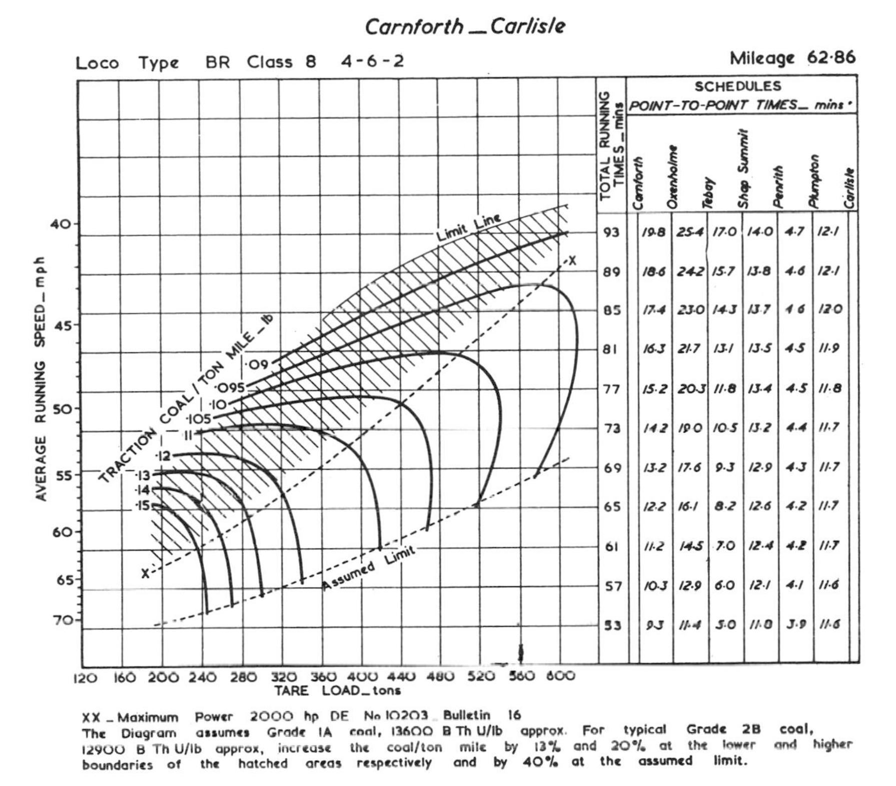

TOTAL RUNNING TIMES _ mins	SCHEDULES POINT-TO-POINT TIMES _ mins: Carnforth–Oxenholme	Oxenholme–Tebay	Tebay–Shap Summit	Shap Summit–Penrith	Penrith–Plumpton	Plumpton–Carlisle
93	19·8	25·4	17·0	14·0	4·7	12·1
89	18·6	24·2	15·7	13·8	4·6	12·1
85	17·4	23·0	14·3	13·7	4·6	12·0
81	16·3	21·7	13·1	13·5	4·5	11·9
77	15·2	20·3	11·8	13·4	4·5	11·8
73	14·2	19·0	10·5	13·2	4·4	11·7
69	13·2	17·6	9·3	12·9	4·3	11·7
65	12·2	16·1	8·2	12·6	4·2	11·7
61	11·2	14·5	7·0	12·4	4·2	11·7
57	10·3	12·9	6·0	12·1	4·1	11·6
53	9·3	11·4	5·0	11·0	3·9	11·6

Fig. 367.—*Performance of* " BR8 " 4-6-2 *engine No.* 71000 *applied to Carnforth-Carlisle line.*

For the cost in fuel consumption, the intersection of overall time and load is referred to the grid of constant coal per ton mile. Thus, the intersection of a 380-ton load with an 80 min. timing falls on the curve marked 13 lb. per ton mile. Total fuel consumption for traction is therefore $13 \times 380 \times 82.54 = 4075$ lb., where 82.54 is the mileage. This is at the rate of 49.4 lb. per mile. Conditions appertaining to maximum built-in efficiency (minimum coal per d.h.p. hour) lie within the hatched band.

To all students of locomotive performance however it is the diagram referring to the mountain section of the West Coast Route that will undoubtedly prove of the greatest interest, in the application of the results of scientific test procedure to the ascent of Shap. The Cost of Energy and Performance diagram for the section from Carnforth to Carlisle is reproduced in (Fig. 367). It is to be noted that the fuel economy of the locomotive is not adversely affected by working the engine heavily on the rising gradients when the speed is low. The diagram shows that a train of 600 tons tare could be worked over Shap Summit at 24 m.p.h. without exceeding the assumed limit of continuous steaming, namely 30,000 lb. per hour.

However the fuel consumptions given on the diagrams assumed ideal operating conditions, which, of course, could not be regarded as the general case. The permanent way, for instance, has to be repaired. Temporary speed restrictions for this purpose, and other minor disturbances to the timetable were expected to be met by adjustment in the running. Recovery, by the locomotive, of time lost in meeting these contingencies results in increased fuel consumption because of the additional energy required, irrespective of whether or not the thermal efficiency is significantly decreased by the heavier working rate entailed in the recovery. Referring to the graph for the southern end of the line it may be seen that the coal consumption for a 400 ton train timed at 84 min. Euston to Rugby non-stop was .12 lb./ton mile in the ideal conditions assumed. If 6 min. were lost and recovered during the journey, thus making the equivalent running time 78 min. the fuel consumption is seen to rise to .13 lb./ton mile, an increase of 8 per cent.

It is interesting to compare the theoretical timings contained in the Swindon diagrams with the actual running times made by this locomotive on a service run with the " Midday Scot ", as between Preston and Carlisle. The schedule times then laid down for that train were 26 min. from Preston to Carnforth, and 73 min. from Carnforth to Carlisle.

On the service run concerned the trailing load was 465 tons, and it will be seen from the graph that this was well within the maximum assumed limit of the locomotive. It will also be seen that while the theoretical timing, corresponding to a constant steam rate corresponded closely to the actual point-to-point timings laid down over the level sections between Preston and Carnforth, there was some variation over the mountain section (Fig. 369), where the service

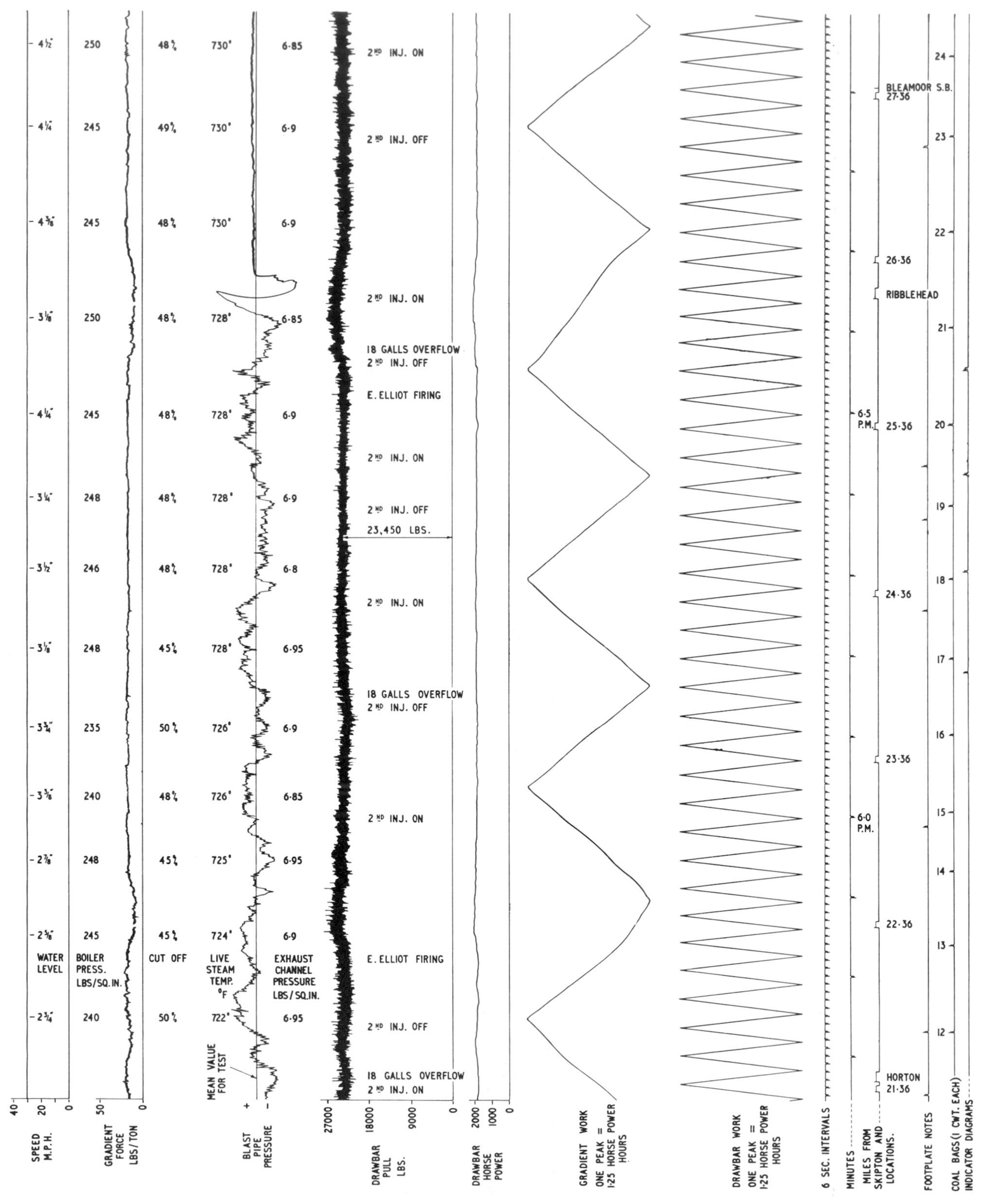

Fig. 368.—*L.M.R.* 4-6-2 *engine* No. 46225*: portion of dynamometer car record between Horton and Blea Moor; maximum output.*

Fig. 369. *Tho* " BR8 " 4 6 2 *No.* 71000 *on up Anglo-Scottish express near Tebay, Westmorland.*

Dist. miles		Schedule laid down		Schedule Swindon diagram	
		min.	Av. speed m.p.h.	min.	Av. speed m.p.h.
0.0	Preston	0	—	0	—
1.3	Oxheys	3	26.0	2.9	26.9
9.4	Garstang	11	60.7	10.6	63.1
21.0	Lancaster	21	69.6	20.3	71.8
27.2	Carnforth	26	74.4	25.6	70.2
40.1	Oxenholme	41	51.6	39.8	54.5
53.2	Tebay	58	46.3	58.8	41.5
58.7	Summit	68	33.0	69.3	31.4
72.3	Penrith	81	62.8	82.5	61.8
77.0	Plumpton	86	55.2	86.9	62.8
90.1	Carlisle	99	60.5	98.6	67.2

timings demanded harder work uphill and evidently provided some recovery margin between Shap Summit and Carlisle. The times made by the locomotive were as opposite, inclusive of a temporary speed restriction in a very awkward location, namely the foot of the Shap Incline.

The point-to-point timings then laid down for this express were disproportionately hard between Preston and Carnforth, and if strictly observed would have required a performance practically up to the assumed limit of the locomotive, and requiring a steam rate of 30,000 lb. per hour, whereas on the mountain section the point-to-point times required only a little more than 24,000 lb. per hour. It is thus not surprising that in the practical observance of his end-to-end schedule the driver should have lost a little time between Preston and Carnforth, and then more than recovered it afterwards. The scientific application of test results to timetable planning was therefore of great assistance in the avoidance of anomalies in point-to-point timings, which had in the first place often been determined by no more than " rule of thumb " methods.

Engine: " *B R* 8 " 4-6-2 *No.* 71000
Load: 465 *tons*

	Theoretical Schedule (point-to-point) min.	Actual times run (point-to-point) min.
Preston	—	—
Oxheys Box	2.9	3.5
Garstang	7.7	8.2
Lancaster	9.7	10.2
Oxenholme	14.2	12.4
—	—	check
Tebay	19.0	15.7
Shap Summit ..	10.5	11.8
Penrith	13.2	12.3
Plumpton	4.4	4.6
Carlisle	11.7	12.5
Total	98.6	96.5

From working over Shap it is a natural transition to the neighbouring mountain route over Aisgill, and here some worthy northern counterparts to the 25-coach test runs with the Western Region " Kings " were made with " Pacific " engines. It is a great pity that difficulties with slipping prevented the maximum capacity of the " Merchant Navy " class engines from being established over that difficult route. For this reason it was not possible to run any trial at a greater steam rate than 28,000 lb. per hour. But with the " Britannia " class " BR7 " 4-6-2, and with the Stanier " Duchess " class 4-6-2 some examples of very high power output were successfully sustained. With these locomotives the L.M.S.R. mobile testing unit was used to provide much of the load, to avoid using a very large number of coaches in the train; and over this mountain route trains equivalent in tractive resistance to 850 and 900 tons of ordinary stock were

Fig. 370.—*Southern Region: dynamometer car test run with rebuilt "Merchant Navy" class 4-6-2.*

hauled in both directions between Carlisle and Skipton over the 1151 ft. altitude of Aisgill summit.

The "Britannia" class engine No. 70005 *John Milton* was steamed at the very high rate, even for an engine of this class, of 36,000 lb. per hour—considerably above the maximum obtained on the Rugby testing plant with the same engine. With the ex-L.M.S. 4-6-2 No. 46225 *Duchess of Gloucester* an even higher output was obtained. The results were:

Engine 4-6-2 No. . . Class	70005 "BR7"	46225 L.M.R. "8"
Equiv. trailing load in ordinary stock, tons	850	900
Duration of effort, min.	29 1/4	43
Steam rate, lb./hr. . .	36,000	40,000
Coal rate, lb./hr. . .	5,600	5,700
Av. equiv. d.h.p. . .	2,000	2,200
Section of line . .	Lazonby to Crosby Garrett	Low House to Kirkby Stephen
Av. speed, m.p.h. . .	47.3	46.2

As an indication of haulage ability on a severe rising gradient it can be added that the "Duchess" class engine sustained a speed of 30 m.p.h. on a 1 in 100 gradient for some miles in the neighbourhood of Kirkby Stephen. It will be noted that the "Duchess" used very little more coal to produce 40,000 lb. of steam per hour than the "Britannia" did to attain 36,000 lb.; but that is understandable because the latter engine was being pressed almost beyond the limit of continuous steaming in order to ascertain maximum output. The graph on page 242 (Fig. 368) shows the dynamometer record of a similar performance on the ascent from Settle Junction to Blea Moor, when the engine was being steamed at 38,500 lb. per hour. These test runs with the ex-L.M.S.R. engine No. 46225 probably represent the greatest continuous outputs of power ever achieved by a British steam locomotive, though of course they have been substantially exceeded for brief periods as the final chapter of this book will relate.

Fig. 371.—*Western Region: Maximum load test with* "BR9" 2-10-0, *leaving Stoke Gifford yard for Reading.*

Chapter Nineteen Last Steam for Overseas

In the years following the end of World War II when railway administrations all over the world were anxious to make good the arrears of locomotive modernisation and replacement occasioned by the war, the British locomotive building industry was responsible for the construction of a great variety of new and improved classes. Among the neatest and most conventional by previous British standards were some 5 ft. 3 in. gauge 4-4-0s for the Great Northern Railway of Ireland. The Class " U " branch line engines (Fig. 373) of which five were built in 1948 by Beyer Peacock and Co. Ltd. were a development of the Glover engines of similar proportions first introduced in 1915. The new engines were of simple and straightforward proportions, and in accordance with the contemporary practice of the G.N.R. (I) they were gaily finished in a livery of bright blue, with crimson and black below the running plate.

Immediately following delivery of these engines Messrs. Beyer Peacock and Co. completed five main line express passenger 4-4-0's (Fig. 374) that were generally similar in appearance to the 3-cylinder compounds of 1932, and had interchangeable boilers. But the new engines were 3-cylinder simples having many points of similarity, so far as principles were concerned to the Southern Railway " School " class 3-cylinder simple 4-4-0s. At the time of delivery of these engines arrangements had been made for nonstop running between Belfast and Dublin for the first time in history, and the working of these nonstop trains was one of the duties to which the new engines were assigned. The dimensions of these new Irish 4-4-0s were:

G.N.R. (I) 4-4-0 Locomotives

Class .. / Service ..	U Branch	VS Express Passenger
Cylinders: number ..	2	3
dia., in. ..	18	$15^1/_4$
stroke, in. ..	24	26
Coupled wheel dia., ft. in. ..	5-9	6-7
Heating surfaces, sq. ft.:		
Tubes ..	757	973.5
Firebox ..	106	160.5
Superheater ..	177	293
Combined total ..	1040	1527
Grate area, sq. ft. ..	18.3	25.2
Nom. tractive effort at 85% boiler pressure, lb. ..	16800	21469

In contrast to these relatively small Irish 4-4-0s the resources of the British locomotive building industry were shown in the production of some very imposing mixed traffic 2-10-0s for the Turkish State Railways (Fig. 375). These engines were a development of a German design previously built by several manufacturers; but while the outline of the design and the arrangement of cab fittings followed the practice previously favoured in Turkey, British methods of construction, and many proprietary fittings of British manufacture were incorporated. The engines were built on bar frames, $3^1/_2$ in. thick, with fully compensated suspension. Straight-air and automatic Westinghouse brake equipment was included, together with the Riggenbach counter-pressure brake apparatus for use in descending long grades with the

Fig. 373.—*Great Northern Railway (Ireland): light branch* 4-4-0 *locomotive* (1948).

Fig. 374.—*Great Northern Railway (Ireland): 3-cylinder simple* 4-4-0 (1948) *built by Beyer Peacock and Co. Ltd.*

engine in reverse. The leading dimensions were:

Cylinders dia., in.	25 5/8
Cylinder stroke, in.	26
Coupled wheel dia., ft. in.	4-9 1/8
Heating surfaces, sq. ft.:	
Tubes	2406
Firebox	169
Superheater	908
Combined total	3483
Grate area, sq. ft.	43
Boiler pressure, lb./sq. in.	228
Nom. tractive effort at 85% boiler pressure, lb.	57560

Two other interesting locomotives for the Middle East may also be mentioned at this point: a new 4-6-0 for Egypt and a 4-8-2 for the Sudan. There was a time, just before World War II when the latest locomotives designed and built for the Egyptian State Railways were distinguished by domeless boilers; but in the new 4-6-0s supplied by the North British Locomotive Company in 1949-50 (Fig. 376) there was a reversion to more conventional lines. These engines were powerful units, having a nominal tractive effort, at 85 per cent. working pressure of 30,610 lb. An interesting feature was the steel all-welded firebox, incorporating a Nicholson thermic syphon. These engines were designed for oil-firing. The firebox walls were lined with firebrick to a suitable depth, and the firebox was similarly lined on the bottom, side, and back plates. The oil burner was located at the front of the firepan above a small primary air door. Modern items of equipment included manganese steel liners for the coupled axlebox guides and Timken roller bearings throughout on the tender.

The Sudan Railways 4-8-2, of 1955 (Fig. 377), was also an oil-burner, and was designed for routes which have a severely restricted axle load. In fact a greater weight was carried on the three coupled axles of the Egyptian 4-6-0 than on the four coupled axles of the Sudanese 4-8-2—62 as against 59.7 tons. The latter was a skilfully designed locomotive providing a

Fig. 375.—*Turkish State Railways: Beyer-Peacock* 2-10-0 (1946).

Fig. 376.—*Egyptian State Railways:* 4-6-0 *passenger locomotive* (1949) *by North British Locomotive Co. Ltd.*

Fig. 377.—*Sudan Railways: oil-burning* 4-8-2 *of* 1955, *by North British Locomotive Co. Ltd.*

Fig. 378.—*Victorian Railways:* 2-8-2 *of* " N " *class* (1950).

nominal tractive effort of 35,940 lb. at 85 per cent. boiler pressure. Like the Egyptian 4-6-0, the 42 engines of this class were built by the North British Locomotive Co. Ltd., and shipped in the fully erected condition from Glasgow docks to Port Sudan. The leading dimensions of these two designs of locomotive, 4 ft. 8 $^{1}/_{2}$ in. gauge for Egypt and 3 ft. 6 in. gauge for Sudan were:

Railway	Egypt	Sudan
Type	4-6-0	4-8-2
Maximum axle load, tons	21	15
Cylinders (2) dia., in.	21	21$^{1}/_{2}$
Cylinder stroke, in.	28	26
Coupled wheel dia., ft. in.	6-0	4-6
Heating surfaces, sq. ft.:		
Tubes	1958	2027
Firebox	233	203
Superheater	472	542
Combined total	2663	2772
Grate area, sq. ft.	31.25	40
Boiler pressure, lb./sq. in.	210	190
Nom. tractive effort at 85% boiler pressure, lb.	30,610	35,940

These latter were designed for high grade express passenger service, with a view to their use eventually on Inter-State expresses working alternatively with engines of the New South Wales railways, between Melbourne and Sydney. They were massively constructed engines having bar frames cut from rolled slab and finished to a thickness of 5 in. A cast steel headstock incorporating buffer beam and drag-box was fitted at the front end, and the rear end of the frame consisted of a one-piece cast steel cradle attached to the main frames at the rear of the coupled wheels. Self-aligning roller bearings were fitted to all coupled axles. The boiler was a large one, in conjunction with a firebox having an all-welded inner shell. Two thermic syphons were included. It is significant of the hard work envisaged for these engines that they were fitted with mechanical stokers.

Particular attention was paid to the design of the cylinders, valves and steam circuit, to provide for free running, and a good feature was the large diameter of piston valves: 11 in. diameter valves in conjunction with cylinders 21 $^{1}/_{2}$ in. diameter by 28 in. stroke. An attempt was made to impart a " new look " to these engines, in keeping with " Operation Phoenix ", and although they were finished in plain black without any lining a dash of brilliant colour was provided by painting the running plate valences, and the small smoke deflecting plates in the same scarlet as the buffer beams. The leading dimensions of these 2-8-2 and 4-6-4 locomotives were:

Fig. 379.—*Victorian Railways: express passenger* 4-6-4 " R " *class, for " operation Phoenix ".*

Locomotives for Australia

Some very large orders for locomotives of new design for the Victorian Railways (5 ft. 3 in.) were completed by the North British Locomotive Co. Ltd. in 1950 51: 50 2-8-2s of Class " N " and 70 4-6-4s of Class " R ". An interesting feature of both designs, foreshadowing the eventual standardisation of rail gauges in Australia, was that although built to operate on the 5 ft. 3 in. gauge they were convertible to 4 ft. 8 $^{1}/_{2}$ in. The 2-8-2 (Fig. 378) can be considered as a development of traditional practice on the Victorian Railways, exemplified by the famous " A2 " class 4-6-0 express passenger engines of 1907. As time went on these engines, in their later developments, began to lose their characteristic Edwardian-British appearance, and acquire stove-pipe chimneys and large smoke deflecting plates. The 2-8-2 was a post-war development of the " N " class of 1925, and together with the new 4-6-4s formed part of " Operation Phoenix ", or the post-war rehabilitation of the Victorian Railways. Being based largely on previous practice they had no new features, and were built on plate frames. The leading dimensions are given in the table alongside those of the new 4-6-4s (Fig. 379).

Victorian Railways
" Operation Phoenix "

Engine Class	N	R
Type	2-8-2	4-6-4
Cylinders (2) dia., in.	20	21 $^{1}/_{2}$
Cylinder stroke, in.	26	28
Coupled wheel dia., ft. in.	4-7 $^{3}/_{16}$	6-0
Heating surfaces, sq. ft.:		
Tubes	1250	1958
Firebox	203	285
Superheater	324	462
Combined total	1777	2705
Grate area, sq. in.	31	42
Boiler pressure, lb./sq. in.	175	210
Nom. tractive effort at 85% boiler pressure, lb.	28,650	32,800

Fig. 380.—*Western Australian Government Railways:* 3*ft.* 6*in. gauge* 4-8-2 *for low grade fuel utilisation* (1951).

The post-war problems of the 3 ft. 6 in. gauge Western Australian Government Railways were of a different kind. A powerful general purpose locomotive (Fig. 380) was required to run over routes with 45 lb. rails, and moreover to use the local low grade coal, which had an ash and moisture content of 8 and 25 per cent. respectively. As might be expected the outstanding feature of a non-articulated locomotive, having a nominal tractive effort at 85 per cent. boiler pressure of 21,760 lb., operating in such conditions was the boiler. The firebox was of unusual shape having a steeply sloping door plate and front, as can be seen from the accompanying photograph. The inner firebox was of steel, all-welded, and its large proportions and very large combustion chamber gave the unusually high volume to grate area ratio of 6 to 1. This was required in order to provide for the slow burning and full-flame length of the low grade fuel employed. The firebox also included two arch tubes and a thermic syphon. The boiler had a relatively short barrel, with a distance of 14 ft. between the tubeplates. The apparent length seen in the broadside photograph is accounted for by the extreme length of the combustion chamber. This extends forward almost to the position marked by the boiler band immediately in rear of the dome. The design features included in these locomotives proved very successful in service, and they hauled loads of 900 tons at 35 m.p.h. on level track.

While the Class " N " 4-8-2 locomotives of 1951 were designed for general service on lightly laid routes using low grade Collie coal, a new design of heavy mineral engine was prepared in 1955 by Beyer Peacock and Co. Ltd., for working coal traffic from the mines themselves (Fig. 381). The lines traversed permitted of a much larger and heavier engine, and the 24 units of " V " class were the most powerful non-articulated locomotives in Australia running on the 3 ft. 6 in. gauge. An axle load up to 14 $^1/_4$ tons was permitted in these engines, as compared with 10 $^1/_4$ tons in Class " W ". In the latter engines, actually the most heavily loaded axle was the trailing truck, carrying 10.3 tons; the four driving axles each carried 9 $^3/_4$ tons. In the " V " class 2-8-2 the general design was similar, though much enlarged over the " W " class, including the same unusual shape of firebox. There were two thermic syphons, and a large combustion chamber. The leading dimensions of these two interesting locomotive designs, both from Beyer Peacock and Co. Ltd were:

Western Australian Freight Locomotives

Class	" W "	" V "
Type	4-8-2	2-8-2
Number built	60	24
Max. axle load, tons	10 $^1/_4$	14 $^1/_4$
Cylinders dia., in.	16	19
Cylinder stroke, in.	24	26
Coupled wheel dia., ft. in.	4-0	4-3
Heating surfaces, sq. ft.:		
Tubes	930	1570
Firebox	184	244
Superheater	305	499
Combined total	1419	2313
Grate area, sq. ft.	27	40
Boiler pressure, lb./sq. in.	200	215
Nom. tractive effort at 85% boiler pressure, lb.	21,760	33,630

In this same period, between 1945 and 1955 Messrs. Beyer Peacock and Co. Ltd. prepared two very fine new designs of Beyer-Garratt locomotives for service in Australia. The first was a 4-8-2 + 2-8-4 for the 3 ft. 6 in. gauge Queensland Railways (Fig. 382) with a maximum axle load of only 9.64 tons, while the second was a very powerful 4-8-4 + 4-8-4 for the 4 ft. 8 $^1/_2$ in. gauge New South Wales Government Railways (Fig. 383). In the former case the operating conditions on this important railway are not generally realised. The new locomotives, of which 30 were built, were intended for general service, but also for handling the crack " Sunshine Express ", working this throughout with one engine over the 1043 miles between Brisbane and Cairns. The northern part of the line takes this express far into the tropics, and so particular attention had to be paid to those features of design subject to the effects of extremes of climate. The characteristic features of the Beyer-Garratt design, combined with the most modern details of equipment resulted in a locomotive of handsome proportions and excellent performance, as befitted machines having a tractive effort nearly 50 per cent. higher than anything previously operating in Queensland.

The New South Wales locomotives were constructed for a railway system having generous loading gauge limits, and the designers thus had a freedom unusual where Beyer-Garratt locomotives are normally

Fig. 381.—*Western Australian Government Railways: Heavy mineral* 2-8-2 *locomotive* (1955) *built by Beyer Peacock & Co. Ltd.*

Fig. 382.—*Queensland Railways:* 4-8-2 and 2-8-4 *Beyer Garratt, for general service on 3ft. 6in. gauge.*

Fig. 383.—*New South Wales Railways:* 4-8-4 + 4-8-4 *Beyer Garratt; the largest and heaviest steam locomotive design ever introduced in Australia.*

concerned. They were more accustomed to working to severe structural limits. With this freedom a magnificently proportioned locomotive was produced, yet one having an axle-load not exceeding 16 tons, and thus able to work over practically the whole of the system. A very powerful engine was specified, and the result was the largest and heaviest locomotive ever introduced on to the Australian railways, having a total weight in working order of 255 tons, and a nominal tractive effort at 85 per cent. boiler pressure of 59,560 lb. Apart from their great size however there were certain very interesting design features in these locomotives. The New South Wales railway authorities had, for example, a long and successful experience of one-piece cast steel bed frames, and this type of frame was specified for the 50 new Beyer-Garratt locomotives. They were the first engines of this type to have this kind of frame.

The generous loading gauge, coupled with the Garratt form of construction made possible a boiler and firebox of exceptionally favourable dimensions. The firebox was no less than 9 ft. 8 in. long, with a grate area of 63.4 sq. ft. and mechanically fired. The boiler barrel was short—only 13 ft. 6 1/2 in. between the tube plates but with an outside diameter of 7 ft. 3 in. The proportions were ideal for free steaming, and the total combined heating surface of 3789 sq. ft. was amply adequate to feed four cylinders 19 1/4 in. diameter by 26 in. stroke. The superheater had 50 elements and provided 750 sq. ft. of superheating surface. Of the 50 engines of this class 25 had two thermic syphons in the firebox while the remaining ones each had five arch tubes. The leading dimensions of these fine articulated locomotives for the Australian railways were:

Railway	Queensland	New South Wales
Gauge, ft. in.	3-6	4-8 1/2
Type	4-8-2+2-8-4	4-8-4+4-8-4
Max. axle load, tons	9.64	16
Cylinders (4) dia., in.	13 3/4	19 1/4
Cylinder stroke, in.	26	26
Coupled wheel dia., ft. in. ..	4-3	4-7
Heating surface, sq. ft.:		
Tubes	1490	2799
Firebox	178	242
Superheater	453	748
Combined total	2121	3789
Grate area, sq. ft.	39	63.4
Boiler pressure, lb./sq. in. ..	200	200
T.E. at 85% boiler pressure, lb.	32,770	59,650
Total weight in working order, tons	136 3/4	255

Locomotives for India

Following upon the valuable experience gained in pre-war years with the first group of standard locomotives the second stage of standardisation saw the production of several important new designs which were based upon a blend of British and American practice. The " WG " 2-8-2 (Fig. 384), of which the first examples were built by the North British Locomotive Co. Ltd. in 1950, had many parts standard with the " WP " class standard passenger engines built in the U.S.A. Interesting departures from previous Indian practice were the provision of greater firebox volume in relation to grate area, by incorporation of a large combustion chamber, one thermic syphon and two arch tubes, and very massive bar frames. The front end and back end of the frames are each composed of a single steel casting, as in the Australian 4-6-4 express locomotives already described. At the front this casting incorporates the buffer beam and drag box, and a similarly comprehensive casting is incorporated at the back end. In common with practice becoming universal at that time the piston valves, actuated by Walschaerts gear, were made of a diameter large in relation to cylinder diameter to facilitate the free flow of steam; the piston valves were 12 in. diameter. The North British Locomotive Company was entrusted with an order for no fewer than 100 of these powerful engines.

Another interesting Indian design which was also built in Glasgow, in large numbers from 1952 onwards was the " YP " standard " Pacific " for the metre gauge lines of the Government Railways (Fig. 385). From the accompanying illustration it is perhaps difficult to appreciate that these handsomely proportioned engines weigh no more than 57 tons; but the overall height is only 11 ft. 2 in. and the coupled wheel diameter 4 ft. 6 in. The locomotive is virtually a reproduction on a miniaturised scale of the standard features incorporated in broad gauge standard types, including bar frames; one-piece steel castings at the front and rear end; very large firebox with combustion chamber, thermic syphon, and arch tubes; roller bearing axle-boxes on the bogie trailing truck and all tender wheels, and 9 in. diameter piston valves, supplying cylinders 15 1/4 in. diameter by 24 in. stroke. The comparative dimensions of the broad gauge 2-8-2 and the metre gauge 4-6-2 are set out in the accompanying table.

Post-War Indian Standard Locomotives Built by North British Locomotive Co. Ltd.

Class	WG	YP
Type	2-8-2	4-6-2
Gauge	5 ft.-6 in.	Metre
Cylinders (2) dia., in.	21 7/8	15 1/4
Cylinder stroke, in.	28	24
Coupled wheel dia., ft. in.	5-1 1/2	4-6
Heating surfaces, sq. ft.:		
Tubes	1962	905
Firebox	275	196
Superheater	683	331
Combined total	2920	1432
Grate area, sq. ft.	45	28
Boiler pressure, lb./sq. in.	210	210
T.E. at 85% boiler pressure, lb. ..	38,890	18,450

African Medium-Power Types

The Indian " YP " metre gauge " Pacific ", with its evidence of American influences in its design can also be studied in conjunction with a group of medium-powered locomotives built for various metre and 3 ft. 6 in. gauge railways in central and southern Africa, on all of which the influence remained almost entirely

Fig. 384.—*Indian Government Railways: " WG " type* 2-8-2 *of* 1950, *built by North British Locomotive Co. Ltd.*

Fig. 385.—*Indian Government Railways: Metre gauge* 4-6-2 *Class " YP ", built from* 1952 *onwards by North British Locomotive Co. Ltd.*

British, even including the Benguela Railway. Four designs can be considered together, as follows:

Nigerian Railways:	2-8-2 of 1948
South African Railways:	2-8-4 of 1949
Benguela Railways:	4-8-2 of 1951
East African Railways:	2-8-2 of 1952

African Medium-Power 8-Coupled Engines

Railway	Nigerian	S.A.R.	Benguela	E.A.R.
Gauge, ft. in.	3-6	3-6	3-6	Metre
Type	2-8-2	2-8-4	4-8-2	2-8-2
Max. Axle load, tons	13	9 1/2	13	
Cylinders (2) dia., in.	18	19	21	18
Cylinder stroke, in.	26	26	26	26
Coupled wheel dia., ft. in.	4-0	4-3	4-6	4-0
Heating surfaces, sq. ft.:				
Tubes	1732	1497	1623	1732
Firebox	146	144	154	146
Superheater	489	380	420	446
Combined total	2384	2021	2197	2341
Grate area, sq. ft.	38	36	40	38
Boiler pressure, lb./sq. in.	200	200	200	200
Nom. tractive effort at 85% boiler pressure, lb.	29,800	31,280	36,200	29,800
Fuel	Coal	Coal	Wood	Oil

The " River " class 2-8-2s of the Nigerian Railways (Fig. 386), of which batches were built both by the Vulcan Foundry Ltd. and by the North British Locomotive Company proved to be a prototype for general adoption in territories then covered by the Crown Agents for the Colonies. The design was used subsequently, and with only the slightest detail alterations, on the East African Railways (Fig. 390) and on the Tanganyika Railway (Fig. 395) examples of both of which are illustrated herewith. The Nigerian and Tanganyika engines were designed to burn low grade coal, that of Nigeria coming from the Udi coalfield and having the relatively low calorific value of 11,350 B.Th.U. per pound. Although having a fairly large grate area of 38 sq. ft. it was not considered necessary to have an unusually large firebox volume, and there was no combustion chamber. The boiler had a length of 16 ft. 4 1/2 in. between the tube plates. The inner firebox was of steel.

The valve gear details were calculated to provide a free-running and economical engine, with 10 in.

Fig. 386.—*Nigerian Railways: " River " Class* 2-8-2.

diameter piston valves, and the following details:

Lap of valves	$1\,^{7}/_{8}$
Lead, in.	$^{1}/_{8}$
Exhaust clearance, in. ..	$^{1}/_{16}$
Travel in full gear, in. ..	$6\,^{13}/_{16}$

The East African variant of this design, first supplied by the North British Locomotive Company in 1951 was arranged for working on the metre gauge, but to be readily convertible to 3 ft. 6 in. gauge if the necessity arose. These latter engines were equipped for oil firing, but in such a way that conversion to solid fuel could be easily done. The arrangement of the firepan was devised skilfully within the general limits of the existing design for Nigeria and Tanganyika, and two oil-burners were mounted at the front of the firebox. As in the early examples of the design the East African engines had bar frames, but a development on the latter was the use of roller bearings on all axles throughout the locomotive and tender.

The South African 2-8-4 (Fig. 387) was of quite a different design, suitable for lines where the maximum permissible axle load was only 11 tons. The general proportions are set out in the table of dimensions but two points of particular interest were the use, for the first time in any British-built locomotive of a one-piece cast steel bed frame, and the adoption by the South African Railways of the Vanderbilt type of tender. The author was able to inspect some of the bed frame castings just as they were being delivered from the General Steel Castings Corporation, of Eddystone, Philadelphia, and found them a fine piece of foundry work, in which the cross stretches, cylinders and valve chests were all incorporated in the one casting. They were nearly 40 ft. long, and weighed about 10 tons each. Much of the machining had already been done in America, and the chief work being done in Glasgow was the fitting of cylinder and valve liners. As mentioned earlier in this chapter the one-piece cast steel bed frame was subsequently specified for the huge Beyer-Garratt engines supplied to the New South Wales railways in 1952.

The use of the Vanderbilt type of tender tank had been made on the South African Railways a few months prior to delivery of the first of the 2-8-4 engines of Class " 24 ", on the main line " 19D " class 4-8-2s. These latter engines were very similar in outward appearance to the " 24 " class 2-8-4s, except that they were generally larger, and had bar frames. The tenders of both " 19D " and " 24 " class engines were carried on 6-wheeled bogies. That of the " 24 " class engine illustrated had a capacity of 9 tons of coal and 4,500 galls. of water, and had a total wheelbase of

Fig. 387.—*South African Railways:* 2-8-4 *lightweight locomotive Class* " 24 ".

Large African Beyer-Garratts: 1953–56

Railway / Type / Class	Benguela 4-8-2+2-8-4 —	Rhodesia 2-6-2+2-6-2 "14A"	Rhodesia 2-8-2+2-8-4 "16A"	Rhodesia 4-8-2+2-8-4 "20"	E.A.R. 4-8-2+2-8-4 "59"	S.A.R. 4-8-2+2-8-4 "GMAM"
Max. permitted axle load, tons	13	13.5	14.5	17	21	15.4
Cylinders (4) dia., in.	18 1/2	16	18 1/2	20	20 1/2	20 1/2
Cylinder stroke, in.	24	24	24	26	28	26
Coupled wheel dia., ft. in.	4-0	4-0	4-0	4-3	4-6	4-6
Heating surfaces, sq. ft.:						
Tubes	2224	1667	2131	2791	3313	2960
Firebox	229	174	212	233	248	237
Superheater	458	374	481	748	747	747
Combined total	2911	2215	2824	3772	4308	3844
Grate area, sq. ft.	51.5	38.6	49.6	63.1	72	63.2
Boiler pressure, lb./sq. in.	180	180	190	200	225	200
Nom. tractive effort at 85% boiler pressure, lb.	52,360	39,170	55,270	69,330	83,350	68,800
Total weight in working order, tons	175	131 1/2	167	224 3/4	252	190.2

24 ft. 7 1/2 in. The tenders of the "19D" class were enormous, with a capacity of 12 tons of coal, 6500 galls. of water, and having a total wheelbase of 34 ft. 9 in. In working order the "19D" tenders weighed no less than 73. 2 tons, against a total engine weight of 79.95 tons.

Last among these medium-powered African locomotives was the 4-8-2 for the Benguela Railway (Fig. 388), built by the North British Locomotive Company in 1951. These engines were designed for burning wood from the eucalyptus trees having a calorific value of 7600 B.Th.U. per pound, and to work passenger trains of 500 tons gross trailing load over a route with a ruling gradient of 1 in 80. In view of the fuel to be used the boiler proportions of these locomotives were interesting. The firebox having a grate area of 40 sq. ft., was not provided with a combustion chamber, and the distance between tube plates was much longer than on any of the other African locomotives considered in this particular group. It is no less than 19 ft. 3 in. against 16 ft. 4 1/2 in. on the Nigerian "River" class. The nominal tractive effort of 36,200 lb. at 85 per cent. boiler pressure was high in relation to the size of the boiler and to the low calorific value of the fuel employed; but on the Benguela Railway the eucalyptus wood, which is grown specially by the administration in belts of trees extending for many hundreds of miles on both sides of the line, has been proved very suitable for rapid combustion in large quantities.

Large African Beyer-Garratts

The Benguela Railway was one of the earliest in Africa to use the Beyer-Garratt type of locomotive for heavy freight service. The need for a very powerful engine will be appreciated from the physical conditions existing over much of the first 250 miles from the coast at Lobito until the interior plateau is reached and the gradients become undulating and not unduly severe (Fig. 389). But in the first 250 miles the ruling gradient is 1 in 40 for about 100 miles between Benguela and Cubal, and 1 in 80 from the latter point for a further 150 miles. The Beyer-Garratt type of locomotive was first introduced in 1926, and in 1952 delivery was taken, from Beyer Peacock and Co. Ltd. of some further examples of a much larger and

Fig. 388.—*Benguela Railway: wood burning* 4-8-2 *locomotive* (1951).

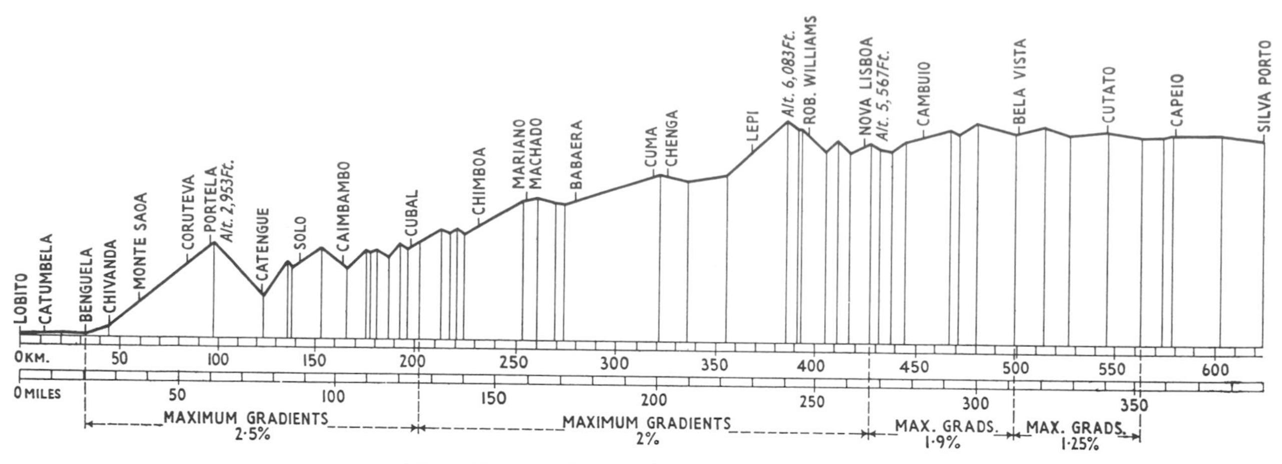

Fig. 389.—*Gradient profile, Benguela Railway.*

improved design. These locomotives incorporate all the specialities developed by this firm in connection with the Beyer-Garratt type of locomotive, but the firebox proportions are of interest in view of the use of wood fuel (Fig. 391). The grate is large and deep, with a grate area of 51.5 sq. ft. The ashpan was specially arranged for the burning of wood, and provided with suitable air and cleaning doors. A dust protection plate was fitted across the back end of the ashpan.

The cab was widened to the limit of the loading gauge, to provide plenty of floor space that is necessary when firing with wood. When working hard up the severe gradients with maximum load trains several men were kept continuously employed feeding logs into the firebox by hand, under the supervision of the "fireman". As these logs were 2 ft. long and 10 in. diameter the human "belt conveyor" from tender to firedoor did not permit of any interruption. The dimensions of these locomotives are given in the general table on page 255.

The Rhodesian Railways have always been extensive users of the Beyer-Garratt type of locomotive, and the dimensional particulars are included in the accompanying table of these new designs introduced between the years 1953 and 1955. They can be summarised thus:

Class 14A	2-6-2 + 2-6-2	Branch line duties.	(Fig. 393)
Class 16A	2-8-2 + 2-8-2	Heavy Freight.	(Fig. 394)
Class 20	4-8-2 + 2-8-4	Heavy Freight.	

The "20" Class were indeed very large and powerful engines for the 3 ft. 6 in., gauge having a nominal tractive effort of 69,330 lb., and were required to handle loads of 1,400 tons over gradients of 1 in 64 on the Kafue-Broken Hill section. They were, of course, mechanically fired.

These large engines were nevertheless considerably surpassed in size and tractive capacity by the outstanding "59th" class of the East African Railways introduced in 1955 (Fig. 396). In this case however the designers had the advantage of a permitted axle load up to 21 tons, and the result was the most powerful

Fig. 390.—*East African Railways:* 2-8-2 *locomotive built* 1951, *by North British Locomotive Co. Ltd.*

Fig. 391.—*Benguela Railway: wood burning Beyer-Garratt* 4-8-2 + 2-8-4.

Fig. 392. *Benguela Railway: oil-burning Beyer-Garratt* 4-8-2 + 2-8-4.

Fig. 393.—*Rhodesia Railways:* " 14A " *Class Beyer-Garratt* 2-6-2 + 2-6-2 *for branch line duties.*

Fig. 394.—*Rhodesia Railways:* " 16A " *Class Beyer-Garratt locomotive:* 2-8-2 + 2-8-2 *for heavy freight work.*

Fig. 395.—*Tanganyika Railway:* 2-8-2 *locomotive.*

locomotive ever put on to metre gauge metals anywhere in the world. The power of these locomotives was commensurate with the work they were required to do. In a distance of 330 miles the line climbs from sea level at Mombasa to an altitude of 5,600 ft. at Nairobi, on a ruling gradient of 1 in 66, uncompensated for curvature, while the summit level, at 9,000 ft. above sea level attained at Timboroa, is 530 miles from the coast. The new engines were of most handsome appearance. They were originally arranged for oil firing, though with facility for mounting mechanical stokers to fire solid fuel at a later date if required.

The last of the Beyer-Garratt locomotives of which dimensions are tabulated on page 255, are the " GMAM " class of the South African Railways, which was not only an excellent design in itself, but which in speed of construction was an altogether extraordinary example of the resource and ingenuity of the British locomotive building industry. In December, 1955 Messrs. Beyer Peacock and Co. Ltd. received an order for 35 huge Beyer-Garratt locomotives on condition that delivery commenced in seven months. *Seven months*! On the face of it the task was impossible, seeing that the engines had to be designed, as well as constructed. Yet the fact remains that the first engine was steamed and under test at Gorton one month *before* the contract date. Such was the brilliant combination of design and constructional organisation behind the production of the South African " GMAM " class (Fig. 397). One could hardly end the long and distinguished history of British locomotive building for overseas railways with a finer achievement.

Mention must be made however of British participation in the production of the remarkable South African Railways Class " 25 " 4-8-4 condensing locomotive, ordered in 1951. This was a joint project between the North British Locomotive Company and Messrs. Henschel and Sohn, the well-known German locomotive builders. The condensing arrangements had been designed by Henschel, but when it came to placing a bulk order for 90 of these very powerful locomotives the British manufacturers were awarded the order for all 90 of the locomotives, and for 30 of the enormous condensing tenders. With the use of these locomotives it was claimed that distances of up to 700 miles could be covered without taking water

Fig. 396.—*East African Railways:* 59*th Class Beyer-Garratt* 4-8-2 + 2-8-4 (1955), *with a tractive effort of* 83,350 *lb. on metre gauge.*

intermediately. The locomotives, which were equipped with roller bearings on all axle boxes, had the following leading dimensions:

Cylinders (2) dia., in.	24
Cylinder stroke, in.	28
Coupled wheels dia., ft. in.	5-0
Heating surfaces, sq. ft.	
Tubes	3059
Firebox	294
Superheater	630
Combined total	4020
Grate area, sq. ft.	70
Boiler pressure, sq. ft.	225
Nom. tractive effort at 85% boiler pressure, lb.	51,300

So far as the condensing arrangements are concerned, exhaust steam from the cylinders was passed into a turbine driving a fan blower in the smokebox which latter replaces the normal blastpipe draught. From the blower turbine the exhaust steam passed along the side of the locomotive, and then through an oil separator to another turbine in the tender which drove the air intake fans. The fans were driven by bevel gears on a line shaft from the turbine. Finally the steam passed to the condensing elements mounted on both sides of the tender. The condensate was collected in a tank fitted underneath the tender frame, and from there it was fed back into the boiler. The tender was considerably longer than the engine, having a total wheelbase of 45 ft. 10 in. against 38 ft. for the engine itself.

Fig. 397.—*South African Railways:* "GMA" *Class Beyer-Garratt locomotive* (1956).

Chapter Twenty Finale

In the foregoing chapters the evolution of the British steam railway locomotive has been traced from the time of the Railway Centenary to the virtual end of steam traction on British Railways at the close of the year 1965. The story has been traced through points of design, through changing traffic requirements to the time when locomotive testing had become an exact science. At the end one comes to the inevitable question: what can be the eventual verdict on steam. Was it like some fruit that had matured to the point where it was over-ripe and fell in disorder? Had it reached a point beyond which no worth-while development was possible? Or was it merely outmoded. None of these verdicts is really complete in itself. The circumstances that led to the sudden demise of steam on British Railways will be debated as long as railways are remembered; but among engineers it can be said with absolute certainty that steam did not play itself out. The development was unfinished. This book is not the place for discussion as to why the story was so abruptly cut short. In this final chapter, rather, it is my wish to examine the last 10 years on British Railways critically, from the viewpoint of an engineer, not a politician, to try and see, in retrospect, the position to which the steam locomotive had finally attained.

Although a number of attempts had been made to break away from the conventional—the last being O. V. S. Bulleid's double-ended tank engine of the " Leader " class—the long line of development ended in extreme simplicity, with the British Standard range of locomotives, each having two outside cylinders, and Walschaerts valve gear. None of the last minute variations, such as Caprotti valve gear, or the Franco-Crosti boiler came to displace the conventional, and in this respect the last phases of British steam traction came to resemble those of the U.S.A. where the many complicated stages of compounding eventually gave way to huge, ultra-simple 2-cylinder simple expansion locomotives. British engineers have been criticised for not persisting with compounding, at a time when its potentialities were being so vividly displayed anew by the work of André Chapelon, in France. Furthermore this development was in its flood-tide of success at a time when all four main line railways of Great Britain were using 3 or 4 cylinders for their top line express passenger locomotives, and the reversion to the ultra-simple had not yet commenced.

There were times when arguments were produced to show, structurally, why high powered compound locomotives could not be designed within the limitations of the British loading gauge, while in other schools of design it was still felt that the compound versus simple controversy had been settled once and for all by the work of Churchward on the Great Western Railway in 1903–06. That great engineer had then found it possible to design a simple engine that was equal in thermal efficiency to the de Glehn compound " Atlantics ". But it was in radical improvement of this very design, as very simply adapted 20 years earlier from the " Atlantics " to the " Pacifics " on the Paris-Orleans Railway, that Chapelon had achieved his most spectacular success; and in so doing he attained levels of overall thermal efficiency that have never been surpassed by steam locomotives.

Nevertheless thermal efficiency is not the only yardstick by which the steam locomotive must be judged as an effective tool to railway management; and the circumstances of personnel and their training in France were undoubtedly far more favourable to the effective utilisation of compound locomotives of highly advanced design than they would have been in Great Britain in the same era. One can therefore accept, without criticism, the reasons why British locomotive engineers continued to introduce machines that were simple to handle, from the viewpoint of the crews, particularly as multiple-manning—in many cases indeed, indiscriminate manning—had become a major point of policy in the utilisation of the motive power stud.

Earlier chapters in this book have traced the rapid advance in thermal efficiency obtained by the use of improved front-ends, and how basic coal consumption, as measured in pounds of coal per d.h.p. hour had been reduced from the general level of 4 to 5 lb. in 1925 down to about 3 lb. in all new designs introduced from about 1927 onwards. A reduction, in round terms, of some 33 per cent. was truly a major achievement, surpassing in its overall effects the improvements obtained in the period 1905–15, by the introduction of superheating. On the London and North Western Railway in 1910–12, C. J. Bowen-Cooke had secured a 25 per cent. saving in coal consumption by superheating, on two brand new and otherwise identical engines—the *George the Fifth* and the *Queen Mary*, of July 1910. But a reference to those two celebrated engines of an earlier period is important in another respect, because it draws attention to the efforts made to sustain the " mint-condition " performance of locomotives at a high level throughout their span of service from one major overhaul to the next. That is a respect in which engines like the *George the Fifth* of 1910 were seriously deficient.

Fig. 398.—*Eastern Region: a Gresley " A3 " 4-6-2 as fitted with Kylchap double-blastpipe and chimney.*

In the last days of steam not only were locomotives running longer mileages between general repairs, but it was usual for high speed express passenger units to remain on their normal work until they were due for shopping. Although certain very exacting duties were the subject of special engine allocation, as will be mentioned later, in general the schedules and loadings were such that any locomotive that could be accepted for traffic could time the trains, no matter how large the mileage since last general repair. Some of the services operated in this manner were the fastest ever to be worked over the routes in question, such as the Liverpool St.—Norwich two-hour expresses, with " Britannia " class " Pacifics "; and the Atlantic Coast Express of the Southern, between Waterloo and Exeter, while the work of " Britannia " class " Pacifics " and rebuilt " Royal Scots " 4-6-0s reached a very high level of attainment on the St. Pancras-Manchester services by the Midland route. On the Great Eastern line the " Britannia " class engines continued to work their fast schedules at a mileage of 100,000 and more, from the previous general repair.

Nationalisation, and the need to consider the British steam locomotive stock as a whole, made possible some interesting comparisons in vital features of performance never previously made public, and of these one of the most interesting was the mileage run between periodical repairs. In the years between 1950 and 1960 the frequency of repairs was governed as much by the mechanical condition of the engine as by the boiler. It was found in many cases that heavy boiler repairs involving the removal from the frame were necessary only at every other periodical repair, whereas axleboxes, tyres, motion and frames required attention on each occasion. For the entire steam stock of British Railways, in 1951, before the introduction of the new standard designs, the average mileage from general to intermediate repairs was 64,268, and the average mileage between general repairs was 101,130. As could well be imagined there was considerable difference between the performance of various regional types, of widely divergent design, and the average mileages between periodical repairs —general to intermediate, or intermediate to general— were quoted in a paper read before the Institution of Locomotive Engineers in March 1953 by R. C. Bond, then Chief Officer (Locomotive Construction and Maintenance) British Railways.

It would have been still more interesting if the various classes could have been subdivided into the different phases of utilisation, but certain points emerge. It is often claimed that one feature of design contributing to trouble-free long mileage working is to have the valve gear and all working parts outside and readily accessible. Against this must be set the mileage records of three Great Western designs with outside cylinders and inside Stephensons link motion—the " Hall " 4-6-0; the series of tank classes, and the " 28XX " class heavy freight engine, all of which were pre-eminent in their respective spheres, except for those ex-L.M.S. engines that were fitted with manganese steel liners on the coupled axleboxes. The inference must be drawn that factors other than design contributed to this high mileage result. Another point is the good mileage records of the Gresley " Pacifics " of classes " A3 " and " A4 ", fitted with

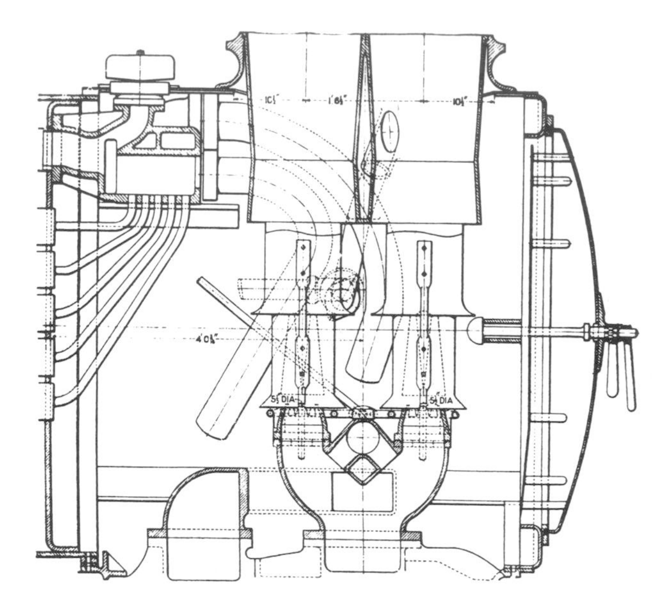

Fig. 399.—*Smokebox arrangement E.R.* 4-6-2 *Class* " A3 ".

the conjugated valve gear, a feature at one time somewhat in disrepute (Figs. 398, 399).

Mileage between periodical repairs is, of course, not the only factor governing the ultimate value of a locomotive in traffic, and a regrettable case was that of the Bulleid air-smoothed " Pacifics " on the Southern Region. Despite the great amount of care and ingenuity involved in the conception and development of these novel designs the incidence of failures in traffic, chiefly from the valve gear and the enclosed motion, was such that the decision was taken to rebuild the whole of the " Merchant Navy " class, and a number of the lightweight " West Country " and " Battle of Britain " engines with an orthodox front end, and three sets of Walschaerts valve gear (Figs. 400, 401). The overall performance of the engines was much improved, though in the modified form and with boiler pressure reduced from 280 to 250 lb. per sq. in. the maximum efforts were somewhat reduced. Thus one of the last attempts in British locomotive practice, to break away from the conventional, ultimately proved unsuccessful.

Average Mileage Between Periodical Repairs

Region	Power Class	Class of Locomotive	Mileage
E./N.E.	8P	4-6-2 " A1 "	93,363
E./N.E.	8P	4-6-2 " A2 "	85,671
E./N.E.	8P	4-6-2 " A3 "	83,574
E./N.E.	8P	4-6-2 " A4 "	86,614
L.M.	8P	4-6-2 " Coronation "	73,188
W.R.	8P	" King " 4-6-0	78,987
S.R.	8P	" Merchant Navy "	75,687
E./N.E.	7P	2-6-2 " V2 "	77,892
L.M.	7P	Converted Royal Scot	70,495
W.R.	7P	" Castle " 4-6-0	87,424
W.R.	7P	" County " 4-6-0	87,588
S.R.	7P	" West Country " 4-6-2	74,650
S.R.	7P	" Lord Nelson " 4-6-0	81,611
E./N.E.	" 5 "	4-6-0 " B1 "	78,396
L.M.	" 5 "	4-6-0 Class " 5 " standard	56,969
L.M.	" 5 "	4-6-0 Class " 5 " with manganese steel liners	97,291
W.R.	" 5 "	4-6-0 " Hall "	87,942
S.R.	—	" King Arthur " 4-6-0	70,995
S.R.	—	" Schools " 4-4-0	69,851
E./N.E.	" 4 "	2-6-2 tank " V3 "	66,821
E./N.E.	" 4 "	2-6-4 tank " L1 "	67,213
L.M.	" 4 "	2-6-4 tank (standard)	55,579
L.M.	" 4 "	2-6-4 tank with manganese steel liners	79,361
W.R.	" 4 "	2-6-2 tanks " 3150-81XX "	71,720
E./N.E.	8F	2-8-0 Class " O1 "	55,616
E./N.E.	8F	2-8-0 " WD " design	62,624
L.M.	8F	2-8-0 Stanier design	50,361
W.R.	8F	2-8-0 " 28XX " class	86,981

Two other notable, though equally unsuccessful ventures must also be mentioned in these last years of steam. The first was the Bulleid double-ended " Leader " class tank engine (Fig. 402), in which one

Fig. 400.—*Southern Region:* " *Merchant Navy* " *Class* 4-6-2 *rebuilt with Walschaerts valve gear.*

Fig. 401.—*Southern Region: "West Country" class 4-6-2 as rebuilt with Walschaerts valve gear.*

Fig. 402.—*Southern Region: Bulleid's double-ender tank engine "Leader" Class.*

Fig. 403.—*B.R. Class* " 9 " 2-10-0 *with Crosti boiler.*

Fig. 404.—*Crosti-boiler* 2-10-0 *under construction.*

Fig. 405.—*Crosti-boiler* 2-10-0; *lowering the boiler unit into position.*

Fig. 406.—*B.R. Class* " 9 " 2-10-0 *with Crosti boiler: view of right-hand side.*

of the designers' aims was to produce a comparatively trouble-free boiler in which firebox legs and flat stayed surfaces were eliminated. The boiler was welded throughout, and included four thermic syphons. The locomotive was articulated, with the main frame carried on two outside-framed six-wheeled bogies. Each bogie was powered by a 3-cylinder engine, with cylinders $12\,^1/_4$ in. diameter by 15 in. stroke. The drive was on to the centre axle of each bogie, with the steam distribution by a cast iron sleeve valve sliding inside a cast iron liner. Separate admission and exhaust ports, equally distributed around the circumference were machined in the sleeve. The valve gear was basically of the Walschaerts type, and as in the Bulleid "Pacifics" the whole of the motion, crank axle, and so on, was totally enclosed and automatically lubricated. A driving cab was provided at either end of the locomotive with all controls duplicated, but the fireman was located in a central cab. It was an ingenious conception, reminiscent both in features of the boiler design and in the use of sleeve valves of the Paget locomotive built at the Derby Works of the Midland Railway, in 1908. The first of the "Leader" class tank engines, produced in 1949, ran no more than a few trial trips. After Mr. Bulleid's retirement work on the remaining four engines was stopped, and the project was virtually stillborn, though it must be added, that in the short period of experimental running of the locomotive much trouble was encountered.

The second departure was the building by British Railways, at Crewe Works, in 1955, of ten of the standard Class "9F" heavy freight 2-10-0 locomotives with the Crosti type of boiler (Figs. 403–406), that had been applied to a considered number of Italian and German locomotives. This device aimed at securing increased thermal efficiency by pre-heating the water passing into the boiler. The pre-heater was mounted beneath the main boiler, and the layout, as applied to the "9F" 2-10-0 is shown in Fig. 407. The exhaust steam from the cylinders, instead of passing directly into the base of the blastpipe, as in a normal locomotive, was carried back in two long pipes. That on the right hand side of the engine passed directly from the cylinders to the blast chamber, G, and that on the left hand side passing back and then under the boiler to join the other pipe just before entering the blast chamber. The "9F" engines thus equipped had a more orthodox look than their Italian counterparts, which had no chimney at all in the usual place. The exhaust was through the final chimney on the right hand side of the boiler, and the chimney on the front smokebox was used only for lighting up.

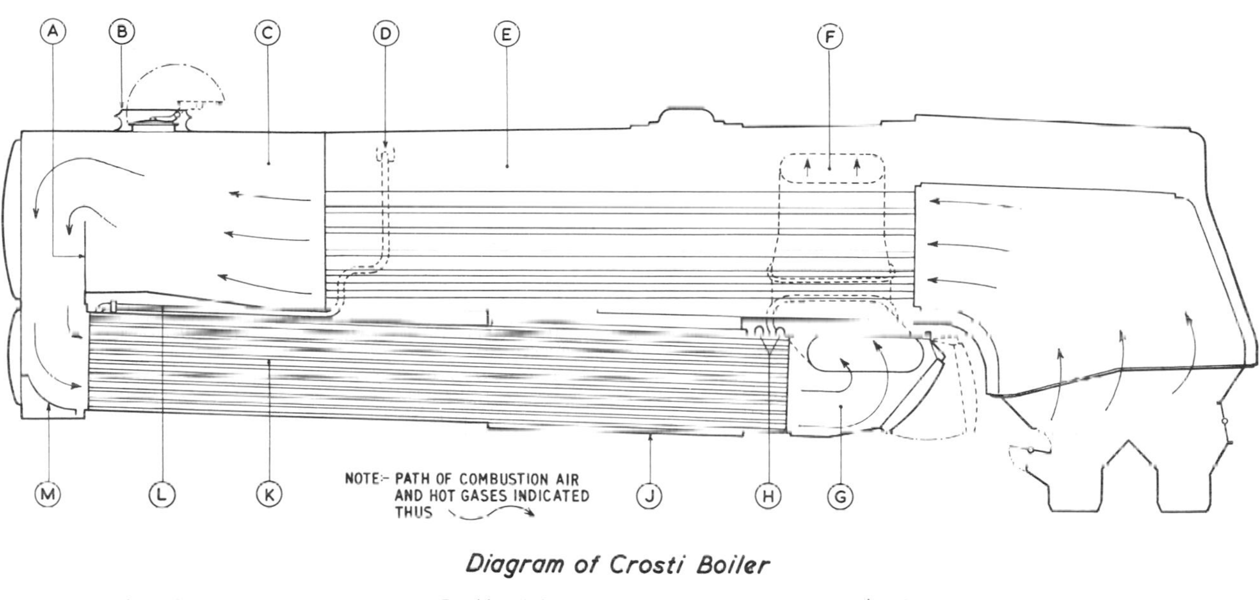

Fig. 407.—*Crosti boiler; diagram of connections on "BR9" 2-10-0 locomotive.*

On the Western Region, experience with British Railways new standard types and particularly with the Class "8" 3-cylinder 4-6-2 No. 71000 (Fig. 408), led to the fitting of double-blastpipes and chimneys to all 30 engines of the "King" class (Fig. 410), and to certain engines of the "Castle" class (Fig. 412). This development, which led to the production of higher i.h.p. at high speed, was more marked in the case of the "Kings" than with the "Castles" though engines of the latter class, stationed at Bristol, did some extremely fast running with the high-speed Bristolian express. The modified arrangement of the

Fig. 408.—*The "BR8" 4-6-2 No. 71000 on a test run near Didcot, Western Region.*

blastpipe and chimney on the "King" is shown in (Fig. 411); it was a simple change, cheaply done, and its effectiveness was shown in a series of trials with the dynamometer car in 1956. Certain details of a run with the up Cornish Riviera Express are shown in the accompanying table, though unfortunately direct comparison cannot be made with a test run made with a similar load and at similar speed with a locomotive having the single blastpipe. On this latter run the tractive resistance was considerably higher on account of a strong headwind which prevailed throughout the journey. Nevertheless, the comparison does show a notable reduction in the basic coal consumption:

Engine No.	6013	6002
Type of blastpipe	Single	Double
Load, trailing tons tare	393 ½	391
Average speed m.p.h.	64.7	64.7
Firing rate lb. per hr.	4210	2800
Net steam rate lb. per hr.	29,835	20,800
Average d.h.p.	1020	799
Coal per d.h.p. hour lb.	4.0	3.5

The foregoing results are more a revelation of the adversity of the weather conditions on the run with the single blastpipe engine rather than a comparison of the two front end designs; but the average coal consumption for the two sets of trials between Paddington and Plymouth, two in each direction, in which the prevailing weather conditions during the first set would to some extent cancel each other out, by being exceptionally favourable one way and equally adverse the other, are more significant. These figures were obtained by the most careful scientific testing and can be accepted as accurate, and truly relevant.

Engine No.	6013	6002
Type of blastpipe	Single	Double
Coal per d.h.p. hour, exclusive of coal theoretically required to raise steam, lb.	3.62	3.34
Water per d.h.p. hour, lb.	25.8	25.7

The double-chimneyed engine thus showed an economy of approximately 8 per cent.—a worthwhile saving. Summary details of the running of engine No. 6002 between Exeter and Paddington during the trials were as follows:

W.R. Engine No. 6002

Section	Load tons tare	Dist. miles	Booked av. sp. m.p.h.	Actual av. sp. m.p.h.	Net av. sp. m.p.h.
Padd.-Taunton (start-to-stop)	449* 413†	142.7	57.9	64.3	65.4
Exeter-Padd. (pass-to-stop)	391	173.5	59.0	59.8	62.4
Padd.-Exeter (start-to-stop)	457* 390†	173.5	63.2	50.9‡	67.6
Exeter-Padd. (pass-to-stop)	391	173.5	59.0	63.6	65.1

* Paddington to Heywood Road Junction (94.6 miles)
† Heywood Road to Exeter
‡ Includes exceptional traffic delay of 37 min. at Patney.

A notable performance with another of the "King" class engines fitted with double chimney was in working the Cornish Riviera Express punctually into Plymouth after delays totalling 21 min. had been experienced en route (Fig. 409). The load conveyed

Fig. 409.—*Western Region; Double-chimneyed " King " Class 4-6-0 on Cornish Riviera Express*

Fig. 410.—*W.R. " King " Class* 4-6-0 *fitted with double blastpipe and chimney.*

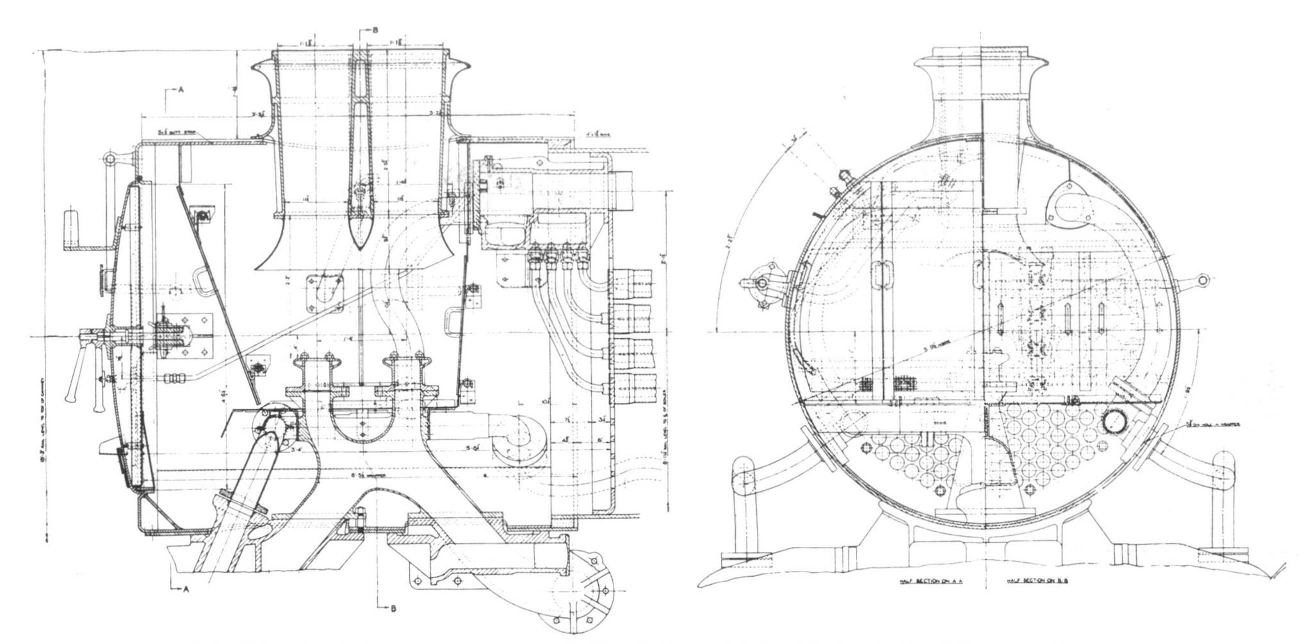

Fig. 411.—*Smokebox arrangement on " King " Class* 4-6-0 *with double blastpipe and chimney.*

was 362 tons to Heywood Road Junction and 327 tons beyond, and the 225.4 miles were completed in $238\,^1/_2$ min., or $218\,^1/_2$ min. net—a net average speed of 61.8 m.p.h. over this difficult route. The author rode on the footplate of the locomotive on this occasion, and noted that the substantial recovery of time was made entirely by fast hill-climbing, and rapid acceleration from sections where speed restrictions were in force.

The first " Castle " class 4-6-0 to be equipped with double-blastpipe gave some interesting results on a series of dynamometer car tests carried out between

Fig. 412.—*W.R. " Castle " Class* 4-6-0 *with double blastpipe and chimney.*

Fig. 413.—*Western Region: Double-chimneyed "Castle" class* 4-6-0 *at* 90 *m.p.h. with the up Bristolian express.*

Paddington and Kingswear. One particular exposition of high output at maximum speed was noted as follows:

Load tare, tons	310
Load gross trailing, tons	330
Speed sustained on level, m.p.h.	86
Drawbar pull, tons	1.6
Drawbar horsepower	825
Boiler pressure, lb./sq. in.	220
Steam chest pressure, lb./sq. in.	205
Cut-off, per cent.	20
Back pressure, lb./sq. in.	3.8
Superheater temperature, deg. fahr.	730

This particular engine, No. 7018 *Drysllwyn Castle*, was subsequently allocated to Bath Road shed, Bristol, and in the course of ordinary duty made some very fast runs on the Bristolian express (Fig. 413). Details of two of these are set out in the accompanying table:

W.R. The "Bristolian"
Engine No. 7018 *Drysllwyn Castle*

Run No.			1		2	
Load tons gross			265		260	
Dist.		sch.	Actual		Actual	
miles		min.	m.	s.	m.	s.
0.0	BRISTOL	0	0	00	0	00
4.8	Filton Junc.	8 1/2	9	40	9	03
17.6	Badminton	21 1/2	22	11	21	15
34.7	Wootton Bassett	34	34	43	32	48
40.3	SWINDON	39	39	26	37	29
64.5	Didcot	59 1/2	56	13	54	15
81.6	READING	71 1/2	68	20	66	11
93.4	Maindenhead	80	77	09	75	03
99.1	SLOUGH	84 1/2	81	10	79	07
108.0	Southall	91	88	01	85	45
111.9	Ealing		90	33	88	11
116.3	Westbourne Park		94	28	91	45
117.8	PADDINGTON	105	97	08	93	50
Max. speed after Badminton m.p.h.			94		100	
Average speed Swindon-Ealing m.p.h.			84		84.8	

These were both excellent examples of sustained free-running, and their end-to-end average speeds of 73 and 75.5 m.p.h. start to stop represented performances up to the highest standards attained with steam traction in this country. They were, of course, the product of a picked engine, reserved for the duty and used on nothing else at the time and the work of keen and experienced crews. But nevertheless, even with this reservation the performances rank very high indeed.

Perhaps the finest, and certainly the most spectacular achievement of the British steam railway locomotive was the regular running during the summer months of the London-Edinburgh non-stop expresses, on the accelerated times in operation up to the time when steam was superseded by diesel traction on the East Coast Route. From the time of its re-introduction after the war the end-to-end time over the 392.7 miles between King's Cross and Edinburgh Waverley was successively reduced from 7 hr. 20 min. in 1948 to 6 3/4 hr. in 1953, representing average speeds of 53.6 and 58.2 m.p.h. respectively. Throughout its post-war existence this service was entrusted exclusively to the Gresley Class "A4" streamlined "Pacifics" despite their increasing age. It is true that they, and some of the still older Gresley non-streamlined "Pacifics", were the only locomotives in Great Britain to be fitted with corridor tenders; but they were preferred on this very long non-stop run because of their very low coal consumption. It would not have been a difficult matter to fit corridor tenders to the more recent L.N.E.R. "Pacifics" of Thompson or Peppercorn design; but then the latter engines with their 50 sq. ft. grates were heavier coal burners on duties requiring a medium power output. With the London-Edinburgh non-stops it was above all mechanical reliability that counted, and by use of selected engines, in first class condition, a very high standard of punctuality was maintained.

The capacity of the Gresley "A4 Pacifics", and the enterprise and skill of their crews was subjected to a very severe test in the late summer of 1948,

Fig. 414.—*East Coast Route: Class "A4" Pacific on the "Elizabethan" express non-stop between King's Cross and Edinburgh.*

when damage to the track following exceptional storm conditions in the Lammermuir hills closed the main line between Berwick and Dunbar for a period of three months, and necessitated the diversion of all through Anglo-Scottish traffic via Kelso and Galashiels. The only water troughs north of Newcastle are at Lucker, Northumberland, 74 miles from Edinburgh by the direct route. But on account of the diversion this distance was increased to 90 miles, over a route involving greater physical handicaps. The heaviest gradient on the normal route is 1 in 96, extending for $4\frac{1}{2}$ miles from Cockburnspath to Penmanshiel Tunnel. But on the diversion route, in addition to some sharp gradients in the immediate neighbourhood of Edinburgh, there was the severe ascent, for southbound trains from Hardengreen Junction to Falahill, including a section of 1 in 70 for $9\frac{1}{2}$ miles continuously.

When the diversion was first needed arrangements were made for the "non-stop" to take water at Galashiels, while rear-end banking assistance was always available from Hardengreen Junction to Falahill. If drivers availed themselves of both facilities—as they were entitled to do—it so happened that the "non-stop" on the southbound run actually made two stops, purely for locomotive purposes. After a little experience of working over this route the enterprising engine crews from Haymarket shed, Edinburgh, essayed the difficult task of getting through from Edinburgh to Lucker troughs without a stop. This meant disdaining any assistance up to Falahill, and despite the heavier work thus involved so spinning out their water supply as to avoid the need for stopping at Galashiels. This remarkable feat was actually achieved on no less than 17 occasions. These runs of 408.6 miles were the longest made non-stop ever to be achieved in Great Britain. The regular summer load of the "non-stop" was about 430 tons tare; with heavier loads rear-end banking assistance was always taken from Hardengreen Junction to Falahill, on the southbound run. Very expert management of the locomotives was necessary on the Falahill ascent, and the speed throughout from Hardengreen Junction to the summit was rarely more than 20 m.p.h. The emergency schedule nevertheless provided ample running time. The main point, purely for prestige purposes, was to run non-stop between London and Edinburgh.

In the year 1953 in honour of the Coronation of Her Majesty Queen Elizabeth II the train was renamed "The Elizabethan", and accelerated to $6\frac{3}{4}$ hr. run. Even this was not the limit of post-war development with steam, for in the following year a further 15 min. was cut from the schedule, thus bringing the overall speed slightly in excess of 60 m.p.h. In working this train the Gresley "A4" streamlined "Pacifics" showed that they had still a comfortable margin in reserve for coping with incidental delays en route. The train ran to the $6\frac{1}{2}$ hour timing during each summer season from 1954 to 1961 inclusive, conveying a standard rake of 11 vehicles, representing a gross trailing load of about 420 tons. What the fastest net running time achieved on any occasion has been cannot be stated for certain, but there were times when more than 20 minutes of incidental delays were recovered leaving a net running time of around 370 min. for the run of $392\frac{3}{4}$ miles. The fastest time regularly scheduled between London and Edinburgh with steam was the level 6-hours of the Coronation streamlined train, which ran between 1937 and 1939. The maximum load of that train was 325 tons behind the tender, and in the down direction the journey included two intermediate stops, at York and Newcastle. Having regard to the difference in load the two schedules were probably about equal in the demands they made for power output on the locomotive concerned. On the "Coronation" although there were intermediate stops one locomotive worked through from end to end. The accompanying log gives details of a particular southbound run on the Elizabethan when the effects of a number of incidental delays were recovered, and the net average speed from end to end was 63.5 m.p.h.

The Elizabethan
Load 11 *cars*, 403 *tons tare*, 425 *tons full*
Engine: Class "A4" 4-6-2, *No.* 60030 *Golden Fleece*

Dist. Miles		Sch. min.	Actual m.	s.
0·0	Edinburgh (Waverley)	0	0	00
			*	
29·2	Dunbar	31	32	22
57·5	Berwick	60	60	10
89·6	Alnmouth	90	88	32
124·4	Newcastle	125	124	08
			*	
138·4	Durham	141	144	40
160·4	Darlington	$163\frac{1}{2}$	165	44
182·3	Thirsk	182	182	16
204·5	York	$201\frac{1}{2}$	200	03
218·3	Selby	$216\frac{1}{2}$	214	15
236·7	Doncaster	237	231	48
254·1	Retford	254	248	51
			*	
272·6	Newark	$270\frac{1}{2}$	265	36
287·2	Grantham	$283\frac{1}{2}$	280	10
292·6	Stoke Box	—	285	44
			*	
316·3	Peterborough	$308\frac{1}{2}$	305	28
			*	
333·8	Huntingdon	$325\frac{1}{2}$	323	58
360·8	Hitchin	$348\frac{1}{2}$	346	01
375·0	Hatfield	362	359	46
			*	
392·7	King's Cross	390	385	10

* Out of course check in this section
maximum speed 96 m.p.h.

In the 140 years of locomotive development covered by Ahrons' work, and my own, there have been many

Fig. 415.—"*B.R.9*" *class* 2–10–0 *No.* 92250 *fitted with Giesl oblong ejector.*

instances of foreign inventions being used to great advantage in British locomotive engineering. One has only to mention the Walschaerts valve gear, the Schmidt superheater and Kylchap blastpipe; other names will also come readily to mind. At the very end of the steam age in Britain another ingenious development, the Giesl oblong ejector, combined with the same inventor's superheater booster might have had a considerable effect in prolonging the usefulness of the steam locomotive at a time when the quality of coal made available to railways in Great Britain was still further on the decline. Two main line locomotives were equipped, a class " BR9 " standard 2-10-0, and an unrebuilt " West Country " 4-6-2 of the Southern. Both applications achieved a degree of success that indicated clearly what could have been achieved with a little development work. But unfortunately for the advancement of the science of steam locomotive engineering in this country the invention came at a time when the decision had been taken to eliminate steam traction, and these two isolated applications were not pursued (Fig. 415).

On the overseas railways which were still depending upon British-built steam locomotives for the major part of their motive power the Giesl oblong ejector and superheater booster has been applied with great success to the improvement of many locomotives that have been described in earlier chapters in this book: in India, in Australia, and on the wood-burning Beyer-Garratts of the Benguela Railway. But perhaps its most notable success, so far as British-built locomotives are concerned has been on the East African Railways, where the capacity of the " 59th " class Beyer-Garratt locomotives has been so enhanced—by 14 per cent. in fact—as to make possible the scheduling in the timetables of the " Giesl Goods " services, between Mombasa and Nairobi, carrying heavier loads and at higher speeds than were previously possible. Furthermore the working efficiency of these locomotives is now such that from figures quoted in the financial returns from that railway their operating costs are now less than for diesel power engaged on the same work. In 1964 the railway administration quoted the following, in its " Motive Power Report 1963 "

East African Railways: " 59th " *class Beyer-Garratt**
Nominal Tractive Effort, 83,350 *lb.*

Description	Cost per engine mile	Cost per 1000 gross ton-miles
	In East African shillings	
Shed labour and material	1·032	1·08
General repair	1·006	1·05
Fuel	3·670	3·84
Train crew	1·550	1·61
Totals	7·258	7·58

* Fitted with Giesl ejector

The corresponding total costs for two 1,850 h.p. diesel electric locomotives that would have been needed to do the same work were quoted as 8.28 shillings per engine mile, and 13.32 shillings per 1,000 gross-ton miles.

However, at this late stage I must not be lured into any participation in the evergreen steam versus diesel controversy. The saga of steam on British Railways is now virtually at an end; a premature end maybe. And in simply recording that fact this book must end too.

Index